T0184984

Palgrave Studies in Educational Futures

Series Editor
jan jagodzinski
Department of Secondary Education
University of Alberta
Edmonton, AB, Canada

The series Educational Futures would be a call on all aspects of education, not only specific subject specialist, but policy makers, religious education leaders, curriculum theorists, and those involved in shaping the educational imagination through its foundations and both psychoanalytical and psychological investments with youth to address this extraordinary precarity and anxiety that is continually rising as things do not get better but worsen. A global de-territorialization is taking place, and new voices and visions need to be seen and heard. The series would address the following questions and concerns. The three key signifiers of the book series title address this state of risk and emergency:

1. **The Anthropocene:** The 'human world,' the world-for-us is drifting toward a global situation where human extinction is not out of the question due to economic industrialization and overdevelopment, as well as the exponential growth of global population. How to we address this ecologically and educationally to still make a difference?
2. **Ecology:** What might be ways of re-thinking our relationships with the non-human forms of existence and in-human forms of artificial intelligence that have emerged? Are there possibilities to rework the ecological imagination educationally from its over-romanticized view of Nature, as many have argued: Nature and culture are no longer tenable separate signifiers. Can teachers and professors address the ideas that surround differentiated subjectivity where agency is no long attributed to the 'human' alone?
3. **Aesthetic Imaginaries:** What are the creative responses that can fabulate aesthetic imaginaries that are viable in specific contexts where the emergent ideas, which are able to gather heterogeneous elements together to present projects that address the two former descriptors: the Anthropocene and the every changing modulating ecologies. Can educators drawn on these aesthetic imaginaries to offer exploratory hope for what is a changing globe that is in constant crisis?

The series Educational Futures: Anthropocene, Ecology, and Aesthetic Imaginaries attempts to secure manuscripts that are aware of the precarity that reverberates throughout all life, and attempts to explore and experiment to develop an educational imagination which, at the very least, makes conscious what is a dire situation.

More information about this series at
http://www.palgrave.com/gp/series/15418

Marc Higgins

Unsettling Responsibility in Science Education

Indigenous Science, Deconstruction, and the Multicultural Science Education Debate

Marc Higgins
Department of Secondary Education
University of Alberta
Edmonton, AB, Canada

Palgrave Studies in Educational Futures
ISBN 978-3-030-61301-3   ISBN 978-3-030-61299-3   (eBook)
https://doi.org/10.1007/978-3-030-61299-3

Cover illustration: © Mihai Popa/Stockimo/Alamy Stock Photo

This Palgrave Macmillan imprint is published by the registered company Springer Nature Switzerland AG
The registered company address is: Gewerbestrasse 11, 6330 Cham, Switzerland

PREFACE

> There is no shortage of epistemologies and ontologies (e.g., feminist, indigenous, African, anticolonial, etc.) which have called attention to and critiqued Western humanism and nature-culture binary thinking; nonetheless, there *is* a shortage of collective organizing, reconfiguring, and reimagining of science curriculum in light of these calls. (Kayumova, McGuire, & Cardello, 2019, p. 212, emphasis in original)

Science education is increasingly being heralded as remedy to a multiplicity of contemporary issues (e.g., scientific literacy) and is actively worked upon in order to be more accessible, inclusive, and empowering. However, the ways in which science education takes up its responsibility in making itself accessible to all learner does not always come to value the ways-of-knowing and -being that diverge from the standard account of Western humanist thought (e.g., nature/culture as binary): be it feminist, Indigenous, postcolonial, queer, or other. As a result, in our contemporary moment in science education, the concepts of justice, equity, and ethics are simultaneously ubiquitous, necessary, yet un- or under-theorized: science education is a *pharmakon* (Derrida, 1976). This is to say, science education is a *panacea* that cannot account for or be accountable to the ways in which it is always already a *poison*, despite there being no shortage of productive critiques. In turn, the oft depoliticized and atheoretical work of inclusion and empowerment continues to house the potential for reproducing and reifying systems of power: be it at the level of policy, curriculum and pedagogy, or research practice.

In response, there is a recent but growing movement within science education that follows the call by Kayumova and colleagues (2019) to move "from empowerment to response-ability". Echoing Elizabeth Ellsworth's (1989) seminal piece and poignant question, *why doesn't this feel empowering?*, Kayumova and colleagues (2019) invite us to attend to the ways in which disempowering logics and practices linger and lurk, even when the intentions are otherwise (e.g., taking up one's responsibility): "'empowerment' discourses are reminiscent of a colonizer narrative that assumes for a dominant to be in a position to 'give' and bestow the 'power' to those who they have subjugated in the first place" (p. 224). However, we are not off the hook for empowerment: *response-ability* is about (re)opening the norms of responsiveness through which the attempt is made.

The shift from empowerment to response-ability is particularly significant for two reasons:

> Children and communities living under socio-spatial and environmental challenges are powerful individuals and collectives, they do not "need" to be, per se, "empowered" into dominant thinking or practices, but what we might need powerful frameworks, which take into consideration and legitimizes the diverse ways-of-knowing, -being, describing the naturalculutral world among diverse socio-cultural groups that are different from the dominant ways in which conventional science education understands and explains it. (Kayumova et al., 2019, p. 225)

The first is, as Kayumova and colleagues explain, is that notions of "empowerment" are more often than not about including marginalized peoples within the practices and processes of dominance without disrupting or displacing these very systems that placed them at the margins in the first place (and, thus, naturalizing and normalizing such frames). Not only does inclusion becomes but a differing and deferred articulation of the problematic norms, practices, and structures through which exclusion previously occurred: the implicit message is that other(ed) ways-of-knowing and -being are not welcome and that assimilation is the desired outcome. The second reason is that differentially situated peoples not already within dominant discourses have always already had rich practices of knowing nature as well as their own priorities, both of which might not align with those of science education: "empowerment" often comes at the expense of further delegitimizing peoples, places, practices, and priorities. In turn, the question is not one of "giving voice" or "power" to those who are marginalized by systems of dominance, but rather work with that which diverse communities of knowers bring to the table to collectively organize otherwise. As Kayumova and colleagues (2019) state, there is no shortage of ways-of-knowing and - being which might provide rich orientations towards differentially conceptualizing and enacting science education should we take them seriously: epistemic plurality as gift rather than a lack that requires repair.

However, the challenge that this presents, and it is one that is pressing and persistent, is the following: *if science education has a responsibility, is it able to respond?* As both science, and in turn science education, are premised upon the Othering of Nature and those "closer to nature" via nature-culture binary thinking as a means of (re)constituting itself as science, this is not a simple question. These binary logics shape not only the ways in which we are always already in co-constitutive relation but also the ability or inability to respond. Accordingly, and more importantly, the question we should be asking is: how might we go about (re)opening the space of responsiveness?

Recognizing that every act of (re)opening the space of responsiveness is partial, situated, and contingent, the work within this book is not a reply to approaches *"reminiscent* of a colonizer narrative" (Kayumova et al., 2019, p. 224, emphasis mine), but rather the ways in which colonial and neo-colonial practices *literally* continue to linger and lurk within science education. Thus, I turn to, and think with,[1] Sami scholar Rauna Kuokkanen (2007) to frame *response-ability*: "an ability to respond, to respond to the world beyond oneself, as well as a willingness to recognize its existence" (p. 39).[2] Within this book, the question of response-ability is taken up as central, particularly articulating it towards the relation between Western modern science and Indigenous[3] ways-of-living-with-Nature: *Is science education able to recognize the gift (and the logic of the gift) that Indigenous ways-of-knowing-in-being offer? If so, can science education responsibly receive it? Can science education learn (and learn to learn) from Indigenous ways-of-knowing-in-being in order to hospitably receive this gift?*

Responding to the above questions, like all questions of decolonizing education, are journeys rather than destinations. In prefacing the journey presented within this book, I provide a map of the networks of paths constituting this book so that the pathways of science education might be wandered anew. In particular, I speak to the general structure of each chapter, as well as the methodological moves I make therein (e.g., positional vignettes, differential articulation of deconstruction). Following this, I outline the overall arrangement of the book as a whole, as well as some recommended strategies for reading. This is supplemented by a short outline of upcoming chapters.

STRUCTURE OF THE PATHS: MOVING WITH THEORY-PRACTICE-ETHICS THROUGH THE CHAPTERS

Within a larger network of paths, each chapter represents a *differential*[4] hike, journey, or outing through *a* path of science education. Each journey is iterative, travelling through, against, and/or beyond a particular path, wherein the learning is enfolded and carried forward into the next trip. Representationally, this requires that each chapter as journey is presented as self-contained, yet interconnected. I regularly speak to the *path* being journeyed upon, as well as the *ways* in which I attempt (and encourage the reader) to "get lost" within said path to find a way back anew (see Cajete, 1994). Accordingly, each chapter generally includes its own positional piece, literature review, theoretical framework, methodology, analysis, and relational findings in a variety of forms such as open-ended questions, future orientations, and/or considerations and applications for situated practices—all of which come together giving the larger project of this book shape as a network of paths towards (re)opening the space of responsiveness towards Indigenous science (to-come) in science education. As the network of paths come to cover a plurality of locations, this differential approach is in line with aforementioned postcolonial and decolonizing approaches to education and educational research that call for attentiveness and responsiveness to the relations that come to constitute these theoretical, methodological, and substantive sites (e.g., Battiste, 2013a, 2013b; Carter, 2010; Smith, 1999/2012; Smith, Maxwell, Puke, & Temara, 2016). As this significantly inflects conventional book norms (e.g., theory and methodology in separate chapters), I quickly speak to what this means for theory, methodology, and positionality.

With regard to theory, I align myself with scholars such as Fikile Nxumalo (e.g., 2016, Nxumalo & Cedillo, 2018), Julie Kaomea (e.g., 2001, 2016), and Rauna Kuokkanen (e.g., 2007, 2010) who simultaneously resist and refuse the (neo-)colonial[5] desire for theoretical purity and transcendentalism—no singular theory could ever account for or be ethically accountable to the complex and contingent relations that are at play in everyday placed encounters on stolen land. As Kaomea (2001) states, engaging plurality is "consistent with the logic of post-colonialism and its declining emphasis on grand theories and narratives, my hybrid methodology, and thus my story, is intentionally eclectic; mingling, combining, and synthesizing theories and techniques from disparate disciplines and paradigms" (2001, p. 68). Complicating and challenging the "rage for unity" (Spivak, 1976, p. xvi) produced by the *clôture* (i.e., [en]closure) of Western modern(ist) metaphysics "demand[s] such theoretical innovation and flexibility" (Kaomea, 2001, p. 69). Further, As Kuokkanen (2007) states, the working of plurality is also a working towards a re-opening of responsiveness: such an approach brings "a confluence of voices" which has "resulted in a practice of reading—or even misreading—that has allowed me to be carried away by the ideas that various theories and approaches represent instead of seeking orthodox interpretations of them" (p. xx). Accordingly, I (give myself permission to) relationally draw from a diverse and often commensurate range of scholars who support the exploration of response-ability towards Indigenous science to-come. This theoretical plurality is purposeful as "many of the arguments against IK [Indigenous Knowledges] inclusion in the curriculum are more of a philosophical nature" (McKinley, 2007, p. 210). Diverse approaches are required to (re)open a seemingly ever-threatened and -shrinking landscape of plurality. I attempt to critically inhabit this process by engaging theoretically with philosophies and arguments that do not typically hold a central position within science education (e.g., postcolonialism, decolonizing, posthumanism).[6] Importantly, this includes (and must include, given the scope of the research) thinking with Indigenous scholarship:

> I also contend that the significance of [I]ndigenous philosophies extends beyond [I]ndigenous communities; these can be employed in various non-[I]ndigenous contexts as well. Indeed, I believe that indigenous philosophies offer a timely alternative paradigm for the entire world, which is increasingly characterized by tremendous human suffering and environmental destruction. (Kuokkanen, 2007, p. 25)

As Kuokkanen (2007) states, Indigenous ways-of-knowing-in-being offer alternative ways of addressing our very real and pressing contemporary problems; more so, to not think Indigenous scholarship is to (differently) engage in the longstanding and ongoing (fore)closure of Indigenous peoples, places, practices, and protocols that has historically shaped Indigenous-settler relations in most institutional places of learning. Additionally, to take theoretical plurality seriously is to also recognize the ways in which Indigenous ways-of-knowing-in-being extends beyond traditional (yet ongoing) knowledge-practices: it also comes to include "deep knowledge of colonizers and the practices and effects of colonization" (Smith et al., 2016, p. 136). Further, in working towards the possibility of a response-able reading, I take seriously Kuokkanen's (2007) call that diverse perspectives might also come to include "the work of (relatively) well known critics":

> Injecting the work of critics who are (relatively) well known – especially within dominant discourses and scholarly circles – into an inquiry dealing with the gift of [I]ndigenous epistemes and the academic responsibility of hospitality, is a way for me to bring closer the two sometimes separate worlds of [I]ndigenous and non-[I]ndigenous scholarship... It is also a strategy for summoning circles which might dismiss considerations on [I]ndigenous issues as either irrelevant to their own fields or, worse, unscholarly. (Kuokkanen, 2007, pp. xx–xxi)

Here, the inclusion of post-humanist theories is of particular significance in this contemporary moment which is often referred to as the ontological turn in education and educational research. In a nutshell, post-humanist theories call for a double(d) decentering of *both* the human *and* Western humanism. Post-humanisms call for a reconfiguration of priorities that brings to the forefront questions regarding the nature of Nature (e.g., time, space, matter) which, importantly, is not an *instead of culture*, but rather an *after* of previous educational turns (e.g., socio-cultural, linguistic, ethical). In other words, the ontological turn is a moment in which we are still called to continue responding to, and be answerable to, education's previous challenges and possibilities, but with ontology as entry point. As Milne and Scantlebury (2019) suggest, this is of particular significance within science education: "it is ironic that a field such as science education has ignored (or has been silent) about the material in learning science given science's focus on understanding matter and materiality" (p. 4). Science education's late rejoinder to this conversation

speaks to the ways that science education is a field that is at once politically and theoretically conservative (see Lemke, 2011), particularly when
many post-humanist theories are rooted in the materiality of science.[7]
Yet, there is nonetheless a growing body of work that is taking up the
posthuman question of how matter matters in science education, as well
as its epistemological, ontological, and ethical consequences (e.g., Bang
& Marin, 2015; Bazzul & Kayumova, 2016; Higgins, 2016; Kayumova,
McGuire, & Cardello, 2019; Milne & Scantlebury, 2019).

However, this taking seriously of the ontological turn is not without
its problems—problems which extend above and beyond challenging
claims of "newness" (i.e., who considered materiality first) despite it
being a recent "turn" in the academy. However, this positioning of
newness and focussing on other-than-humans, but not always other-
than-(Western-)humanism, risks subsuming or suturing over the ways
in which Indigenous ways-of-knowing-in-being have been thinking and
practicing co-constitutive relations to other-than- and more-than-human
worlds since time immemorial, and thus differentially (re)producing
(neo-)colonial relations (see Bang & Marin, 2015; Cole & O Riley,
2017; Jones & Hoskins, 2016; Patel, 2016; Todd, 2016; Tuck, 2010;
Watts, 2013). As decolonizing scholar Leigh Patel (2016) suggests, that
"understanding that knowledge is inseparable from materiality does not
necessarily move to a less colonial stance" (p. 56). It is perhaps for this
reason that Métis scholar Zoe Todd (2016) states explicitly, "ontology"
might come to be "just another word for colonialism" if these dynamics
go unmarked and unchallenged. Nonetheless, there is also a growing
body of scholarship that is engaging in the interface between Indigenous
and post-human theories and practices which recognizes and productively
labours these tensions (e.g., Kerr, 2019; Rosiek, Snyder, & Pratt, 2019;
Nxumalo & Cedillo, 2017; Zembylas, 2018).

On the subject of methodology (i.e., the interconnection of theory-
practice-ethics), most chapters use a deconstructive approach (see
Derrida, 1976).[8] As deconstruction is always already in relation to the
context in which it is being applied, articulations and enactments of
deconstruction differ from chapter to chapter, often building upon one
another.[9] In turn, it is more apt and useful to describe and situate these
concepts with/in the proximal relations by which they are co-constituted.
However, as a nod to Derridean approaches, (near-)homonyms are
frequently used throughout to productively defer and differ meaning-
making practices. For example, *response-ability* is a central deconstructive

concept which retains the significance of and need for responsibility while simultaneously working to trouble some of the received assumptions that might be inherited (e.g., that responsibility is something that one takes up, and that one is, in turn able to respond).

Further, taking seriously deconstructive writing practices, there are multiple footnotes located throughout the text. These are, in a way, a form of writing *under erasure*. To put something under erasure "is to write a word, cross it out, and then print both word and deletion. (Since the word is inaccurate, it is crossed out. Since it is necessary, it remains legible)" (Spivak, 1976, p. xiv). Because textuality is always already complicitous to its otherness,[10] footnotes present themselves as traces of a main body that either once was or could have been. They are the ~~main body of the text~~, at once useful and unnecessary, as well as multiplicitous in their form and purpose (e.g., productive tangents, elaborations, definitions, stories).

Similarly, while it is commonplace and of importance to position oneself within decolonizing work (e.g., through identity), Carter (2004) offers that:

> Postcolonialism's ability to delve into these processes, and into the deeper ravines of referents like modernity, identity, representation, and resistance underpinning many theorizations of culture and difference including those used, but underexplored, within science education, can open spaces to generate different discussions about what science education is, and could be. (p. 821)

I take up this invitation to consider the ways in which identity, representation, modernity, and (neo-)colonization are inevitably intertwined through differential articulations of metaphysics by using and troubling "position" (see also Spivak, 1988; 1993/2009). Thus, rather than offer *a* (i.e., singular) positional piece here in the introduction, you will find small and partial positional vignettes at the beginning of each chapter. This approach is by design and aligns with my understanding that positionality, too, is always already in relation (and is a differential articulation of *all my relations*). Such positionality cannot be disclosed through the self-sameness of identity, least not in the essentialized, stable, and singular understanding of the term (i.e., position rather than positionality). Rather, positionality is always contingent, partial, plural, and emerges in relation to the "scene of address" of the account that is being represented (see Butler, 2005). Furthermore, drawing on the work of the late Vine

Deloria, Wildcat (2005) reminds that the very concepts we hold are exceeded by lived experience and shared experiences can become a site of shared meanings across difference (see also Bohm, 1996). In turn, the vignettes are also stories of why the very concepts under exploration are always already deconstructing, require *deconstructive* engagement, and/or are productive sites of relational *reconstruction*.

NAVIGATING THE NETWORK OF PATHS: OR, ON HOW TO READ THIS BOOK

The book invites a process of wandering the pathways of science education, journeying the paths *anew* so that new path*ways* (e.g., Indigenous science to-come) might come to constitute the network of paths (see also Cajete, 1994; Kuokkanen, 2007). On pathways, Tewa science educator Gregory Cajete (1994) states that,

> In travelling a pathway, we make stops, encounter and overcome obstacles, recognize and interpret signs, seek answers, and follow the track of those entities that have something to teach us. We create ourselves anew. *Path* denotes a structure; *way* implies a process. (p. 54, emphasis in original)

What might appear as a sedimented and stratified *path* is inseperable from its enactment, its journeying, its *way*. Such is important in differentially enacting the double(d) closure (i.e., enclosure [noun], enclosing [verb]) of modernity's metaphysics that seek to make the ends (i.e., the path) congruent with the means (i.e., the way). However, when *the* path is revealed as but *a* path, a multiplicity of possibilities opens up.

As the human, other-than-human, and more-than-human worlds that constitute a(ny) path "move in never ending cycles of creation and dissolution" (p. 43), it is necessary to engage in what Cajete (1994) refers to as "creative acts of perception. A free play of thought and an opening of the field". (p. 19). This process requires realizing that there are tacit, ever shifting infrastructures (i.e., like paths) that frame what *is*, *is not*, as well as *becoming* possible within a field. Rather than engage in the destruction of a path (if such were even a desirable possibility)[11] or complicit journeying on a path as is (upholding the status quo in which Indigenous science is to-come), the task as one of coming-to-know the "nature" of many of the paths that lay before me and others in science education, their possibilities and problematics, and to look for different *ways* (and potentially (re)open new *paths*) to journey through this field. In other words, the goal is to wander these pathways *deconstructively*, "turn[ing] the *outside in* and the *inside out*" (Donald, 2012, p. 544, emphasis in original) while recognizing that each pathway presents its own challenges and possibilities. As the book is not knowledge made but knowledge in the making (see Ellsworth, 2005), reading the book invites engaged, active, and relational reading.

Wandering pathways anew can be described as a process of "getting lost" within science education. Such strategic straying off the beaten path or taking 'the path' in unintended ways is positioned "not as 'losing *one's* way' but as losing *the* way—as losing any sense that just one 'way' could ever be prefixed and privileged by the definite article" (Gough, 2006, p. 640, emphasis in original; see also Lather, 2007; Patel, 2016). Furthermore, as Gough (2006) states on "getting lost" in science education:

> ...to "wander" away from the semiotic spaces of science education text-books and scientific media reports, and to experiment with making passages to hitherto disconnected systems of signification, is neither "haphazard" nor "careless" but a deliberate effort to unsettle boundary distinctions and presuppositions. (p. 640)

In reading this book, I invite the reader to tactically wander within, against, and beyond the sedimented spaces of knowing in/of science education, because "Western science and science education also tend to be written from a sedentary point of view" (Gough, 2006, p. 640). Importantly, it is important to recognize that coming-to-know otherwise in science education requires being vulnerable and attuned to what can be known through the process of *not knowing*. In addition, this tactical wandering, or science education as *on-the-move* (literally and towards Indigenous science to-come), is not about wholly stepping away from the pathways of science education, but rather about wandering them anew. Importantly, such wandering is and must not be without aim: "to liberate thinking about science education from the sedentary points of view and judgmental positions that function as the nodal points of *Western* academic science education discourse. What happens when we encourage random, proliferating and decentred connections?" (Gough, 2006, p. 628, emphasis mind). Particularly, as a reader, you are encouraged to tactically wander within, against, and beyond the *sedimented* and *stratified* locations which continue to (re)centre the West as the standard against which other ways-of-knowing-in-being are measured.[12]

Further, while not wanting to be (wholly) prescriptive about how the book is to be read, one of the ways in which a tactical wandering might be engaged with is through *slow science* (Stengers, 2018). As science education is often caught up in efforts to rapidly increase pedagogical and curricular effectiveness, the complex question of Indigenous science's place and how we might go about meaningfully incorporating it cannot be and is not productively met with a certain degree of expediency. Questions of how the sciences interact with society are often perceived as non-scientific and are rarely taken up in a "fast science"[13] curriculum whose image of creativity, as critical science studies scholar Isabelle Stengers (2018) offers, is one of the sleepwalker:

> The... current image of scientific creativity, that of the sleepwalker walking on a narrow ridge without fear or vertigo because he is blind to the danger. Asking creative scientists to be actively concerned about the consequences of their work would the equivalent of waking the sleepwalkers, making them aware that the world is a long way from obeying their categories. Struck by doubt, they would fall from the ridge into the morass of turbid opinions. They would, that is, be lost for science. (p. 114)

Stuck in a form of "fast science" whose "imperative not to slow down" (Stengers, 2018, p. 115) dictates which pathways it travels upon, as well as how, scientists and science educators are often channelling attention away from the very landscape it traverses, only attending, albeit momentarily to the obstacles that present themselves. In contrast,

> Slow science is not about scientists taking full account of the messy complications of the world, it is about them facing up to the challenge of developing a collective awareness of the particularly selective character of their own thought-style. (Stengers, 2018, p. 100)

In turn, the *slow* in slow science invites not only a different style of perception and awareness to that which we inherit and enact as science educators, but also a different pace and temporality. As "colonial projects have shaped technology, knowledge, and connection", (Patel, 2016, p. 1) the act of slowing down is one that is not inconsequential to the central exploration in this book:

> Perhaps one of the most explicit decoloni[zing] moves we can make, at this moment, is to sit long enough to see clearly what we need to reach beyond. This stillness should not be confused with doing nothing. Without pause, it's difficult to ascertain what structures, what inequitable structures, are enlivened by narratives, even and perhaps especially the progressive narratives. Our pauses, actions, and revisiting should be answerable to a constant desire for material transformation, repatriation, and rectification. (Patel, 2016, p. 88)

Without slowing down or pausing, it might be impossible to slow down or pause that which we are working against, or notice the ways in which (neo-)colonial logics are always already at play when we, like the sleep-walker, are continuing on our pathway dictated by the mandate of fast science (see also Higgins, Wallace, & Bazzul, 2019).

Along these lines, it has been productively suggested by some readers that this work is enjoyed most when read a chapter or two at a time. Like when one approaches a network of pathways covering a large expanse of territory, only the most experienced hikers should attempt a multi-path hike. Furthermore, as each pathway (as process and product) builds upon the previous one, it may even be worth revisiting a path a second time before continuing on.[14] To assist in this process, the remainder of the book is divided into four two-chapter arcs (discussed below) that are strongly connected to one another. Furthermore, because this book is not meant to be read in a single sitting (and importantly not the *only* text read on the subject), you will find "trail markings" throughout that provide orientations through signalling preceding and forthcoming work. These may serve as a welcome reminder of where one has journeyed and where one is going.

The book is broken into four two-chapter arcs as a suggested reading pace.[15] The first section, *Unsettling the Metaphysics of Responsibility*, introduces and troubles the relationship between Indigenous and Western modern metaphysics and its consequences for science education, with a focus on science education's responsibility and (in)ability to respond to Indigenous ways-of-living-with-Nature. The second section, *Critical Possibilities and Possible Critiques through Deconstructive Play in/of the Multicultural Science Education Debate*, works within and against the tradition of WMS-based science education that (re)produces science educator and science education as a field by critically inhabiting the multicultural science education debate and critically questioning the potency of critique when enacted through the *clôture* of metaphysics. In the third section *Tinkering with Ontology with/in the Multicultural Science Education Debate*, the insight that ontology matters in the (re)construction of norms and practices through which Indigenous science is yet-to-come is pursued. The fourth and final section, *Towards a Curriculum for Indigenous Science To-Come*, uses insights from previous sections and works imperfectly towards leveraging these differential (re)openings towards *a* re(con)figuring of science education knowledge-practices.

A short synopsis of each chapter that appears in the remainder of the book follows.

Overview of Arc 1: Unsettling the Metaphysics of Responsibility

Within Chapter 1, the simultaneously co-constitutive and othering relation between Western modern science and Indigenous ways-of-knowing-in-being is introduced, with specific attention the metaphysics of this relationship as well as how it is taken up with science education, both as curriculum and research. Unsettling science education is presented as a double(d) approach to address the ways in which settler colonial logics linger and lurk within sedimented and stratified knowledge-practices (Bang & Marin, 2015; Bang, Warren, Roseberry, & Medin, 2013). As a more nascent approach to the question of Indigenous science within science education, this is expanded upon by drawing from decolonizing (e.g., Aikenhead & Elliot, 2010; Chinn, 2007), and postcolonial (e.g., Carter, 2004; McKinley, 2007) approaches to science education. Specifically, drawing across the two, deconstruction is highlighted as a (meta-)methodological approach that is employed across the two approaches, to bear witness to the ways in which settler coloniality often manifests as absent presence and to (re)open the space of response *within* science education towards Indigenous ways-of-knowing-in-being.

Within Chapter 2, attention is brought to taken-for-granted notions of responsibility as the ways in which science education's responsibility towards Indigenous ways-of-living-with-nature (IWLN) and traditional ecological knowledge (TEK) is often and inadvertently over-coded by the (neo-)colonial logics that it sets out to refuse and resist. To animate this, I revisit a significant personal pedagogical encounter in which this distinction between responsibility and the ability to respond made itself felt and known. Thinking with the work of Sami scholar Rauna Kuokkanen (2007, 2008, 2010), this narrative provides a platform to explore and deconstruct practices of epistemic ignoranceignorance its (co-)constitutive relation to knowledge, as well as "the homework of response-ability" required to (re)open the norms of responsiveness towards the possibility of heeding the call of Indigenous science from within the structure of science education.

Overview of Arc 2: Critical Possibilities and Possible Critiques through Deconstructive play in/of the Multicultural Science Education Debate

Within Chapter 3, the multicultural science education debate around *how, when,* and *if* TEK and IWLN are included within science education (re)presented as (a) play. However, as the debate comes to be shaped by what Moulton (1983) refers to as the "adversary method", an invitation to (mis)read the dialectic negationnegation of conversation as dialogue is posited in an attempt to open a different space in which cross-culturalists and universalists might come to shared meanings. A significant question emerging from shared meanings within this chapter, and revisited in the chapters following, is *How might considering scientific knowledge as knowledge-practice assist us in collectively working towards the shared goal of working against scientism in science education?* This is significant to the overall scope of this book as scientism often comes to be a central mechanism (alongside exclusivity and Eurocentrism; see McKinley & Stewart, 2012) through which Indigenous science is excluded, differing, and deferred.

Within Chapter 4, the taken-for-grantedness, yet centrality, of metaphoric visuality in science is utilized as an opening to re(con)figure critique in science education. Specifically, three different optical metaphors are offered to inform gazing critically otherwise within science education: the mirror, the prism, and the diffraction grating. Drawing from the work of Latour (a critique of the mirror; 1993, 2004), Foucault (the prism; 1977, 1979, 1997), and Barad (the diffraction grating; 2000, 2007, 2010) each metaphoric visual technology and their relation to critique are respectively explored to inform new lines of critical questioning. Particularly, these critical metaphors are employed to ask theoretical, methodological, practical, and ethical questions of the multicultural science education debate presented within the previous chapter.

Overview of Arc 3: Tinkering with Ontology with/in the Multicultural Science Education Debate

Within Chapter 5, Cobern and Loving's (2008) call to consider scientific knowledge-practice as ontologically situated (articulated in Chapter 3) is used and troubled. As they refer to this consideration as *uncommon*, I explore the ways in which ontology is utilized to make common a science education status quo. In turn, I tinker with/in the common/uncommon dichotomy to work within and against this problematic, yet productive, statement. Particularly, I draw from "uncommon" (to science education) work at the ontological turn to explore how "ontological alignment" might be re(con)figured when ontology is no longer a singular affair that pre-exists scientific meaning-making.

Chapter 6 extends the work of Chapter 5 by positing Cartesianism as *an* ontology, as well as what it might mean to account for and be accountable to the ways in which this classical Western metaphysics comes to (co-)constitute knowledge-practices of WMS and science education. An interview with Dr. Frédérique Apffel-Marglin grounds this exploration (see also Apffel-Marglin, 2011). It is put into conversation with Cobern and Loving's (2008) statement that cites "common sense" in the assertion that WMS offers an epistemologically privileged position from which to know Nature as it best aligns with ontology. Apffel-Marglin's interview gives what she refers to as a "thumbnail account" of the birth of modernity (as an entanglement of economic, political, social, and religious forces and flows). This account elucidates how "common sense" has become common. Within the interview, she unpacks some of the taken-for-granted notions that are naturalized in making such a statement (e.g., values entangled with/in statements that are further entangled within a Cartesian ontology). These insights are then diffractively read through the practices of the multicultural science education debate to produce new insights, differently.

Overview of Arc 4: Towards a Curriculum for Indigenous Science To-Come

In Chapter 7, I draw insights from the previous chapters to work towards responding to Indigenous science to-come, with attention to the dynamics of the cultural interface (Nakata, 2007a, 2007b). Extending the concept of response-ability, I (imperfectly) re(con)figure science education curriculum and pedagogy by opening it to its constitutive otherness. Arguing that scientific literacy presents itself as a rife location to promote response-ability towards Indigenous science to-come, I engage in deconstructive (mis)readings to posit an(other) ontology. Notably, I think with and consider points of convergence and divergence between Karen Barad's (2000, 2007, 2010) quantum philosophy-physics and Gregory Cajete's (1994, 1999, 2000) Indigenous science in order to re(con)figure scientific literacy as ecologies of relationships and agential literacy. Lastly, the consequences of teaching and learning with and across these forms of literacies are explored: troubling how Cartesian modes shape how we know Nature (i.e., space, time, matter)(e.g., other-than-human agency, non-linear space-time).

In Chapter 8, I conclude with a summary and synthesis of the book to highlight the ways in which Indigenous science is *(yet-)to-come*. Notably, each chapter is revisited to (re)articulate the significance of their contributions with the triple(d) understanding of *to-come* explored within the book in mind: (a) Indigenous science, in the context of science education, has not yet (wholly) arrived; (b) where and how science education might be (re)opened towards hospitably receiving Indigenous science; and (c) the types of deconstructive practices that support this work. The chapter, and in turn the book, ends with an affirmative message that the potentiality of Indigenous ways-of-living-with-Nature in science education remains, even if not fully actualized; thus, an invitation.

Towards Response-Ability in Science Education

When applied to notions of justice in education, response-ability invokes the right to epistemological and ontological heterogeneities, all the different ways by which we, communities of different ethnicities, languages, and locations understand ourselves, our relations, and the world around us. (Kayumova et al., 2019, p. 227)

Once again, returning to the initial statement within this preface, there is no shortage of ways-of-knowing and -being which might provide rich orientations towards differentially conceptualizing and enacting science education should we take them seriously. However, in recognizing the gift that is "epistemological and ontological heterogeneities" (Kayumova

et al., 2019, p. 227) requires that we make the shift from (only) empowerment to response-ability.[16] As Kayumova and colleagues (2019) remind,
the shift towards response-ability would, by necessity, address the systems
which render the ability respond lesser or null. In turn, (re)opening
the norm of responsiveness entails: refusing and resisting the systems of
dominance that have Othered human, other-than-human, and more-than-
human beings through the scientific construction of knowledge. Further,
it entails disrupting and displacing the logics which come to constitute it
(e.g., teleology of progress, binary logics). However, to shift to response-
ability is not a move away from empowerment, it is a differential recom-
mitment: it is to labour towards empowerment while (post-)critically
attends to our own complicities and the ways in which empowerment
is conceptualized and enacted *always already* becomes disempowering
to those it attempts the response (see Ellsworth, 1989). This is to say
that the shift towards response-ability requires of "collective organizing,
reconfiguring, and reimagining" (Kayumova et al., 2019, p. 212) that
begins from criticality accounting for and being accountable to science
education's (co-constitutive) margins rather than an a ubiquitous yet
somehow depoliticized and under-theorized notion of empowerment that
is formulated within its centre.

In order to account for and be accountable to difference, figurations
of response-ability must (and will) always take shape differently as they
are always already in relation to the particular relations that co-constitute
them. This will require that the orientations, theories, practices, and ethics
with which we approach the problems and possibilities anew in critical
and creative ways: "to make science education more relevant, response-
able, and reflexive in its efforts to be equitable, inclusive, and robust
requires an integration of 'new' and alternative frameworks" (Kayumova
et al., 2019, p. 212). Herein, as the project is working towards recognizing and hospitably receiving Indigenous science, and particularly the
ways in which it continues to be excluded as policy mandates its inclusion,
such a figuration of response-ability must be a double(d) movement that
ruptures (neo-)colonial systems and also leverages these openings towards
making space for honouring Indigenous peoples, places, practices, and
priorities. In turn, this messy work of creating alternative frameworks
for approaching old questions anew braids together Kuokkanen's (2007)
conception of response-ability, as well as decolonizing, posthuman, and
postcolonialtheories to critically inhabit spaces of science education which
(fore)close the space of responsiveness. We must still attempt a response

within this relation of responsibility, even if response-ability may never be (fully) achieved. Working with purpose, but without guarantee, is par for the course when it comes to such enactments of response-ability; yet, there is (home)work to be done.

Edmonton, Canada   Marc Higgins

NOTES

1. The work herein is in part indebted to the work of Alecia Jackson and Lisa Mazzei (2012, 2017) and the rich framework offered by *thinking with theory*. This framework can be succinctly described as "reading-the-data-while-thinking-the-theory" or "as a moment of plugging in, of entering the assemblage, of making new connectives" (2012, p. 4). It is a research stance that actively works against decontextualization of knowledge-making by encouraging the explicit, transparent and intentional engagement with the theories with which we think: a stance that is deeply relevant given the ways in which the universalization of Western modern science forecloses the ability to respond to Indigenous ways-of-knowing-in-being (see Higgins & Kim, 2019). Significantly, rather than engage in knowledge-practices that (re)produce what is already known and knowable through the frameworks we already hold (which then hold us), Jackson and Mazzei (2012, 2017) invite us to move beyond thinking with scholars who reflect back to us what and how we already know. Instead, they suggest turning to those who offer a "productive provocation: theorists who open up thought rather than foreclose it" (2012, p. 5).

2. Whereas response-ability is making an appearance most frequently in educational research via the work of feminist technoscience scholars Karen Barad (e.g., 2010) and Donna Haraway (e.g., 2016), I want to note the significance of response-ability productively taking differing shapes should we approach it via other theoretical approaches such as postcolonialism (e.g., Spivak, 1994) or Indigenous ways-of-knowing-in-being (e.g., Kuokkanen, 2007).

   Further, there is a growing body of scholarship in science education that is making the move from *empowerment to response-ability* (e.g., deFreitas, Lupinacci, & Pais, 2017; Higgins & Tolbert, 2018; Wallace, Higgins, & Bazzul, 2018).

3. Importantly, Indigenous is capitalized throughout the book not only to recognize the sovereignty of Indigenous nations or to signal the hard-fought political designation (e.g., the United Nations Declaration on the Rights of Indigenous Peoples, but also to differentiate from and trouble

the ways in which the lower-case *indigenous* gets leveraged as a move to innocence, most often by settlers:

> The concept of [I]ndigeneity (at least as it is used by [I]ndigenous peoples themselves) is grounded in and inseparable from the legacy and aftershocks of colonialism.... To claim, that – as is often done – that "we are all indigenous" is either to be blind to this contemporary reality or to refuse to recognize the ways in which colonial history continues to affect not only [I]ndigenous peoples but also relations between states and [I]ndigenous peoples. The statement "we are all indigenous" reflects a reluctance to take responsibility or engage with those issues which, far from belonging to the past, continue to keep [I]ndigenous peoples in a subordinate position and to deny them rights that peoples who belong to nation-states can and do take for granted. (Kuokkanen, 2007, pp. 10–11)

This is not to deny the *veracity* of the statement that we are all indigenous (lower-case) to somewhere, but rather to call attention to what colonial productions it makes possible. Particularly as it often works against a more response-able science education towards Indigenous ways-of-knowing-in-being (as the statement is rarely accompanied with the caveats that would still come to recognize the importance of Indigeneity): namely, as explored in the first chapter, it becomes a way of "masking [colonial] power with innocence" (McKinley, 2001; see also Tuck & Yang, 2012).

4. Differential here signals difference without *a priori* mutual exclusivity (i.e., *A* is not *B*). Rather, differential signals the ways in which difference is co-constitutive: the part enfolds the whole and is articulated as degrees of difference from with/in that whole (see Deleuze, 1994).

5. While coloniality and neo-coloniality are often framed as historically distinct (i.e., past and present), these respective projects of territorial and economic imperialism are bound by a relation of co-constitutive exclusion (Spivak 1999). I signal this assumption through the use of the term *(neo-)colonial* throughout.

6. Since, as Carter (2004) suggests, there is a "paucity of this type [(i.e., post-colonial and decolonizing)] of inquiry in science education" (p. 833), many related concepts may be foreign to some science education readers. Nevertheless, terms and concepts are defined as they appear within the text, and the ways in which they are leveraged towards decolonizing goals.

7. For example, Darwinian evolutionary processes in Elizabeth Grosz's (2011) *Becoming Undone*; quantum mechanics in Barad's (2007) *Meeting the Universe Halfway*.

8. This bears significance as decolonizing and deconstruction share an intimate and (co-)constitutive relationship (Battiste, 2013b): there is

critically subtle work to engage in (re)opening taken-for-granted (i.e., settled) concepts which uphold settler- and/or neo-colonial logics (e.g., representation, see Kayumova, Zhang, & Scantlebury, 2018).

9. For example, in Chapters 5 and 6, the deconstructive metaphor is one of tinkering as it entails misusing science's tools, or using those foreign to science education, to (re)open its structure (see Higgins, Wallace, & Bazzul, 2018). This builds upon deconstruction as the possibility, as well as intentional use, of substitution (or misreading) as deconstructive methodology as used in Chapters 3 and 4 respectively.

10. This is to say that for any and every instance of textuality, the word as signifier has both an irreducible relation to that which *is* signified (if such a signaling could ever be said to be stable), but also to that which *is not* (be it oppositionally other(ed), or otherwise). Deconstruction, which is always already happening, differs and defers the possibility of *a* meaning (in the singular sense, but not the possibility of meaning).

11. Sami scholar Rauna Kuokkanen (2007) reminds of the tenuous path of critically inhabiting spaces—that one must always be critical and complicit—as to engage in *destruction* rather than *deconstruction* results (re)produces similar results albeit differently. There requires, in her words, a "subtlety and responsibility in the process of transforming the [institution of learning], ... proceeding in any other way would eventually backfire and merely too tight[ly] reinforce existing structures and discourse [through]... 'irresponsibilizing destruction'" (p. xx; see also Spivak, 1994).

12. Further, there is a certain degree of pedagogical side-stepping or "wandering" of the pathways of science education required to respond (and to be able to respond) to Indigenous science to-come. In part, this is because of the ways in which WMS-based traditional "school science" frames Nature *a priori*: by taking for granted that Nature is separate, separable, static, and passive (i.e., Cartesianism), Indigenous ways-of-living-with-Nature are presented as lesser means of reflecting *that* reality, by design. Wandering offers a pedagogical form of "getting lost" works within/against many of the nodes of science education, towards the possibility of other-than-Cartesian ontologies: such furtive and deconstructive movements displace and disrupt rather than destroy the structure of science education.

13. Stengers (2018) does not wholly dismiss fast science, but rather highlights its limitations in terms of what it can and cannot engage with:

> What characterizes fast science is not isolation, but rather working in a very rarified environment, and environment divided into allies who matter and those who, whatever their concerns and protests, have to recognize that they are the ultimate recipients of [scientific

  creativity], and therefore should not disturb the progress of science.
  (p. 116)

  Which is to say that the ways in which it frames itself in binary terms
  in relation to that which it deems unscientific makes it rather difficult to
  engage with those very same things.

14. This advice may also be considered for individual sentences which utilize
  the theory-practice of deconstruction to deploy ontological indeterminacy in locations that are problematically singular(izing). Much more
  than a stylistic choice, such deconstruction leverages open a singularizing
  (neo-)colonial center, which itself is always on the move (see Spivak,
  1993/2009, 1999). This means that there are often multiple *ways* to
  wander the *path* that is *a* sentence and that it is productive to linger
  and explore its possible possibilities.

15. Note that differing chapters might be of particular importance for diverse
  audiences. For example: readers firmly familiar with the multicultural
  science education debate and its largely adversarial workings may want
  to forego Chapter 3; readers already familiar with the post-structural turn
  and the ontological turn in educational research could potentially skip
  Chapter 4; readers most interested in pedagogical design and delivery may
  be most interested in Chapter 7. However, also note that each chapter
  is more than its substantive content as it also presents and builds upon
  methodology (i.e., theory-practice-ethics).

16. Furthermore, as Battiste (2013b) suggests, decolonizing education invites
  the nurturance of a diversity of teaching and learning gifts: these are assets
  to be celebrated and lifted rather than liabilities to be addressed through
  a deficit lens.

References

Aikenhead, G. S., & Elliot, D. (2010). An emerging decolonizing science education in Canada. *Canadian Journal of Science, Mathematics and Technology
Education, 10*(4), 321–338.

Apffel-Marglin, F. (2011). *Subversive spiritualities: How rituals enact the world.*
New York, NY: Oxford University Press.

Bang, M., & Marin, A. (2015). Nature–culture constructs in science learning:
Human/non-human agency and intentionality. *Journal of Research in Science
Teaching, 52*(4), 530–544.

Bang, M., Warren, B., Rosebery, A. S., & Medin, D. (2012). Desettling
expectations in science education. *Human Development, 55*(5–6), 302–318.

Barad, K. (2000). Reconceiving scientific literacy as agential literacy. In R. Reed
& S. Traweek (Eds.), *Doing Science+Culture* (pp. 221–258). New York, NY:
Routledge.

Barad, K. (2007). *Meeting the universe halfway: Quantum physics and the entanglement of matter and meaning*. Durham, NC: Duke University Press.

Barad, K. (2010). Quantum entanglements and hauntological relations of inheritance: Dis/continuities, spacetime enfoldings, and justice-to-come. *Derrida Today, 3*(2), 240–268.

Battiste, M. (2013a). Deconstruction and reconstruction: Roles, responsibilities and implications of a decolonizing framework. Retrieved from November 22, 2019, http://www.indigenouseducation.educ.ubc.ca/transformation/indigenous-perspectives/.

Battiste, M. (2013b). *Decolonizing education: Nourishing the learning spirit*. Saskatoon, SK: Purich Publishing.

Bazzul, J., & Kayumova, S. (2016). Toward a social ontology for science education: Introducing Deleuze and Guattari's assemblages. *Educational Philosophy and Theory, 48*(3), 284–299. https://doi.org/10.1080/00131857.2015.1013016.

Butler, J. (2005). *On giving an account of oneself*. New York, NY: Fordham University Press.

Cajete, G. (1994). *Look to the mountain: An ecology of indigenous education*. Durango, CO: Kivaki Press.

Cajete, G. A. (1999). *Igniting the sparkle: An Indigenous science education model*. Durango, CO: Kivaki Press.

Cajete, G. (2000). *Native science: Natural laws of interdependence*. Santa Fe, NM: Clear Light Books.

Carter, L. (2004). Thinking differently about cultural diversity: Using postcolonial theory to (re)read science education. *Science Education, 88*(6), 819–836.

Carter, L. (2010). The armchair at the borders: The 'messy' ideas of borders, border zones and epistemological diversity in multicultural science education. *Science Education, 94*, 1–20.

Chinn, P. (2007). Decolonizing methodologies and Indigenous knowledge: The role of culture, place and personal experience in professional development. *Journal of Research in Science Teaching, 44*(9), 1247–1268.

Cobern, W. W., & Loving, C. C. (2008). An essay for educators: Epistemological realism really is common sense. *Science & Education, 17*, 425–447.

Cole, P., & O'Riley, P. (2017). Performing survivance: (Re)storying STEM education from an Indigenous perspective. *Critical Education, 8*(15), 24–40.

de Freitas, E., Lupinacci, J., & Pais, A. (2017). Science and technology studies × educational studies: Critical and creative perspectives on the future of STEM education. *Educational Studies, 56*(6), 551–559.

Deleuze, G. (1994). *Difference and repetition*. New York, NY: Columbia University Press.

Derrida, J. (1976). *Of grammatology* (G. C. Spivak, Trans.). Baltimore, MD: John Hopkins University Press.

Donald, D. (2012). Indigenous Métissage: A decolonizing research sensibility. *International Journal of Qualitative Studies in Education, 25*(5), 533–555.

Ellsworth, E. (1989). Why doesn't this feel empowering? Working through the repressive myths of critical pedagogy. *Harvard Educational Review, 59*(3), 297–325.

Ellsworth, E. (2005). Places of learning: Media, architecture, pedagogy. New York, NY: Routledge.

Foucault, M. (1977). *Discipline and punish: The birth of the prison.* London, UK: Allen Lane.

Foucault, M. (1979). Truth and power. In M. Morris & P. Patton (Eds.), *Power, truth, strategy* (pp. 29–48). Sydney, NSW: Feral Publications.

Foucault, M. (1997). *The politics of truth.* New York, NY: Semiotext(e).

Gough, N. (2006). Shaking the tree, making a rhizome: Towards a nomadic geophilosophy of science education. *Educational Philosophy and Theory, 38*(5), 625–645.

Grosz, E. (2011). *Becoming undone: Darwinian reflections on life, politics, and art.* Durham, NC: Duke University Press.

Haraway, D. J. (2016). *Staying with the trouble: Making kin in the Chthulucene.* Durham, NC: Duke University Press.

Higgins, M. (2016). Decolonizing school science: Pedagogically enacting agential literacy and ecologies of relationships. In C. Taylor & C. Hughes (Eds.), *Posthuman research practices* (pp. 267–289). Basingstoke, UK: Palgrave Macmillan.

Higgins, M., & Tolbert, S. (2018). A syllabus for response-able inheritance in science education. *Parallax, 24*(3), 273–294.

Higgins, M., Wallace, M., & Bazzul, J. (2018). Disrupting and displacing methodologies in STEM education: From engineering to tinkering with theory for eco-social justice. *Canadian Journal of Science, Mathematics and Technology Education, 18*(3), 187–192.

Higgins, M., Wallace, M., & Bazzul, J. (2019). Staying with the trouble in science education. In C. Taylor & A. Bayley (Eds.), *Posthumanism and higher education: Reimagining pedagogy, practice and research* (pp. 155–164). Basingstoke, UK: Palgrave Macmillan.

Jackson, A. Y., & Mazzei, L. A. (2012). *Thinking with theory in qualitative research: Viewing data across multiple perspectives.* New York, NY: Routledge.

Jackson, A. Y., & Mazzei, L. A. (2017). Thinking with theory: A new analytic for qualitative inquiry. In N. K. Denzin & Y. S. Lincoln (Eds.), *The sage handbook of qualitative research* (5th ed., pp. 717–727). Thousand Oaks, CA: Sage.

Jones, A., & Hoskins, T. K. (2016). A mark on paper: The matter of Indigenous-settler history. In C. Taylor & C. Hughes (Eds.) *Posthuman research practices in education* (pp. 75–92). Basingstoke, UK: Palgrave Macmillan.

Kaomea, J. (2001). Dilemmas of an indigenous academic: A Native Hawaiian story. *Contemporary Issues in Early Childhood, 2*(1), 67–82.

Kaomea, J. (2016). Qualitative analysis as ho 'oku 'iku 'i or bricolage: Teaching emancipatory indigenous research in postcolonial Hawai'i. *Qualitative Inquiry, 22*(2), 99–106.

Kayumova, S., McGuire, C. J., & Cardello, S. (2019). From empowerment to response-ability: Rethinking socio-spatial, environmental justice, and nature-culture binaries in the context of STEM education. *Cultural Studies of Science Education, 14*(1), 205–229.

Kayumova, S., Zhang, W., & Scantlebury, K. (2018). Displacing and disrupting colonizing knowledge-making-practices in science education: Power of graphic-textual illustrations. *Canadian Journal of Science, Mathematics and Technology Education, 18*(3), 257–270.

Kerr, J. (2019). Indigenous education in higher education in Canada: Settler re-education through new materialist theory. In C. Taylor & A. Bayley (Eds.), *Posthumanism and higher education* (pp. 313–328). Basingstoke, UK: Palgrave Macmillan.

Kuokkanen, R. J. (2007). *Reshaping the university: Responsibility, Indigenous epistemes, and the logic of the gift.* Vancouver, BC: UBC Press.

Kuokkanen, R. J. (2008). What is hospitality in the academy? Epistemic ignorance and the (im)possible gift. *Review of Education, Pedagogy, and Cultural Studies, 30*(1), 60–82.

Kuokkanen, R. (2010). The responsibility of the academy: A call for doing homework. *Journal of Curriculum Theorizing, 26*(3), 61–74.

Lather, P. (2007). *Getting lost: Feminist efforts toward a double(d) science.* New York, NY: State University of New York Press.

Latour, B. (1993). *We have never been modern.* Cambridge, MA: Harvard University Press.

Latour, B. (2004). Why has critique run out of steam? From matters of fact to matters of concern. *Critical inquiry, 30*(2), 225–248.

Lemke, J. (2011). The secret identity of science education: Masculine and politically conservative? *Cultural Studies of Science Education, 6*(2), 287–292.

McKinley, E. (2001). Cultural diversity: Masking power with innocence. *Science Education, 85*(1), 74–76.

McKinley, E. (2007). Postcolonialism, Indigenous students, and science education. In S. K. Abell & N. G. Lederman (Eds.), *Handbook of research on science education* (pp. 199–226). Mahwah, NJ: Lawrence Erlbaum.

McKinley, E. & Stewart, G. (2012). Out of place: Indigenous knowledge in the science curriculum. In B. Fraser, K. Tobin, & C. J. McRobbie (Eds.), *Second international handbook of science education* (pp. 541–554). Dordrecht, NL: Springer Netherlands.

Milne, C., & Scantlebury, K. (Eds.). (2019). *Material practice and materiality: Too long ignored in science education* (Vol. 18). Cham, Switzerland: Springer.

Moulton, J. (1983). A paradigm of philosophy: The Adversary Method. In S. Harding, & M. B. Hintikka (Eds.), *Discovering reality: Feminist perspectives on epistemology, metaphysics, methodology, and philosophy of science* (pp. 149–164). Dordrecht, NL: Reidel.

Nakata, M. (2007a). *Disciplining the savages: Savaging the disciplines. Exploring inscriptions of Islanders in Western systems of thought.* Canberra, Australia: Aboriginal Studies Press.

Nakata, M. (2007b). The cultural interface. *The Australian Journal of Indigenous Education, 36,* 7–14.

Nxumalo, F. (2016). Towards 'refiguring presences' as an anti-colonial orientation to research in early childhood studies. *International Journal of Qualitative Studies in Education, 29*(5), 640–654.

Nxumalo, F., & Cedillo, S. (2017). Decolonizing place in early childhood studies: Thinking with Indigenous onto-epistemologies and Black feminist geographies. *Global Studies of Childhood, 7*(2), 99–112.

Patel, L. (2016). *Decolonizing educational research: From ownership to answerability.* New York, NY: Routledge.

Rosiek, J. L., Snyder, J., & Pratt, S. L. (2020). The new materialisms and Indigenous theories of non-human agency: Making the case for respectful anti-colonial engagement. *Qualitative Inquiry, 26*(3–4), 331–346.

Smith, L. T. (1999/2012). *Decolonizing methodologies: Research and Indigenous People (2nd Ed.).* London, UK: Zed Books.

Smith, L. T., Maxwell, T. K., Puke, H., & Temara, P. (2016). Indigenous knowledge, methodology and mayhem: What is the role of methodology in producing Indigenous insights? A discussion from mātauranga Māori. *Knowledge Cultures, 4*(3), 131–156.

Spivak, G. C. (1976). Translator's preface. In J. Derrida (Ed.), *Of grammatology* (G. C. Spivak, Trans.) (pp. ix-lxxxvii). Baltimore, MD: Johns Hopkins University Press.

Spivak, G. C. (1988). Can the subaltern speak? In C. Nelson & L. Grossberg (Eds.), *Marxism and the interpretation of culture* (pp. 271–313). Urbana, IL: University of Illinois Press.

Spivak, G. C. (1993/2009). *Outside in the teaching machine.* New York, NY: Routledge.

Spivak, G. C. (1994). Responsibility. *boundary 2, 21*(3), 19–64.

Spivak, G. C. (1999). *A critique of postcolonial reason.* Cambridge, MA: Harvard University Press.

Stengers, I. (2018). *Another science is possible: A manifesto for slow science.* Cambridge, MA: Polity.

Todd, Z. (2016). An indigenous feminist's take on the ontological turn: "Ontology" is just another word for colonialism. *Journal of Historical Sociology, 29*(1), 4–22.

Tuck, E. (2010). Breaking up with Deleuze: Desire and valuing the irreconcilable. *International journal of qualitative studies in education, 23*(5), 635–650.

Tuck, E., & Yang, W. (2012). Decolonization is not a metaphor. *Decolonization: Indigeneity, Education & Society, 1*(1), 1–40.

Wallace, M., Higgins, M., & Bazul, J. (2018). Thinking with nature: Following the contours of minor concepts for ethico-political response-ability in science education. *Canadian Journal of Science, Mathematics and Technology Education, 18*(3), 199–209.

Watts, V. (2013). Indigenous place-thought and agency amongst humans and non-humans (First woman and sky woman go on a European world tour!). *Decolonization: Indigeneity, Education & Society, 2*(1), 20–34.

Wildcat, D. R. (2005). Indigenizing the future: Why we must think spatially in the twenty-first century. *American Studies, 46*, 417–440.

Zembylas, M. (2018). The entanglement of decolonial and posthuman perspectives: Tensions and implications for curriculum and pedagogy in higher education. *Parallax, 24*(3), 254–267.

ACKNOWLEDGMENTS

I would like to begin by expressing my gratitude and appreciation for the scholarship and friendship of the many peers, colleagues, and mentors whose presence differentially lingers in (and in-between) the lines of this work.

Learning alongside Marie-France Berard, Elsa Lens Kothe, Brooke Madden, Heather McGregor, and Sam Stiegler—our own (self-styled) *Thinking with Theory* reading, writing, and research group—has been and continues to be a space of hospitality in which the difficult work of faltering, failing, and getting lost in educational research is met with serious playfulness.

Connecting, co-conspiring, and collectivizing with my science education colleagues continues to nurture both my research and teaching. Scheming with Jesse Bazzul, Sara Tolbert, and Maria Wallace ever remains a life-giving form of critical collective organizing as we work together to disrupt and displace the normative logics of science education. I am ever indebted to the work of and ongoing support from Glen Aikenhead whose thoughtful lines of questioning make space for the complexities, contradictions, compromises, and complicities of working within and against disciplinary spaces towards decolonizing goals in science education. Further, I continue to be deeply appreciative the ongoing conversations of approaching science education from the "posts" had with my extended (post-)critical science education kin: Rouhollah Agasaleh,

Lars Bang, Shaknoza Kayumova, Eun-Ji Amy Kim, Blue Mahy, and Kate Scantlebury.

Offering a warm welcome, I am ever thankful for my new University of Alberta colleagues. Thank you to those with whom I am still working towards making mathematics and science education a more hospitable curricular space for local Indigenous learners and knowledges: Florence Glanfield, Steven Khan, Mijung Kim, and Kerry Rose. To the Aboriginal Teacher Education Program team and extended family who push me to be a better relative in Treaty 6 territory and lift me in the task of being pedagogically relevant, responsive, and responsible: Trudy Cardinal, Brooke Madden, Sharla Peltier, Evelyn Steinhauer, Noella Steinhauer, Patsy Steinhauer, Angela Wolfe, Misty Underwood, as well as ATEP students. To my curriculum colleagues whose invitation to (re)think how we think about the thing we call curriculum is always generous and generative: Dwayne Donald, Kent den Heyer, Cathryn van Kessel, David Lewkowich, Jason Wallin. A special thanks goes to jan jagodinski, who encouraged me to write the book that is between your hands (or on your screen).

Mentoring me during my time at the University of British Columbia, I continue to be profoundly grateful to my academic mentors. I give special thanks for Dónal O Donoghue's generous criticality and attentiveness to the ways in which concepts differentially live alongside one another as they travel across and are translated into diverse contexts; for the lively and fierce community of decolonizing scholarship that Michael Marker fosters; and for Cynthia Nicol's attentiveness to the complicated space of cross-cultural ethics that entails accounting for and being accountable to both the institutions and communities with whom we work. Also, I would like to thank Dr. Frédérique Apffel-Marglin for her scholarship on, and lived practice of, keeping the spirit of the gift in circulation, and the time, energy, care, and curiosity shown during a summer institute at the Sachamama Center for Biocultural Regeneration in Peru.

Uplifting my own lines of personal and professional engagement, I wish to recognize the contributions of multiple and occasionally overlapping communities of scholars. Here, I would like to acknowledge (in alphabetical order): Cash Ahenakew, Vanessa Andreotti, Jo-Ann Archibald, Andrea Belczewski, Michael Cappello, Heather Commodore, Karlee Fellner, Alecia Jackson, Lisa Korteweg, Jeannie Kerr, Lisa Mazzei, Susan Nordstrom, Fikile Nxumalo, Julia Ostertag, Amy Parent, Marcelina Piotrowski, Nikki Rotas, Alexa Scully, Carol Taylor, Jasmine Ulmer, and Alannah Young. Particularly, I am indebted to Masayuki Iwase's indomitable spirit, and thankful for our frequent discussions on visuality, the dark hope of futurity, and post-coloniality.

For their patience, flexibility and support amidst the final stages of (re)writing (which, unfortunately, coincided with a global pandemic), I am indebted to the editorial team at Palgrave MacMillan, particularly Linda Braus and Milana Vernikova.

Fostering an ethic of hard work and perseverance and for continuously checking-in, a thank you goes out to my family.

Lastly, I would like to thank (again) Brooke Madden whose multiple forms of partnership differentially manifest as ongoing and unwavering support and love, peppered with spirited reminders that the journey within, against, and beyond institutional pathways may not be simple but that it is a journey worthy of pursuit.

CONTENTS

Unsettling the Metaphysics of Responsibility

CHAPTER 1

Unsettling Metaphysics in Science Education

This 'beginning,' like all beginnings, is always already threaded through
with anticipation of where it is going but will never simply reach and of
a past that has yet to come. It is not merely that the future and the past
are not 'there' and never sit still, but that the present is not simply here-
now. Multiply heterogeneous iterations all: past, present, and future, not
in a relation of linear unfolding, but threaded through one another in
a nonlinear unfolding of spacetimemattering, a topology that defies any
suggestion of a smooth continuous manifold. (Barad, 2010, p. 244)

As critical science studies scholar Karen Barad (2010) reminds us, "the
present is not simply *here-now*" (p. 244, emphasis mine). Rather, it is
also a dis/continuous enfolding of heterogenous *there-thens*.[1] This is
to say that the central process of this book—accounting for and being
accountable to the uneven and unequal relation between Indigenous
metaphysics and classical Western metaphysics by way of quantum metaphysics, and the ethical, epistemological, and ontological implications for
science education—has and will have already begun *elsewhere* and *elsewhen*
(both past and futures to-come).

Such is significant in science education where often, or perhaps *too
often*, the work is framed in way that asks *where do we begin to engage
the question of including Indigenous knowledges or perspectives in science
education?* While there is usually an intent of being in relation in a good
way, the language and the practices they signal are fraught. There is

© The Author(s) 2021
M. Higgins, *Unsettling Responsibility in Science Education*,
Palgrave Studies in Educational Futures,
https://doi.org/10.1007/978-3-030-61299-3_1

often a forgetting (be it one that is individual or systemic) that science education is *always already* in relation to Indigenous ways-of-knowing-in-being.[2] As Ngāti Kahungunu ki Wairarapa and Ngāi Tahu scholar and science educator Liz McKinley (2001) states, this has much to do with the ways in which dominance operates in science education: its response to difference is often a form of "masking power with innocence" (see also Kuokkanen, 2007). Primarily, McKinley (2001) suggests that a lack of knowledge (or a positional stance of "not knowing") often serves to (re)produce the norms of power; in turn, "we need to challenge the mask of innocence and ask ourselves how relations of domination and subordination regulate encounters in classrooms" (p. 76). This is not only significant because science education has a responsibility towards Indigenous ways-of-knowing-in-being that goes beyond a responsibility *for* the Other, but it is also a responsibility as well as an indebtedness *to* the Other.[3] If responsibility is an "an incarnate relation that precedes the intentionality of consciousness" (Barad, 2010, p. 265) the question is no longer whether or not we are responsible but rather if we are able to respond. Importantly, "we no longer have any excuse, only alibis for turning away from this responsibility" (Derrida, 1994/2006, p. 14).

Then again, the question of already having begun is substantially relevant. Both quantum and Indigenous metaphysics "caus[e] trouble for the very notion of 'from the beginning'" (Barad, 2010, p. 245; see also Cajete, 2000; Kawagley, 2006). They are *un-settling*. Nonetheless, because the ability to respond is always situated, this inquiry must begin some-where and some-time (as well as given over to someone; see Butler, 2005), even though these spacetime coordinates (what are conventionally referred to as history and geography, as separate and separable; see Barad, Barad 2010) cannot be torn asunder from their co-constitutive otherness (see also Cajete, 2000; Kawagley, 2006; Kirby, 2011; Kuokkanen, 2007).

This book's intended purpose is to take seriously this simultaneously co-constitutive and othering relation between Western modern science and Indigenous ways-of-knowing-in-being; dialogically engaging with the field of science education[4] to so that it might practice "an iterative (re)opening up to, an enabling of responsiveness" (Barad, 2010, p. 265) to the possibility of Indigenous metaphysics (to-come). Yet, before "beginning" if one could make such a proposition, the purpose of this introductory chapter is to put forward the relationships between metaphysics, decolonizing, and post-colonial approaches to science education, and deconstruction that are central to the work to-come within this book. The framing of these relationships is done by attending to the double(d) meaning of *unsettling science education* in two parts to provide orientations for the reading journey. The first provides an overview of some of the pathways explored with/in decolonizing science education: decolonizing and post-colonial science education in response to the metaphysics of modernity. The second unpacks deconstruction in relationship to decolonizing methodologies as well as decolonizing science education as a (meta-)methodological approach to (re)open the metaphysics of modernity. In this chapter, as in those that follow, I initiate the work with *a* positional vignette that give glimpses of the curiosities and questions that motivate and guide my explorations and give shape to the inquiries to come.

My Relation to Indigenous Metaphysics, the Metaphysics of Modernity, and Science Education

Because we need to "begin" some-where and some-time, let's "begin" in Calgary, Alberta on 1 June 2016. Blackfoot Elder and scholar Leroy Little Bear is giving a *Big Thinking* address at the annual Canadian *Congress of the Humanities and Social Sciences*. The title of his talk: "Blackfoot Metaphysics is Waiting in the Wings". Playing off the title as he walks onto the main stage, Little Bear jokes "[waiting in the wings] just like I was a few minutes ago". This calls to mind(-body-heart-spirit) the importance of making connections through humour (see also Little Bear, 2000). Leroy Little Bear invites those in the crowded auditorium to (re)consider what Blackfoot and other differentially articulated Indigenous metaphysics (i.e., the co-constitutive space of axiology, epistemology, ontology, ethics, and cosmology) continue to offer: ways-of-knowing-in-being premised on ethics, relationality, process, flux, and renewal centering a sense of place. Already, the talk is rife with significance: metaphysics is "classically" understood as a philosophy of *being* or *what is* (i.e., ontology) or the nature of Nature which comes to describe multiple concepts and enactments such as space, time, matter, causality, agency, identity, among others. Little Bear implicitly calls it out as but *one* metaphysics amidst many. Furthermore, he articulates a need to consider Indigenous metaphysics in response to the metaphysics that are already and often in operation within educational spaces: "What are the metaphysics of our schools?

Where are those metaphysics taking us?" Articulating a metaphysics of modernity as taken-for-granted, unquestioned, and unstated, Little Bear subtly shifts the statement that metaphysics *is* (i.e., singular and a priori) to one wherein metaphysics *are*[5] and are *in relation* (i.e., plural and entangled in the world's ongoing becoming). Furthermore, he motions that the metaphysics of modernity continues to provide some comforts (e.g., material goods), "but at what price? Is our metaphysics making us better? Happier?" he asks. In referring to this metaphysics as *ours*, Little Bear signals that metaphysics is not strictly a binary either/or affair, as in Indigenous *or* Western. Rather, metaphysics is always both/and. In other words, metaphysics are neither separate nor separable, but rather always co-constituted and co-constitutive. In turn, responsibility for the metaphysics of modernity is also shared, albeit, and importantly, not in the same way. Making his concluding remarks, Little Bear suggested that it is time to move Blackfoot and other Indigenous metaphysics from the wings to the main stage where their contributions might significantly come to bear in generative ways.

During this address, Little Bear only hinted at the ways in which the knowledge-practices of Indigenous metaphysics come to be positioned in the wings where they have been waiting for a long time. He signalled dialogues that began over 20 years ago[6] between Indigenous Elders and scholars (e.g., Leroy Little Bear, Chickasaw and Cheyenne scholar Sakej Youngblood Henderson) and Western scientists and linguists (e.g., quantum physicists David Peat and David Bohm). During these dialogues, they met "to discuss the underlying principles of the cosmos, not from an adversarial point of view, but from one of mutual respect and deep listening" (Parry, 2008, p. 37). The purpose of the *Science Dialogues* was not to work towards knowledge, but rather understanding (see Little Bear, 1994; Parry, 2008; Peat, 2002, 2007). This certainly was not the first time, nor would it be the last that such an initiative towards cross-cultural understanding would take place. Yet, despite such efforts, Indigenous metaphysics *still* waits in the wings of science education.

When I began graduate studies in decolonizing science education in 2008, the very first book I read was Peat's (2002) *Blackfoot Physics*.[7] While I was on the lookout for Indigenous science,[8] David Peat (2002) reminds that Indigenous metaphysics and Indigenous science are differential articulations of one another that cannot be separated (see also Cajete, 1994, 2000; Kawagley, 2006):

As a science, [Indigenous science] is a disciplined approach to understanding and knowing, or rather, to the processes of coming to understanding and knowing. It has supporting metaphysics about the nature of
reality, deals in systems of relationship, is concerned with the energies and
processes of the universe, and provides a coherent scheme and basis for
action. On the other hand, *it is not possible to separate Indigenous science
from other areas of life such as ethics, spirituality, metaphysics, social order,
ceremony, and a variety of other aspects of daily existence.* This it can never
be a "branch" or a "department" of knowledge, but rather *remains inseparable from the cohesive whole,* from a way of being and of coming-to-know.
(p. 241, emphasis mine)

Blackfoot Physics was a powerful early read for me as it discusses the
"points of resonance" between Indigenous metaphysics and quantum
physics that emerged from the *Science Dialogues.* Holding the complexity
of difference without subsuming it into sameness, *Blackfoot Physics*
explored these two systems that diversely articulate flux and relationality
concurrently, providing me with a hopeful potentiality for science education to be constituted and enacted otherwise. It could be stated that this
was *an* example of what Yupik science education scholar Oscar Kawagley
and his settler ally Ray Barnhardt (2005) meant when they suggested
that "there is a growing appreciation of the complementarity that exists
between what were previously considered two disparate and irreconcilable systems of thought" (p. 12).[9] Notably, from this exploration
of putting Indigenous and Western science into proximal relation and
productively their *similar differences* and *differing similarities* (see Bohm,
1994), there was and continues to be rich potentiality. Notably, the possibility for respectful, relevant, and responsive science education whose
pedagogical potency is enriched from cross-cultural diversity resonated
with my own professional experiences of working as a fourth generation white Euro-settler of Irish ancestry working as an informal science
educator in First Nations, Métis, and Inuit communities across Canada.
I had witnessed and worked towards contributing to science education
as plurality; science education shaped by cross-cultural understandings
through similar, yet different, practices; and science education that draws
strength from cultural and placed locations, instead of treating difference
as an individual problem located with the one who diverges from the
norm (e.g., Higgins, 2014; see Chapter 2). In my experience, and at the
time, there was no lack of respectful, reciprocal, and relational models
for cross-cultural science education drawing from Indigenous traditions

(e.g., Aikenhead, 1997, 2006b; Cajete, 1999; Barnhardt & Kawagley, 2005, 2008). Yet, despite science being a fruitful location for cross-cultural "points of resonance" (Peat, 2002), to my surprise Indigenous metaphysics was *still* waiting in the wings when it came to most science education spaces. *Blackfoot Physics* illuminated the ways in which Indigenous science was yet-to-come and productive locations to bring about that potentiality.

Now, over ten years later, at the time of writing this book, Indigenous metaphysics' status can still be referred to as "waiting in the wings", of not yet having (fully) arrived, in relation to school science. This, in part, has a great deal to with the ways in which Indigenous knowledges are generally approached from within spaces of science education. Above and beyond the aforementioned treating of Indigenous ways-of-knowing-in-being as if it were *a new* relation rather than approaching the relation *anew*, there is often an attempt to come-to-know the defining character-istics of Indigenous metaphysics (i.e., knowing about) without accounting for (or being accountable to) the ways in which Western modern meta-physics has always and continues to dialectically subsumes, sublates, or sutures over Indigenous metaphysics. This becomes problematic when there a positional stance of "not knowing" that operates in tandem (see McKinley, 2001).

To dig deeper into the various ways in which science education is seemingly unable to *know* Indigenous science, I turn to Sami scholar Rauna Kuokkanen (2007), whose scholarship centres Western modern educational and institutional responsibility towards Indigenous ways-of-knowing-in-being. Specifically, with respect to the question of knowledge and learning, Kuokkanen (2007) asks us,

> Why is the academy, the supposed generator of knowledge, so disinclined to engage with [I]ndigenous ontologies and philosophies? Considering the endless number of studies on, and the voluminous information about, practically every imaginable topic dealing with the world's various [I]ndigenous peoples, how can this general ignorance of [I]ndigenous epistemes continue to be so pervasive? (p. 56)

How is it that science education continues to engage from a place of "know-nothing-ism" (Kuokkanen, 2007) or "sanctioned ignorance" (Spivak, 1999)[10] despite there already beings so many, and an ever-growing quantity of, resources describing, defining, and documenting

Indigenous metaphysics? Is this more than simply being "commonplace 'moves of innocence'" through which educators "claim the right to not know" (Kuokkanen, 2010, p. 65)? While these are questions that are approached from multiple angles over the course of this book, it is important to immediately dispel the easy, yet problematic answer: "the problem is not that that there are no [or not enough] books on Indigenous peoples by Indigenous peoples" (Kuokkanen, 2007, p. 102), even on the topic of Indigenous science.

As a white settler scholar, I see it as both my role and responsibility to take up the task of accounting for and being accountable to the ongoing dialectic negation of Indigenous metaphysics, and the ways in which this process of negation cannot be separated from colonial referents which linger and lurk in science education. Simultaneously, I recognize that an *a-political* documenting of Indigenous metaphysics cannot be disassociated from forms of differentially (re)producing contemporary and historically ongoing colonial relations: here, shoring up my privilege as a settler academic through describing and defining Indigenous Others in a contemporary moment in which the will-to-know is driving a great deal of educational change. However, the work of addressing the ways in which science education is produced by and reproducing colonial logics is not a straightforward or simple task. As Mik'maq educational scholar Marie Battiste (2008) states, "what is becoming clear to educators is that any attempt to decolonize education and actively resist colonial paradigms is a complex and daunting task" (p. 508): not only because they are strictly complicated tasks, but also because colonial structures and systems also come to shape our ability to respond to them and even imagine something beyond them (see also Ahenakew, 2017; Battiste, 2005). For example, even within Peat's (2002) articulation of the rich potentiality of points of resonance, there are subtle and not-so-subtle traces of the ways in which the relationship Indigenous and Western science is *always already* one that cannot not be colonizing (even when working towards something else)[11]:

> It is at this point that a tantalizing paradox presents itself. On one hand it seems that the very activity and busy-ness of our analytic, linear Western minds would obstruct us from entering into Indigenous coming-to-knowing, yet, on the other, scientists who have been struggling at the cutting edges of their fields have come up with concepts that resonate with those of Indigenous science. (p. 6)

While, as Peat (2002) suggests there are rich points of resonance between the Indigenous and Western metaphysics (e.g., around questions of quantum physics), there are still patterns of difference that matter. To uniquely focus on commensurability here becomes an act of mirroring sameness elsewhere, dialectically subsuming into or sublating through sameness patterns of difference, as well as making it difficult to account for and be accountable to the enactment of difference. Relationships of commensurability become all the more complicated between Indigenous and Western metaphysics as equivocation often becomes a move to mask colonial relations of power between the two (Carter, 2004; McKinley, 2001; Tuck & Yang, 2012). Further, as Kuokkanen (2007) states, "the demand and desire that [I]ndigenous cultures and epistemes be translated into forms recognizable by the dominant colonial society is at least as old as colonialism itself" (p. 75). Colonial logics always come to *differ* and *defer* what can be said, to whom, and for what purposes. Here, the move the make space for Indigenous metaphysics simultaneously (re)centres scientist as subject as a form of listening-as-imperial-benevolence in which the Indigenous other can only be heard in the spaces in which it relates to and potentially benefits the Western scientist (Kuokkanen, 2007). It is for this reason that Ngāti Kahungunu ki Wairarapa and Ngāi Tahu scholar Liz McKinley and Maori scholar Georgina Stewart (2012) state, "the aspiration of defining and understanding IK [Indigenous knowledges] (in order to place it in the science curriculum) can be likened to chasing the pot of gold at the end of the rainbow, which remains permanently out of reach" (p. 551).

However, rather than reverse the logics and wholly adopt a position of *incommensurability* when considering the space between Indigenous and Western metaphysics, Tewa scholar Gregory Cajete (2000) suggests that we account for and are accountable to the diverse ways in which they come to be constituted and enacted:

Native science is a product of a different creative journey and a different history than that of Western science. Native science is not quantum physics or environmental science, but it has come to similar understandings about the workings of the natural laws through experimentation and participation with the natural world. The groundwork for a fruitful dialogue and exchange of knowledge is being created. (p. 14)

While both may have reached similar destinations when it comes to notions of flux, uncertainty, and co-constitutiveness, Cajete (2000) reminds us that this destination is shaped by "different creative journey[s]" (p. 14) both in terms of where they have come from as well as where they are going. These currently intersecting pathways are differentially produced, and produce ways-of-knowing-in-being differently. Furthermore, such patterns can become rich locations for dialogue: points of resonance become an entry point towards what one can learn from the others' points of divergence. Thus, rather than sameness and its constitutive other of difference as oppositional negation, we can consider the ways in which Indigenous and Western metaphysics are in ever-shifting, situated, and differential relations.

Despite its "to-come" status, the potential of Indigenous metaphysics (and its differential articulation as Indigenous science) within and beyond cross-cultural science remains a central motivation that continues to drive the inquiry herein that is grounded in an ethical commitment towards a future[12] in which Indigenous metaphysics is no longer "waiting in the wings".

Barad (2010) suggests, "we inherit the future, not just the past" (p. 257). In considering the future, we not only inherit *the* future (*avenir*) that is the most possible possibility; one that prolongs and replicates the present condition, albeit differently, by restituting a foreclosed past that has yet to happen (again). We also inherit futures that are yet-to-come (*à-venir*); those unexpected arrivals that produce a (re)opening of difference whereby possibilities and consequences are not (fully) knowable (see also Smith, 2005). However, "there is no inheritance without a call to responsibility" (Derrida, 1994/2006, p. 114); a responsibility that is not only an epistemological and ontological accounting *for* but also an ethical accountability *towards* that which is yet-to-come (see also Kuokkannen, 2007; Spivak, 1994). The Otherness that has yet-to-arrive (e.g., a future to-come where Indigenous metaphysics is no longer "waiting in the wings" of science education),[13] whose arrival cannot be anticipated, is entangled in what Barad (2010) refers to as co-constitutive "relations of obligation":

Othering, the constitution of an 'Other,' entails an indebtedness to the 'Other,' who is irreducibly and materially bound to, threaded through, the 'self'—a diffraction/dispersion of identity. 'Otherness' is an entangled relation of difference (différance). Ethicality entails noncoincidence with oneself. (p. 265)

As the future (*avenir*) and the to-come (*à-venir*) are not one and the same (without being mutually exclusive),[14] the present of science education is irreducibly bound to and ethically indebted to Indigenous science to-come. Articulated otherwise, it is a responsibility that is not *for* the other but *to* the Other whose labelling as "non-scientific" allows that which is "scientific" to persist and thrive (see Wallace, 2018). Indigenous metaphysics is already entangled within the production of science education. Rather, science educators are tasked with (re)opening the ability to respond to towards Indigenous ways-of-knowing-in-being. This potentiality that has yet-to-(fully-)come, whose arrival is unforeseeable, invites "*the continual reopening and unsettling of what might yet be, of what was, and what comes to be*" (Barad, 2010, p. 264, emphasis in original).

In order to engage in the work of *un-settling* (i.e., what I have come to understand and will present as deconstructing and decolonizing) what might yet be, what was, and what will come to be science education, the central question that guides the inquiry presented within this book is: *How is Indigenous science to-come with/in the context of science education?* The central question is understood and explored in this book through three guiding inflections that are inseparably entangled. First, *to-come* signals that Indigenous metaphysics, in the context of science education, has not yet (wholly) arrived. This precipitates the questions: *How is it that Indigenous science is still to-come? How do the structures of science education—the assumptions, terms, categories, practices, and beliefs—contribute to exclusion of Indigenous science, as well as inclusion that disciplines, differs from, and defers Indigenous science to-come?* Secondly, *to-come* signals ethical indebtedness; this invites the question *How might the structure, culture, and discipline of science education be (re)opened and re(con)figured to receive Indigenous science to-come, on its own terms, and in ethical relation?* Thirdly, *to-come* entails a responsibility for and towards that which is *to-come*. Yet, modes, practices, and enactments of responsibility cannot be prescribed when that which is *to-come* is never (fully) knowable and distorted by the current frames of science education. Stated otherwise, responsibility requires the occasion and ability to respond. Accordingly, I wonder: *What types of practices*[15] *might allow for and nurture the possibility of Indigenous science to-come?* This final query recognizes that potentiality need not require actualization for it to be worthy of consideration.

To situate the engagement with the central questions and its different inflections, the remainder of this introductory chapter provides a double(d) orientation to guide the reader through the book: *unsettling science education*. To animate the two ways in which unsettling can be read, first, decolonizing and post-colonial approaches to science education will be introduced as these are the disciplinary spaces within which I situate this work. Secondly, in addressing the ways in which meaning and matter become stratified or sedimented in science education, deconstruction as a (meta-)methodology will be explored. However, as the metaphysics of modernity is often entangled with/in enactments of WMS and science education, (re)producing Indigenous science as to-come, this relation is also explored as (co-constitutive) site to unsettle.

On Unsettling Science Education: Decolonizing and Deconstructing

In our view science education is a key site in which nature–culture relations are defined, enacted, brought-to-life, expanded, narrowed and legislated. The manifestations of *nature-culture relations*, from the very constructions of subject matter, to focal content, to the configurations of practice, engaged in science learning environments are often deeply unreflective of the most pressing scientific questions—rather they focus on "settled" phenomena as well as "settled" perspectives and relations to phenomena. (Bang & Marin, 2015, p. 531, emphasis in original)

As Ojibwe scholar Megan Bang and Choctaw scholar Ananda Marin's state in their seminal 2015 piece on *unsettling science education* (see also Bang, Warren, Rosebery, & Medin, 2012), they state that the science education is a key site for addressing and the ways in which relations between Nature and Culture are produced through a double(d) *settling*. Accordingly, unsettling science education is a practice that is first about addressing the ways in which *settler colonialism* (i.e., the structure rather than event through which settlers continue a project of Indigenous erasure; see Tuck & Yang, 2012) manifests within science education by refusing and resisting the logics and structures through which the colonial project remains ongoing. As Bang and Marin define, "the fundamental tenant of settler-colonial societies is the acquisition of land as property, followed by the establishment of settler lifeways as the normative benchmark from which to measure development" (p. 532). In addition, *unsettling* also has a second, subtler, but no less significant meaning:

attending to the ways in which science education draws from stratified and sedimented knowledges, phenomena, histories, pedagogies, and other practices which complicate questions of making space for and responding to Indigenous ways-of-living-with-Nature.

Yet, working towards unsettling science education is a task that is already trouble(d) from its very beginning (see Higgins, 2014; McKinley & Stewart, 2012; Sammel, 2009).[16] Notions and enactments of decolonizing are often already overcoded by the colonial logics that we attempt to work within, against, and beyond. Working towards a science education that is able to respond to Indigenous ways of Indigenous ways-of-knowing-in-being is often marked by a de/colonizing relation. Like Subreenduth (2006),

> I use the slash (/) in 'decolonizing' as visual demonstration of the incompleteness of the process of de/colonizing. What I suggest in this use is the impossibility (at least at this historical juncture) to speak of a totalizing decolonizing discourse or imagination. (p. 618)

As with others (e.g., Madden & McGregor, 2013; Rhee & Subreenduth, 2006), I employ the term de/colonizing as a post-colonial inflection to decolonizing theories and practices to consider the ways in which decolonizing and colonizing discourses cannot be wholly framed in opposition, particularly within spaces like educational institutions (see Higgins & Madden, 2017, 2019) or in fields such as science education (see Higgins & Kim, 2019; Higgins, Mahy, Agasaleh, & Enderle, 2019; Higgins & Tolbert, 2018). As a result, "the process and acts of de/colonizing are not *only always* an antithesis of colonialism ... but rather a convoluted, complex and paradoxical one" (Subreenduth, 2006, p. 619). De/colonizing invites an ongoing and hyper-vigilant examination of the ways in which (neo-)colonial logics seep (even) into decolonizing efforts, through engaging with the following questions:

> How should we rethink and rearticulate the conceptualization and practice of education and research when we situate them within contemporary imperialism and the history of pervasive colonialism? What theoretical and practical possibilities can be retrieved by analyzing de/colonizing educational practices through the history of imperialism? What alternative ideas of educational theorizing can be articulated in relation to local/global responsibility, equality and justice? How does local/global mobility and changing demographics impact on such knowledge production and consumption? (Rhee & Subreenduth, 2006, p. 546)

For example, and importantly (particularly as it is the primary form of response), if after Barad (2007), "responsibility must be thought of in terms of what matters and what is excluded from mattering" (p. 220), there is a need to attend to the ways in which the logics of exclusion continue to operate through the ways in which Indigenous ways-of-knowing-in-being are included. As Bang and Marin (2015) remind, the curricular inclusion of Indigenous perspectives is differentially problematic if we cannot also attend to the taken-for-granted and naturalized epistemological, ontological, and axiological commitments and enactments of what we are including perspectives into. As Bang and Marin (2015) state, if science education continues to "focus on 'settled' phenomena as well as 'settled' perspectives and relations to phenomena" (p. 531), which rely on and reinforce settler privilege while simultaneously dismissing, diminishing, and denying Indigenous ways-of-living-with-nature, presence, and futurities, it will remain but a tokenistic inclusion which serves to distract from the more unsettling demands of this work (namely, Land) and is often primarily an effort to reconceptualize and recenter the subject of dominance (see also Ahenakew, 2017).

Importantly, science education centring of "settled" phenomena through "settled" perspectives matter and materialize beyond the classroom as well (see also Kayumova, McGuire, & Cardello, 2019). Science education's (pre)dominant conceptualization of Nature(-Culture) makes palatable and possible the ongoing dispossession and devastation of Indigenous Land:

> The maintenance of settler normativity requires the structuration of time-space relations in ways that make the inseparable dynamics of acquisition of land, [I]ndigenous erasure, and the domination of black people appear as an inevitable, unconnected, and natural course of development rather than socio-politically engineered to support and foster white entitlement and privilege. (Bang & Marin, 2015, p. 532)

Also, through this double(d) settling, Indigenous peoples have been and continue to be the *objects* of science rather than its *subjects* (see also TallBear, 2013). In turn, echoing Bang and Marin, I argue that science education is a "key site" to unsettle the relationship between Nature and Culture.

As unsettling's double(d) meaning and practice is underexplored in science education, it will be framed through an exploration of ways in which (neo-)colonial logics are responded to in science education (i.e., decolonizing and post-colonial approaches) as well as the ways in which meanings and practices are productively destabilized and decentered (i.e., deconstruction).

First Orientation: An Introduction to Decolonizing and Postcolonial Science Education and Their Relationships to Metaphysics

As [Derrida] develops the notion of the joyful [i.e., play-full] yet laborious strategy of rewriting the old language—a language, incidentally, we must know well—Derrida mentions the "cloture" of metaphysics. We must know that we are within the "cloture" of metaphysics, even as we attempt to undo it. It would be an historicist mistake to represent this "closure" of metaphysics as simply the temporal finishing-point of metaphysics. It is also the metaphysical desire to make the end coincide with the means, create an enclosure, make the definition coincide with the defined, the "father" with the "son"; within the logic of identity to balance the equation, close the circle. Our language reflects this desire. And so it is from within this language that we must attempt an "opening." (Spivak, 1976, p. xx)

I begin this section by asking: what does metaphysics (i.e., the co-constitutive space of epistemology, ontology, ethics, among others) have to do with science education and Indigenous science to-come? Recall that Indigenous science is always already an articulation of Indigenous metaphysics and an inseparable part of the whole (see Cajete, 1994, 2000; Little Bear, 2016; Peat, 2002). However, what of Indigenous metaphysics within the Western modern science (WMS) which largely comes to inform most of science education's school-based curricula? As Derrida (1976) offers, we are always already within the *clôture* (i.e., enclosure) of metaphysics: there is no outside of metaphysics (see also Spivak, 1976). Also, as stated earlier, there is no outside of the metaphysics of modernity (see also Apffel-Marglin, 2011; Carter, 2010; Kuokkanen, 2007; Little Bear, Bear 2016; Spivak, 1976, 1993/2009, 1994, 1999).

WMS and science education too must also be within, and have, a metaphysics. Sciences, in all shapes and forms, are premised upon the ways in which Nature's enactments (i.e., ontology) are understood through and in relation to Culture (i.e., epistemology) (see Barad, 2000; Cajete, 2000; Kirby, 2011; Latour, 1993). However, as Little Bear (2016) enunciated in his keynote address, the metaphysical relation between Nature and Culture enacted by Western modernity that informs and produces WMS is often one that assumed, presumed, and/or taken-for-granted. Thus, adding to Little Bear's questions, "What are the metaphysics of our schools? Where are those metaphysics taking us?," I ask: *What are the metaphysics of science education?* In response, I offer that WMS and by extension science education are (mis-)articulated as transcending metaphysics (Barad, 2000, 2007; Cajete, 1994, 2000). This (self-)perceived metaphysical exclusion becomes a criticism that is levied against other ways-of-knowing-Nature (e.g., Cobern & Loving, 2001), becoming one of the ways in which Indigenous science is (yet-)to-come (McKinley, 2007). Answering (and being answerable as form of responsibility; see Patel, 2016; Spivak, 1994) to the metaphysics of science education then becomes a question of (mis)reading science education for its subtle and lingering colonial referents and enactments (Carter, 2004, 2005; McKinley & Aikenhead, 2005): the process of "joyful [i.e., play-full] yet laborious strategy of rewriting the old language" (Spivak, 1976, p. xx) that is deconstruction. (Re)opening science education to Indigenous science-to-come labours the structure of education between what it *is*, *is not*, and *could be(come)*, particularly in instances when meanings (and matter) are sedimented and stratified (e.g., through knowledge-practices such as science as metaphysically transcendent).

Regarding metaphysics as they relate to science education, Derrida (1976) offers that the metaphysics of Western modernity are both the process and product of *clôture*: at once being *an* enclosure and a clos*ing*. This double(d) normative process can never be wholly separated from "the metaphysical desire to make the end coincide with the means" (Spivak, 1976, p. xx). In other words, the *closing* is naturalized, rendering the process an *absent presence* whose partial erasure (but irreducible presence) gives the appearance of stable, unitary, separate, and seperable epistemological and ontological units (see also Apffel-Marglin, 2011; Bang et al., 2012; Bang & Marin, 2015; Barad, 2007; Cajete, 1994, 2000; Latour, 1993). However, how the metaphysics of modernity are always already entangled within science education, how this entanglement is produced, as well as what it produces, and what is producible with/in are undertakings engaged within this book. If we are to (re)open science education to Indigenous science to-come, "it is from within this language that we must attempt an 'opening'" (Spivak, 1976, p. xx), we must do so with/in science education, "a language, incidentally, we must know well"

(p. xx). Thus, in the next section of this introduction onto science education, Indigenous science to-come, and metaphysics, I outline a few more of the features of science education and its relationship to Indigenous science. **Understanding School Science and its Relation to Indigenous Science To-Come.** Generally speaking, within science education, "the conventional goal" is one "of thinking, behaving, and believing like a scientist" (Aikenhead & Elliot, 2010, p. 324). Currently, through the two predominant methods of teaching and learning science, this entails: coming to know what scientists know (i.e., cognitivism, intra-personal learning, scientific knowledge as representation of nature) and/or enculturation into how scientists come-to-know (i.e., socio-constructivism, inter-personal learning, scientific knowledge as representation of culture) (Aikenhead, 2006a; Erickson, 2000). Untroubled, both approaches collude and coalesce around the construction and reification of the subject position of "Scientist". It has been argued that this subject position is emblematic of the masculine, Eurocentric, and anthropocentric subject of Western modernity through modes that enact and uphold its metaphysics (e.g., representationalism, universalism, nature/culture divide) (see Barad, 2000, 2007). This (re)produces science as a modern(ist) practice through which nature is knowable and representable (i.e., quantifiable, generalizable, and predictable; see Aikenhead & Michell, 2011; Aikenhead & Ogawa, 2007), and in which neither the culture of science nor the agency of nature can be (wholly) accounted for or be held accountable. Furthermore, this type of scientific literacy and its entangled culture of "school science" potentially produce experiences of cultural *assimilation* and *acculturation* rather than *enculturation* for the vast majority of students.[17]

In other words, rather than a harmonious interfacing of cultures (i.e., enculturation), encounters of school science are more likely to house potential for dialectical negation that is either actualized (i.e., assimilation) or remains un-actualized through students' complex and complicated curricular navigation (i.e., acculturation). Such dialectic negation occurs at the level of the individual, as well as the system. In reviewing literature on science education in diverse school settings, Aikenhead and Elliot (2010) state that "most students (about 90%) tend to experience school science (Grades 6–12) as a foreign culture to varying degrees, but their teachers do not treat it that way" (p. 323; see also Bang et al., 2012; McKinley, 2001, 2007; McKinley & Stewart, 2012).[18]

For students whose daily lived experiences continue to be negatively impacted by Eurocentrism[19] (re)produced with/in (and beyond) science education, learning with/in the cultural practice of "school science" largely continues to be a form of epistemic violence. As such, assimilation is overwhelmingly identified as a common barrier to engagement (Aikenhead, 2006b; Bang et al., 2012; Barnhardt & Kawagley, 2005; Canadian Council on Learning, 2007; McKinley, 2001, 2007).

There are various ways in which this systemic problematic manifests at the level of individual students and groups. For Indigenous, diasporic, and other post-colonial students[20] these include, but are not limited to: (a) under-representation within science and technology occupations, (b) under-representation within formal education and training that paves pathways to such occupations, (c) gaps in achievement on standardized international assessment such as the *Programme for International Student Assessment*, and (d) lower rates of graduation (Barnhardt & Kawagley, 2005; Canadian Council on Learning, 2007; MacIvor, 1995; McKinley, 2007). For Indigenous, diasporic, and post-colonial students who succeed in spaces of WMS despite the odds that are stacked against them, it is often at a cost: learning science is often at the expense of one's cultural being and belonging, becoming otherwise in the process (see Bang et al., 2012; Cajete, 2000; Marker, 2019; McKinley, 2005, 2007). Furthermore, as local Indigenous ways of coming-to-knowing the natural world continue to be underrepresented, misrepresented, misunderstood, and undervalued, WMS tends to be overrepresented and misrepresented (Aikenhead, 1997, 2006b; Aikenhead & Michell, 2011; Aikenhead & Ogawa, 2007). As a result, many students come away from science education with an understanding of WMS that is shaped by myth (e.g., science as culturally neutral, unbiased, and thus ethical), alongside under-appreciation of what other ways-of-knowing-nature might have to offer.

This not only has an impact upon students, but also their teachers: "stereotypical views of [I]ndigenous students [and their knowledge-practices] have led to assumptions of teaching and learning *for* them" (p. 214, emphasis mine). In a study with science teachers of Indigenous students, Aikenhead and Huntley (1999) documented four ways that deficit thinking manifests:

1. Teachers generally viewed Western science as course content or as a way of exploring nature, not as a foreign culture as experienced by many of their students; 2. Aboriginal knowledge was respected by science teachers, but only a token amount was added onto, but not integrated with, school science; 3. Teachers thought that the act of learning science was unrelated to their students' [Indigenous] worldviews; 4. Students' disinterest in pursuing science careers was either unexplainable by the interviewees or was blamed on student deficits. Few teachers blamed their curriculum and teaching. (pp. 162, 164)

If science education is to be (re)opened to Indigenous science to-come, it is important to recall that the ethical imperative of education is "a responsibility *to* the Other (as answerability or accountability) and not *for* the Other (as the burden of the fittest)" (Andreotti, 2007, p. 74, emphasis mine), as well as recognition of the ongoing (re)construction, enactment, and productions that result from such positioning (see also Andreotti, 2011; Kuokanen, 2007, 2010; Patel, 2016; Spivak, 1994).

The ways in which Indigenous ways-of-living-with-nature (IWLN) come to be under- and misrepresented signals how dialectic negation plays out at systemic, cultural, and discursive levels.[21] The very topic of IWLN in science education is itself subsumed within wider concepts such as multiculturalism and equity that fail to wholly account for the complexities of Indigenous-Western relationships (Carter, 2004, 2010; McKinley & Stewart, 2012). Furthermore, the term traditional ecological knowledge (TEK) that is regularly employed potentially (re)centres a Western modern(ist) notion of knowledge as a discrete unit that exists outside of and beyond the knower and its ecology of relationships (Aikenhead & Ogawa, 2007; Aikenhead & Michell, 2011; Kim, Ashgar, & Jordan, 2017; McKinley, 2007). However, McKinley and Aikenhead (2005) state that while these concepts and conceptual locations have been problematic, they nonetheless provide productive locations to critically inhabit science education and gain leverage (see also McKinley & Stewart, 2012). Whether Indigenous science should be included or not within science education, as well as how, where, and when, has become:

...one of the largest (in terms of literature) debates in the field of culture and science education... [which has centered around] the nature of knowledge.... The relevance of this literature to schools is that a universalist understanding of science informs the assumptions implicit in school curricula about the nature of science and how science should be taught. (McKinley, 2007, p. 206)

Here, what is called *the multicultural science education debate* is of central relevance to the ways in which Indigenous science is to-come; it is not only part of our collective inheritance as science educators but rather part a pivotal one, shaping how and who we can be(come).[22] As such, it is a central node that is explored within this book. Following its more fulsome introduction in Chapter 3, it is differentially revisited and explored in further depth in the chapters following (i.e., 4–6).

"Universality" (i.e., transcendental knowledge) is "achieved" when metaphysics of modernity come to mark IWLN, TEK, and WMS through systems of *clôture* (e.g., as either strictly *similar* or *different*), as well as when WMS reasserts itself as the ("neutral") norm and standard against which other knowledge systems are to be judged through Eurocentrism (see Carter, 2004; Lewis & Aikenhead, 2001). Significantly, this does not strictly mean that Indigenous ways-of-knowing-in-being are compared *to* Western modern practices, but also *through* Eurocentric concepts and practices. As Kuokkanen (2007) explains, this includes:

> Eurocentric arrogance of conscience... the simplistic assumption that as long as one has sufficient information, one can understand the "other".... By assuming that epistemologies are universal and that any episteme or system of knowledge can be accessed, this view reflects the Eurocentric claim that Western or modern intellectual traditions are more sophisticated than are other kinds (assuming that the latter even exists). (p. 99)

The result of such "universalism," at both the level of knowledge and knowledge-making practices, are forms of *sameness* and *difference* that are irreducibly bound to the ways in which (neo-)colonial power dynamics manifest in knowledge production (see also Andreotti, 2007, 2011). Within science education, McKinley (2007) states that the relationship between IWLN and WMS can be generalized into four categories: (a) where Indigenous science can be explained within WMS; (b) where Indigenous science *could be* explained through WMS, but the explanation has yet to be developed; (c) where there is a link between Indigenous science and WMS's knowledge claims, albeit through different knowledge principles and practices; (d) where WMS cannot accept aspects of Indigenous science (e.g., spirituality, animism). The extent to which if and how Indigenous science is to be included within school science curriculum depends highly upon the type of Indigenous knowledge (IK) being brought in, as well as science education's ability to ethically respond to difference (see also Kuokkanen, 2007; Marker, 2006); some forms of Indigenous science are more to-come than others. As the relations of power between IWLN, TEK, and WMS are uneven and unequal, it is often the case that "those opposing the inclusion [of IK] argue that there is no place for IK unless it has been subsumed into the body of knowledge referred to as WMS, that is, unless it is made the same as WMS, in which the status quo continues" (McKinley, 2007, p. 208). Alternately, some who uphold the universality of WMS (e.g., Cobern & Loving, 2001; El-Hani & de Ferreira Bandeira, 2008) argue that the inclusion of Indigenous science is a non-issue so long as it is neither called science nor included within the science classroom (but rather as a separate subject, like art, literature, or history). However, such "inclusion" fails to redress the dialectic negation of Indigenous science marked by sublation, subsumption, or suturing over; further, it masks the colonial relations of power that produce these moves (see McKinley, 2001, 2007). Further, as Kuokkanen (2007) explains, "inclusion" is not a move marked by innocence, removed from dynamics of power, nor one that is new:

> The demand and desire that [I]ndigenous cultures and epistemes be translated into forms recognizable by the dominant colonial society is at least as old as colonialism itself. Colonizers have always used translation against indigenous peoples in an attempt to manipulate and displace them and thereby dispossess them of their land. (p. 75)

Inclusion is not the remedy to exclusion when the structures into which inclusion happen continue to (re)produce and uphold settler-colonial ways-of-knowing-in-being: be it complicity in the devaluation and erasure of Indigenous ways-of-knowing-in-being or participating in interrelated delegitimizing claims to Land in the larger project of dispossession (see also Bang & Marin, 2015). As mentioned early within this chapter, meaningful and respectful dialogue between Indigenous science and WMS is in a perpetual state of im/possibility as they are not and never will be (fully) commensurate; Indigenous science will always be *to-come* but the ethical responsibility is ever-present and irreducible.

Decolonizing and Post-Colonial Responses in Science Education. There are growing bodies of work within science education that address Western modernity's Eurocentric legacies that are often referred to as: decolonizing science education (e.g., Aikenhead, 2006c; Aikenhead & Elliot, 2010; Belczewski, 2009; Chinn, 2007; Higgins, 2014) and post-colonial[23] science education (e.g., Carter, 2004, 2005, 2006, 2010;

McKinley, 2001, 2007; McKinley & Aikenhead, 2005; McKinley & Stewart, 2012). Battiste (2013a) describes decolonizing education as is a "two-prong process". It entails *deconstruction* of (neo-)colonial[24] structures and strategies, and *reconstruction* that centres and takes seriously Indigenous, diasporic, and other post-colonial ways-of-knowing and ways-of-being towards reshaping the place-based processes and priorities of education and educational research (see also Donald, 2012).[25] Similarly, post-colonial approaches to science education[26] seek to (re)open (neo-)colonial structures strategies in order to ethically respond to the Otherness of Indigeneity.[27] As both draw from diverging theoretical lineages and enactments of educational practice (e.g., critical pedagogy and post-structuralism respectively), there are productive points of resonance and tension between the two. Of the latter, and of particular relevance to this book, are: (a) the centrality of land as beyond human cultural understandings of it, and (b) whether ethics is a possible possibility or not. Herein, regarding the first statement, I align with decolonizing theories who suggest that post-colonial theories' focus on cultural hybridity, flow, and porousity do not strongly enough consider the ways in which coloniality operates and circulates beyond an anthropocentric (inter-)textuality. The critique is levied to bring attention to the ways in which (neo-)coloniality comes to problematically shape not only human cultural relations, but also those of other-than-humans, and more-than-humans who, together, come to collectively constitute the ecology of relationships that is signified by an Indigenous concept of place (Donald, 2012; Grande, 2004, 2008; Marker, 2006; Smith, 1999/2012). With respect to the latter statement, I align herein with post-colonial notions of ethics as im/possibility to push forth my own decolonizing scholarship; the discursive practices of decolonizing approaches *can* (but do not always) come to mask colonizing tendencies (see Carter, 2004, 2010; Subreenduth, 2006; Rhee & Subreenduth, 2006; Smith, 2005; Smith, Maxwell, Puke, & Temara, 2016; Spivak, 1993/2009). Nonetheless, ethical im/possibility need not be paralyzing; Spivak (1988a, 1993/2009, 1994) reminds of the importance of persistent critical and complicit enactments that work towards "transforming the conditions of impossibility into possibility" (Spivak, 1988b, p. 201), even if/as they are never achieved.

Within science education, this call has been primarily taken up by extending the openings produced through treating both science (e.g., Haraway, 1989; Latour, 1993; Traweek, 1992; see also Shapin

& Schaffer, 1985) and science education (e.g., Nadeau & Désautels, 1984; O'Loughlin, 1992; Pomeroy, 1994)[28] as problematic cultural spaces to be examined through sociological, anthropological, and cultural studies approaches. In particular, a two-pronged approach to decolonizing science education[29] focuses primarily on addressing the ways in which Eurocentrism (re)produces science education as a space of cognitive and cultural imperialism (Aikenhead, 2001, 2006c; McKinley, 2001, 2007; Sammel, 2009) in order to make space for learning that is epistemologically diverse and pedagogically pluralistic (i.e., which recognizes that there are diverse pathways to learning about and with Nature; Aikenhead, 2006a, Barnhardt & Kawagley, 2005, 2008; McKinley, 2007; Sammel, 2009). In Canada, there have been some successes in this area. For example, there are increasingly more policy-mandated curriculums that include Indigenous perspectives on science (e.g., British Columbia Ministry of Education's 2005 *Science K to 7* and 2008 *Science and Technology 11*),[30] general frameworks for school-based integration in place (e.g., Manitoba Education and Youth's 2003 *Integrating Aboriginal Perspectives into Curricula*), as well an overall commitment from Deans of Faculties of Education to prepare teachers accordingly (Association of Canadian Deans of Canada, 2010).

However, given the capillary pervasiveness of Eurocentrism and its co-constitutive mechanisms (e.g., (neo-)colonialism), decolonizing science education is not simply a process of desiring it to be decolonized. Rather, it is (over-)written in a contradictory, conflicted, and contingent space in which the very processes and practices that explicitly seek to dismantle colonial logics often implicitly uphold and reinforce that which they seek to challenge (Carter, 2004, 2005, 2010; Higgins, 2014; McKinley, 2001; McKinley & Stewart, 2012; Sammel, 2009). On this, Carter (2004) states:

> The inclusion of Other's science has the potential to trouble the categories of Western science, but the processes of cultural representation and translation [i.e., differing and deferring Indigenous science] ensure Western science remains authoritative in most settings. These processes simultaneously work to separate, domesticate, and subsume, regulating the boundaries and preserving the integrity of Western science and science education. Hence, the inclusion of the Other's science in school curricula risks an empty form of pluralism implicated... in restorationist agendas to reassert Western cultural control. (p. 832)

In other words, there needs to be a constant vigilance and (re)evaluation of decolonizing goals and processes, as they are always in co-constitutive relation with (neo-)coloniality. As these discussions have primarily and almost exclusively focused on (a particular) epistemological grounds or locations (see Cobern & Loving, 2008; van Eijck & Roth, 2007), one problematic production is the lack of attention to ontology in science education.

On the topic of considering ontology within science education, Sammel (2009) states that "given the pervasiveness of assimilationism in Western science education" (p. 653), to only address the colonial episteme leaves the systemic strategies and structures that "push for assimilation of students into Western science ontology" (p. 653) to continue functioning implicitly (see also Carter, 2004, 2005). This is to say that to treat science education uniquely as a culture potentially masks the ways in which Culture's Other (i.e., Nature) is implicated with/in these processes (see Barad, 2000, 2007; Latour, 1993). Again, this begets the question, What are the ways in which the absent presence of the metaphysics of modernity operate in science education? (e.g., representationalism, Nature/Culture binary; see Apffel-Marglin, 2011; Carter, 2004). While there is space for diverse ways-of-knowing through cultural critique, Sammel (2009) invites us to consider how science pedagogies and curriculums often work to make science accessible and responsive to all learners, but then *differ* and *defer* these goals by falling back on concepts-as-usual (e.g., scientific literacy) to achieve these goals: smuggling back in (neo-)colonial ways-of-knowing-in-being. Such a practice of deferred and differing science-education-as-usual positions diverse ways-of-knowing-nature that are not WMS as but different, and often lesser, ways to attain the same goal of knowing nature with/in the ontology of Western modernity (Carter, 2004, 2005; see also Latour, 1993).[31] The underlying and problematic message is that ontology is a *singular* affair (Barad, 2007).[32]

Cartesianism, the classical Western ontological process through which meaning and matter are individuated through separation from that which co-constitutes them (e.g., mind/body, nature/culture; Apffel-Marglin, 2011; Barad, 2007; Cajete, 2006), often becomes *the* (only) ontology onto which diverse ways-of-knowing differentially map. This tends to differentially re-centre WMS as the metre stick against which all ways-of-knowing and ways-of-being are measured. When Cartesianism is *the* (only) ontology, it only makes sense that the epistemology of WMS that co-constitutes Cartesianism is best suited to work with/in this ontological configuration (see Cobern & Loving, 2008). However, to forget that it is *an* ontology rather than "ontology" (read: singular) when doing cross-cultural and comparative work is to position other-than-Western-modern ways-of-knowing at a taken-for-granted disadvantage, even when the intent is to make space for both positions that extends beyond inclusion and tolerance towards dialogue and collaboration. Accordingly, this also complicates the entangled relationships held with/in school science for those enacting other-than-Cartesian ways-of-being, such as Indigenous ways-of-knowing-in-being, as they continue to be perceived as alternative but lesser ways of "reflecting" Nature as it is understood and enacted through (the singularity of) Cartesianism (see Chapters 4–6).

When reaching and reading singularity, particularly singularities (e.g., Cartesianism) that impede the possibility of Indigenous science to-come, it is productive to consider Cajete's (1994) deconstructive invitation: "Indigenous thinking honors the reality that there are always two sides to the two sides. There are realities and realities. Learning how they interact is real understanding" (p. 31). (Re)opening and re(con)figuring science education to be able to respond to and receive Indigenous science to-come might entail considering co-constitutive relations between what seems separate and separable quantities marked (wholly) by relations of difference (e.g., Nature/Culture as nature-culture, decolonizing/colonizing as de/colonizing, possibility/impossibility as im/possibility).

In the next section, I detail the methodological approach to decolonizing science education and research that guides process of reopening the ability to respond in science education: deconstruction.

Second Orientation: (Re)Opening Science Education to Indigenous Science to-Come Through Deconstruction and Reconstruction

Deconstruction has been developed by the French philosopher Jacques Derrida and, very broadly, involves a critique of Western knowledge or thought. Derrida ... showed how anthropological knowledge is governed by a philosophical category of the center (named Eurocentrism). The argument contends that in the last few hundred years Europe has constituted and consolidated itself as sovereign and subject by constructing the colonized according to the terms of the colonizer's self-image. Deconstruction is the decentralization and decolonization of European thought... Hence, deconstruction is a deconstruction of the concept, the authority, and the assumed primacy of the category of "the West." (McKinley & Aikenhead, 2005, p. 902)

Methodologically, this book works towards the "the decentralization and decolonization of European thought" (McKinley & Aikenhead, 2005, p. 902) through deconstruction as an over-arching meta-approach to work towards Indigenous science to-come. This is in line with Battiste's (2013a, b)[33] and Donald's (2012)[34] conception of decolonizing education as a "recursive process of deconstructing and then reconstructing" (Donald, 2012, p. 547). Simultaneously, I heed the warnings of post-colonial theorists and theory that the potentiality of *deconstruction* and *reconstruction* lay in recognizing them as more-than deconstruction and reconstruction as forms of taking apart and putting together (see Derrida, 1976; Jackson & Mazzei, 2012; McKinley & Aikenhead, 2005). Taking apart (i.e., destruction) through criticism, as McKinley and Stewart (2012) suggest, is a "seemingly potent but ultimately counter-productive strategy" (p. 545) in science education. Rather, Battiste (2013b) argues that approaches to decolonizing education "first and foremost must be framed within concepts of dialogue, respect for educational pluralities, multiplicities, and diversities" (p. 107). However, when criticism is perceived and enacted as taking apart, colonial logics are replaced in but one sense of the word: displaced but not always disrupted (see Kuokkanen, 2007; Spivak, 1994; see Chapters 2 and 3). Deconstruction works against not only the *Euro*-centred through Eurocentrism, but also the *centering* properties of Eurocentrism through endeavouring to dismantle its logics of either/or. In turn, as post-colonial theory presents Indigenous science to-come as a persistent ethical im/possibility, this book tilts more heavily on the deconstructive side, (re)considering *reconstruction* as inseparable from deconstruction and as a form *re(con)figuring* (see Carter, 2005). As structure in science education is not (fully) reached and, simultaneously, will never achieve a state of being "deconstructed", re(con)figuring is a continued deconstruction, labouring between what a structure *is, is not,* and *could be(come)* in response to an otherness who is yet-to-come (e.g., Indigenous science to-come).

Common Approaches to Cross-Cultural Methodologies in Science Education. Prior to tracing how deconstruction as methodological approach creates space for and supports wandering the pathways of science education anew, it is important to touch on the *ways* in which these problematic paths are usually journeyed upon. Just as teaching and learning in science education are increasingly considered through (socio-)cultural approaches (see Aikenhead, 2006a; Erickson, 2000), so too are its cross-cultural methodologies. As McKinley (2007) states regarding approaches to cross-cultural science education,

> Dominating the field are approaches derived from anthropology, such as worldviews, collateral learning, and border crossing. The anthropological approach is a seductive one because it focuses on the culture and cultural practices of different groups and treats science as a cultural activity. (p. 220)

However, at the same time, "science educators are seldom also trained in associated disciplines, such as cultural studies" (McKinley & Stewart, 2012, p. 545). In turn, as McKinley (2001, 2007) and Carter (2004, 2005, 2010) state, culture comes to be perceived and enacted in ways that often come to reify colonial constructs that they are working against, albeit differently. For example, considering school science as having a culture does not necessarily "critique the Eurocentrism inherent in stable and unitary ideas of culture, identity, and context still to be found in some of science education's more traditional comparative and cross-cultural studies" (Carter, 2004, p. 824). Modes such as worldview theory and border crossing might be apt for considering the experience of a student navigating between cultural spaces, but might not account for the power relations in place between these knowledge systems which occurs beyond the individual learner which produce the very borders they must cross (Carter, 2004; McKinley, 2007). However, this is not to state that *culture* should be jettisoned (thus reinforcing a status quo of science as acultural; see Kirby, 2011). Rather, as *culture* offers both methodological possibility and problematic (see Carter, 2010), it is important to use and trouble this central referent to cross-cultural and multicultural science education. Deconstruction, states McKinley and Aikenhead (2005), provides such means to use and trouble culture within decolonizing and post-colonial science education methodologies as it accounts for both *process* and *product* of Eurocentrism and Cartesianism.[35]

Deconstruction of/in Cross-cultural Science Education. Within this book, I take a deconstructive stance that might best be described as an "impossible 'no' to a structure which one critiques, yet inhabits intimately" (Spivak, 1993/2009, p. 316). Critically inhabiting science education entails refusing to inhabit it *like that* without refusing to inhabit it altogether: deconstruction is at once critical and complicit (see Chapter 3). In offering a succinct "how-to" for deconstruction,[36] Spivak (1976) suggests:

> Deconstruction in a nutshell...[is] to locate the promising marginal text, to disclose the undecidable moment, to pry it loose with the positive lever of the signifier; to reverse the resident hierarchy, only to displace it; to dismantle in order to reconstitute what is always already inscribed. (Spivak, 1976, p. lxxvii)

Spivak (1976) describes the process of bearing witness to undecidability as being on the lookout for snags in meaning when it stops working as intended, in which the absence of unified meaning might come to threaten the very structure which it occupies. In short, this entails paying attention to, and making use of, concepts and categories whose meanings vacillate between a meaning and a constitutive otherness; intentionally (mis)reading them by tinkering with meanings otherwise unintended but potentially signalled by that which is there (see Biesta, 2009; Derrida, 1976; Spivak, 1976; St. Pierre, 2011). Echoing Spivak (1976), McKinley and Aikenhead (2005) state of deconstruction in decolonizing science education that:

> ...the key to deconstruction is not the identification of the dichotomy and the inversion, (although that work is necessary and we do not wish to underestimate it), but the *displacement* of such thinking. In other words, how does one re-think these fundamental ideas? How does one displace those assumptions that make "natural" meaning possible? Furthermore, can deconstruction as a critique lift itself off the page to have any practical application? (p. 903, emphasis in original)

Importantly, McKinley and Aikenhead (2005) remind that deconstruction should not strictly be theory for theory's sake. Rather, as Lather (2007) states, it is important to "[put] theory to work" by using theory (e.g., deconstruction) towards and without losing sight of the critical goals that one sets out to achieve (e.g., decolonizing).[37] Deconstruction must be always already be deconstructing the theory/practice that keep the two separate and separable (e.g., producing practice as "atheoretical", and theory as a practice of "armchair philosophy"): "the production of theory is also a practice; the opposition between 'pure' theory and concrete 'applied' practice is too quick and easy" (Spivak, 1988a, p. 275).

There are three inseparable binary relations that feature strongly within this book: Self/Other, Nature/Culture, and ethical possibility/impossibility.

Deconstructing Self/Other. Given that cultural (re)constructions of Otherness continue to be problematic within science education such that they (re)centre colonial logics and subjects, a prevalent (but not unproblematic) solution is often to reverse the gaze onto the Self (i.e., the (neo-)colonial subject) of colonizing relationships (see Pillow, 2003). As Tuck (2009) articulates, researchers do not need, nor should they use the suffering of Indigenous, diasporic, and other post-colonial students as evidence of colonial violence and as ethical motivation for research. Above and beyond providing positive representations of these students, there is always the possibility to look back at the culture of power that produces this violence. However, to (too simply) displace the gaze by reversing the hierarchy does not always disrupt it (particularly if the gaze continues to operate similarly, albeit with a different target). Here, Lather (2007) suggests a double(d) reversal of the ethnographic gaze. Such a double(d) reversal entails both the literal reversal of studying those who do the studying (i.e., in order to reverse the direction of the ethnographic gaze), as well as the study of the ways in which those who do the studying study (i.e., in order to reverse the way in which the ethnographic gaze is produced). Such deconstructive Self-reflexivity might allow for the possibility of thinking without the thing with which you think (when the thing with which you think is part of the problem), producing the possibility for alternate ways of being and becoming science educator and researcher.

For example, inverting the production of the gaze also entails resisting a simple displacement of colonial violence and, further, houses the potential to disrupt it. I recognize there is something important in extending a genuine invitation a relationship-to-come, even if its potentiality is not enacted (see also Kuokkanen, 2007). Herein, I work to not negate the work science educators who might disagree with the very premise of this book. Rather, I extend an invitation to dialogue across difference towards them.

Further, in considering the Self/Other binary, I recognize that it is important to move "the postcolonial critique of Eurocentrism beyond identity politics, to the level of an epistemic challenge to science" (McKinley & Stewart, 2012, p. 551). Spivak (1993/2009) states that to reduce scholarship to identity politics can be a way which the workings of power are (re)produced:

> I have long held that in the arena of decolonization proper, the call to a complete boycott of so-called Western male theories is class-interested and dangerous. For me, the agenda has been to stake out the theories' limits, constructively to use them. (p. x)

In other words, deconstructing the colonial Self/Other binary does not preclude any one identity from participating in the workings of power, even if the circulation of power is uneven and unequal across different identity positions (see also Spivak, 1988a). Furthermore, as Spivak (1994) states,

> It seems more responsible that, instead of falling back on the deceptive simplicity of a proposition [(e.g., "a complete boycott of Western male theories")] and taking that as sufficient fulfillment of ... philosophical responsibility, ... [we could] philosophize with all stops pulled out, without denegating [our] complicity, to present [such proposition] as pharmakon, what could have been medicine turned into poison. (Spivak, 1994, p. 34)

To strictly operate from an identity politics position in which "Western male theorists" (such as myself) are excluded runs the risk of stating that those excluded are "inherently" Eurocentric (i.e., being) rather than shaped with/in Eurocentrism (i.e., becoming). This risks foreclosing the space of possibility to not be Eurocentric *like that* (and invariably, leaving particular individuals "off the hook"; see Kuokkanen, 2007).

In turn, the theory-practices that I employ throughout this book are selected (but not "validated" as non-Eurocentric or unproblematic) for their ability to displace and disrupt the metaphysics of modernity and (re)open science education towards disrupting and displacing the Self/Other binary as it prevents Indigenous science to-come (e.g., Carter, 2004; McKinley & Aikenhead, 2005). Significantly, science's and science education's Self constructs Nature as it's Other as well.

Deconstructing Nature/Culture. As Spivak (1993/2009) states, "if the lines of making sense of something are laid down in a certain way, then you are able to only do things with that something which are possible within and by the arrangement of those lines" (p. 34). Science education often dialectically subsumes, sublates, and sutures over many of the pluralities, multiplicities, and diversities called for in decolonizing and post-colonial science education. As detailed in previous (sub)sections, these enactments are achieved through an implicit and often taken-for-granted centering of Cartesianism,[38] while simultaneously working to erase other ontologies and Cartesianism's own workings by presenting itself as *the* (only) ontology. Recall that Cartesianism is an ontological enactment through which the Nature/Culture binary is (re)produced and producible.

I endeavour to work within and against this problematic structure that (re)constitutes science education; labouring within and against the *clôture* of metaphysics that is (and is always becoming) singular, stable, and subsuming (see Chapter 5 for a fulsome exploration of this).

Deconstructing possibility/impossibility. To reiterate, deconstruction and reconstruction are not deconstruction and reconstruction (see Jackson & Mazzei, 2012); this process is not destroying and then rebuilding. Deconstruction and reconstruction invariably share a relation of co-constitution (see Chapter 7). As Spivak (1994) outlines, the very possibility of reconstruction as ethical response to a call of otherness such as Indigenous science to come is premised on responsibility. In turn, ethical responsibility is inevitably premised upon the ability to respond:

> It is that all action is undertaken in response to a call (or something that seems to us to resemble a call) that cannot be grasped as such. Response here involves not only "respond to," as in "give an answer to," but also the related situations of "answering to," as in being responsible for a name (this brings up the question of the relationship between being responsible for/to ourselves and for/to others); of being answerable for … It is also, when it is possible for the other to be face-to-face, the task and lesson of attending to her response so that it can draw forth one's own. (Spivak, 1994, p. 22)

In its multiplicity, responsibility for Spivak calls upon the ability to respond in the moment, to take responsibility for the (inevitable) inability to respond, and to continuously be responsible towards the very (im)possibility of responding to the other whose experiences, ways-of-knowing, and ways-of-being sit outside of the register of what we can know. The ability to respond is always, at best, partial as the Other to whom response is granted is, as Spivak (1988a) reminds, "irretrievably heterogeneous" (p. 284) and hence "non-narrativisable" (p. 284): that which is to-come can never (fully) be known as it is always already within the co-constitutive exteriority of that which can be known and responded to.

However, working with purpose but without guarantee is par for the course when it comes to deconstruction: "the philosophy of [decon-struction] cannot be used to ward off accountability, answerability, responsibility ... It can only ever be a reminder of its open-ended and irre-ducible risk" (Spivak, 1994, p. 27). While working towards reconstructing science education with Indigenous peoples, places, and protocols in mind, I remain hyper-vigilant: the very frames through which recognition of Indigenous science to-come occurs are differential articulations of that which makes it such that this call "cannot be grasped as such" (Spivak, 1994, p. 22). It follows that the reconstruction herein focuses largely on the ability to respond on a continued deconstruction and (re)opening of the space of response *within* science education towards Indigenous ways-of-knowing-in-being.

Conclusion: Towards a Metaphysics of Response-Ability in Science Education

> Responsibility implies response-ability, but responsibility of [science educa-tion] must also go beyond innocence. ... innocence takes a special form of choosing not to know in order to exercise patterns of systemic privilege. (Kuokkanen, 2007, p. 154)

Returning to the question of *where do we begin?*, it is important to take head of the ways in which Leroy Little Bear (2016) signalled in his talk, *Blackfoot Metaphysics is Waiting in the Wings*, that there is no metaphysics that exists outside of its relationship to others. This is to repeat and reiterate that Western modern science and Indigenous ways-of-knowing-in-being are *always already* in relation. The task at hand is not to begin a new relation but rather to engage the relation *anew*. Specifically, it is significant to not only attend to but also be answer-able to the the troubled and troubling relationship between Indigenous and Western ways-of-knowing-in-being which contribute to the ways in which Indigenous peoples, practices are rendered absent presences within science education, or how Indigenous metaphysics continues to "wait in the wings". Importantly, this work includes accounting for and being accountable to the ways in which science education has been masking power with innocence, such as decontextualized (e.g., ahistorical, acultural) accounts of difference that make it such that Indigeneity comes to be presented as a dichotomously deficient otherness.

While the question is no longer, if it ever was, whether or not science education has a responsibility, the question of *(how) is science education is able to respond?* lingers and persists. Science education has made great strides in the past few decades in treating science and science education, respectively, as cultural practices which could ethnographically be investigated: revealing the ways in which these spaces continue to (re)centre Western ways-of-knowing and erasing others in the process. However, if we take seriously the double(d) task of unsettling science education as the simultaneous processes of decolonizing and deconstructing, it becomes important to recognize the sedimented notions which uphold settler-colonial logics and structures, even when these "settled" beliefs are leveraged towards critical aims as they shape both ability and inability to respond to Indigenous science to-come. For example, the Cartesian belief, and colonial import, that Culture as dichotomously opposed to Nature comes to partially reproduce the colonial structures worked against. It is for this reason that I have stated elsewhere that, within spaces of science education, "it might be time to begin thinking of decolonization as de/colonization" (Higgins, 2014, p. 265) by taking seriously the ways in which (neo-)colonial logics unavoidably come to shape even the ways in which they are responded within our theories, concepts, and practices.

It is not enough to *take* responsibility for the ways in which science education dialectically negates Indigenous science to-come unless such action engages with the "difficult and contradictory nature of de/colonization" (Subreenduth, 2006, p. 628). Moving towards a more responsible science education also engages in "an iterative (re)opening up to, an enabling of responsiveness" (Barad, 2010, p. 265) by attending to the ways in which the ability to respond is always already shaped by (neo-)colonial logics. Recognizing that some nodes are more pressing and productive than others, this work of *response-ability* (see Chapter 2) entails attending to the absent presences of science education which complicate and complexify the ability to respond: What cultural practices are masked when science is presented as or functions as acultural (see Chapter 3)? What theories inhabit scientific practice when they are treated as atheoretical (see Chapter 4)? What metaphysical commitments are rendered invisible when science is defined in opposition to the metaphysical (see Chapter 5)? What shared histories (and futures yet-to-come) haunt science education when the relation between Indigenous and Western ways-of-knowing-in-being are presented as but a new one (see Chapter 6)?

NOTES

1. Within this chapter, such *there-thens* entangled with the *here-now* might notably include the following SpaceTime coordinates: Calgary, Canada 2016 [Leroy Little Bear's *Big Thinking Address* at the Canadian Congress of the Social Sciences and Humanities]; Kalamazoo, US 1992 [first "Science Dialogues"]; diffracted through Stony Nakoda Nation (west of Calgary), Canada 1989 [Native science conference, Little Bear meets Peat and makes arrangements to meet David Bohm]; Albuquerque, US 1999 [Science dialogues continue where original funds from Fetzer institute ran out]; Ottawa, Canada 1994 [David Peat writes the introduction to the first edition of *Blackfoot Physics*]; Thunder Bay, Canada 2008 [I read Blackfoot Physics for the first time]; Iqaluit, Canada 2009 [I am delivering my first cross-cultural science education research project]; Edmonton, Canada, 2019 [where/when I am (re)writing this introduction].

2. I use the expression ways-of-knowing-in-being throughout this book for three reasons. First, it is a nod to the notion that knowing and learning are always already processes. As Aikenhead and Elliot (2010) suggest, "the expression *Indigenous knowledge* is problematic because the word *knowledge* is embedded in a Eurocentric epistemology" (p. 322, emphasis in original; see also Aikenhead & Ogawa, 2007). Rather, within many Indigenous languages, knowing and learning are not expressed as a product (i.e., knowledge, a noun) but rather as a process (i.e., coming-to-knowing, a verb): as Peat (2002) states, "coming-to-knowing means entering into relationship with the spirit of knowledge, with plants and animals, with beings that animate dreams and visions, and with the spirit of the people" (p. 65). Second, coming-to-know is inseparable from coming-to-being. They are ongoing and interconnected epistemological and ontological processes that are deeply relational and holistically interwoven into the fabric of everyday life (Aikenhead & Michell, 2011; Bang & Marin, 2015; Cajete, 1994, 2000, 2015). Lastly, it is to signal

plurality with a reminder that plurality does not entail a form of relativism (McKinley, 2007).

Furthermore, as Aikenhead and Michell (2011) state, "reading a book is not adequate for understanding specific Indigenous practices (e.g., berry picking or fishing), which invariably require experiential learning" (p. xii). Rather than seeking to reach the problematic closure and containment of *knowledge*, they propose that *appreciating* might be a more apt way of approaching Indigenous ways-of-knowing-in-being (while not in itself unproblematic, even if strategic; see Kuokkanen, 2007). Not only this, as Cajete (1994) suggests of the textuality of his own work on Indigenous ways-of-knowing-in-being, that writing for an academic audience shapes what can be said, what cannot, as well as how. In other words, translation with/in, as well as for, academic traditions differentially produces how Indigenous ways-of-knowing-in-being are (re)presented. This differential ecology of relations (partially shaped by the vales of the academy) matters.

3. Importantly, Other is used throughout the book not as a statement of being (e.g., identity), but as a process of *othering* which says more about the colonial Self from which this representation is projected: "the process of containing the 'other' for colonial, imperial purposes...; it involves domesticating an incommensurable and discontinuous 'other' in order to consolidate the imperialist self" (Kuokkanen, 2007, p. 101).

4. The addressee to whom the account of this work is given to is of no small significance, both in terms of what can and cannot be enunciated, as well as its possible possibilities and problematics. As Butler (2005) states,

> There can be no account of myself that does not, to some extent conform to the norms that govern the humanly recognizable, or that negotiate these terms in some ways, with various risks following from that negotiation. ... No account takes place outside the structure of address, even if the addressee remains implicit and unnamed, anonymous and unspecified. The address establishes the account as an account. (p. 36)

Whereas I generally address my work to others already and actively engaged in processes of decolonizing education, the intended addressee in this work is, more generally, those situated in the field of science education whose consideration of decolonization and/or post-coloniality is to-come (as an unactualized potentiality and a *tout-autre* [wholly other] whose voice has gone unheard because it cannot yet be heard; see Spivak, 1993/2009). This entails that, as a means of (ethically) accounting for and being accountable to this implicit addressee, I often draw from epistemic resources that may be intelligible as such (e.g., canonical science education literature and well-recognizable scholars) in my efforts to work at the limits of intelligibility (i.e., to (re)open science education to the possibility of Indigenous science to-come).

However, as Butler (2005) states, the productive necessity of giving an account over to an addressee is not without risk. Here, the most significant risks posed by this are that, first, there risks a too-easy and representational (ist) reading which produces a perception that Indigenous science wholly yet-to-come (see next footnote) or that this is a space that has not been and continues to be laboured by Indigenous and ally scholars. Secondly, working within this space of un/intelligibility nonetheless defers and differs intended meanings of Indigenous science; but how does one articulate the unarticulable within the frames that render them such (see Ahenakew, 2016; Higgins & Kim, 2019; McKinley & Stewart, 2012)? For example, an Indigenous "sense of place" (Cajete, 1994) and other lived concepts are differentially articulable and intelligible when accounting for and being accountable to this double(d) relation to science education and Indigenous science.

5. Interestingly, the common sense and taken-for-granted assumption that metaphysics *is* in a singular sense is so prevalent that even auto-correct suggests that metaphysics as plural is a grammatical error (see Chapter 5).

6. These dialogues began in 1992 (see Little Bear, 1994; Parry, 2008; Peat, 2002, 2007) but have seemingly ceased in the last few years.

7. The first edition, entitled *Lighting the Seventh Fire*, drew from and was released two years following the 1992 *Science Dialogues*.

8. Importantly, throughout the book I use the Indigenous science to not only signal the ways in which Indigenous peoples have practiced ways-of-living-with-Nature since time immemorial, but also the ways in which Indigenous peoples critically and creatively engage with science, such as the emergent field of Indigenous Science, Technology, and Society studies (STS)(e.g., TallBear, 2013). This is significant as:

> Indigenous knowledge includes knowledge of imperialism from the West, the East or even from the neighbours, deep knowledge of colonizers and the practices and effects of colonization, of different religions that were imposed, of nation states formed by different conceptions of a state, western democratic, socialist or communist, and of the institutions of the state. (Smith et al. 2016, p. 136)

Where colonial logics attempt to perpetually relegate Indigeneity and what it has to offer as fixed, particular, and a "tradition": it is important to attend to the ways in which Indigenous peoples continue to engage with the natural world *as well as* with science as conventionally defined.

9. There are also many already present spaces that have been masked through their dialectic negation. Here are but three examples that I am familiar with. First, a more situated naming of *Maslow's hierarchy of needs* might be *Maslow's (mis)interpretation of a hierarchy of Blackfoot needs* from his time spent with the Blackfoot when he was "stuck" on his working developmental theory (Blackstock, 2011). Secondly, aspirin is the synthetic simile of a willow-bark-based traditional medicine that was "discovered" by the Bayer pharmaceutical company (Snively & Corsiglia, 2001). Third, as Cajete (2000) states, Native Americans had a central role in "establishing uses for asphalt and other petroleum products" (p. 190) such as petroleum jelly as a salve for treating burns and open wounds and asphalt as a waterproofing material.

10. Spivak (1999) suggests that *sanctioned ignorance* is not only a practice of strategically not knowing other(ed) ways-of-knowing and -being, but also a collective forgetting of oppressive structures and practices that are (re)produced by dominant groups: "the mainstream has never run clean… part of mainstream education involves learning to ignore this absolutely, with a sanctioned ignorance" (p. 2).

Importantly, such as stance is well documented in spaces of Indigenous education. Notably, Potawatomi-Lenapé scholar Susan Dion (2007) refers to it as the "perfect stranger," a stance simultaneously marked by an avoidance of taking up colonial complicities and claiming to know little to nothing about Indigenous peoples, places, and practices (see also Higgins, Madden, & Korteweg, 2015).

11. This can be stated not only for the substantive content, but also in relation to the ways in which the medium is the message (see Cajete, 2015; Higgins & Madden, 2018): specifically, Peat's (2002) book is an early text written for an audience that might not even consider the existence of Indigenous ways-of-living-with-Nature. However, as Kuokkanen (2007) states, inviting an audience towards an appreciative stance is not without consequence:

> Developping an understanding and appreciation for the "other" is not only an inadequate response, but also an irresponsible one. It reflects a specific type of racism that enables the dominant to occupy the position of universality while consigning the "other" to a partial and particular one. Through distancing, the dominant takes the position of privilege and is able to disassociate itself from any active commitment to a relationship, to reciprocation. (p. 109)

Importantly, the distancing produced through such a stance removes readers in positions of power from being implicated: presenting them with only the lovely knowledge of such an inquiry and sparing them from the difficult knowledge (e.g., inheriting colonial legacies).

12. Directly related to the notion of making a different future occur in the context of colonial institutions is the concept of futurity. As Patel (2016) describes it,

> Futurity is the imprint, the scent, the murmur of what is in the future. In that sense, it is actually unknowable in the immediate, as its discrete details are not available through current lenses. You can't map futurity; you can only map possible futurities. Learning, similarly, is an act of letting go of what one knows for what one does not yet know.... As long as coloniality has been in existence, so has learning, and it's important to remember that not only has learning predated and will survive coloniality, but that it has existed despite, because of, and in defiance of coloniality. (p. 95)

13. This is not to state that Indigenous science is flatly yet-to-come as it is a practice of living with Nature that Indigenous peoples have been enacting with/in place since time immemorial (see Cajete, 1994, 2000; Kawagley, 2006). Rather, it is to state that it is still (partially) "waiting in the wings" of science education (see McKinley, 2007; McKinley & Stewart, 2012).

14. This distinction bears significance both to the ways in which Indigenous science is othered within science education, but also the role of metaphysics. Analogously to science education, which simultaneously strives to be for *all* students but always already fails in the effort, Derrida (1994/2006) suggests that democracies are (by definition) shaped by "the gap between fact and ideal" (p. 80) which is marked by "failure, inadequation, disjunction, disadjustment, being 'out of joint'") (p. 81). Stating that this is not only the case for older forms of government, but also contemporary ones, Derrida (1994/2006) offers that the promise of democracy is always deferred and differing from what can be done and even imagined in the present:

> ... we always propose to speak of a democracy *to come*, not of a *future* democracy in the future present, not even of a regulating idea... – at least to the extent that their inaccessibility would still retain the temporal form of a *future present*, of a future modality of the *living present*. (p. 81, emphasis in original)

This is to state that, for Derrida, the call to democracy cannot be fully heeded within a present moment, or even a *future present* which problematically displaces a current imaginary *elsewhen*. *To-come* signals a not yet which must be worked towards, which must be received hospitably, but whose arrival cannot be anticipated. This is perhaps even more so the case in science education where the very process of becoming scientific is framed in opposition to Indigenous ways-of-knowing-in-being, rather than strictly a failure to achieve an ideal (e.g., science education for all). It is perhaps for this reason that Plains Cree scholar Cash Ahenakew (2017) states, "the work of decolonization is not about what we do not imagine, but what we cannot imagine from our Western ways of knowing" (p. 88): Indigenous science is to-come within the context of science education.

15. Practice, as enacted and discussed herein, is not strictly understood in the conventional sense (e.g., institutional teaching and learning). Rather, the practice prominently articulated and employed herein is that of decolonizing and post-colonial science education scholarship-as-practice. This is in line with decolonizing and post-colonial science education scholars who articulate that theory too is a practice (Carter, 2005; McKinley & Aikenhead, 2005), as well as recent calls in science education there is too much focus on empiricism (and in turn too much data) and not enough scholarship-as-practice (see Carter, 2010). Furthermore, it is also in line with conceptions of decolonizing and educational research which advocate for attentiveness to the practices one is already engaged in, as well as the norms through which attention is deferred elsewhere and differed: paying attention to the process without relegating its justification to the product (see Higgins & Kim, 2019; Smith et al., 2016).

16. One of the reasons for this, as Tuck and Yang (2012) offer, is that "decolonization in a settler context is fraught because empire, settlement, and internal colony have no spatial separation" (Tuck & Yang, 2012, p. 7); coloniality is always a proximal relation in settler contexts.

17. *Potential* is significant to highlight here. As Aikenhead (2006a) points out, "students and many teachers react to being placed in the political position of having to play school games" (p. 28). In turn, they often creatively subvert this positioning by playing what is called "Fatima's game" in science education "to make it appear as if significant science learning has occurred even though it has not" (p. 28).

    Importantly, it is worth noting the larger, beyond-school context in which these logics plays out. Notably as there are, most significantly, ongoing practices in which Indigenous peoples continue to be the *object* of science (see TallBear, 2013).

18. See Aikenhead and Elliot (2010) for the various qualitative and quantitative science education studies that come to inform this figure.

19. In short, Eurocentrism is a discursive force which (re)centres Western modern(ist) culture, people, places, and histories as the normative standard against which other ways-of-knowing are judged, usually as lesser and deficient (see Battiste, 2005, 2013b). It is not only the "colonizer's

model of the world" (Blaut, 1993, p. 10), but also a colonizing model of the world. Operating through diffusionism, a forced spread of culture, it erases or assimilates non-Eurocentric knowledge systems establishing "the dominant group's knowledge, experience, culture and knowledge as the universal norm" (Battiste, 2005, p. 124).

20. I use "post-colonial student" here as a general category and concept to include other-than-Indigenous and other-than-diasporic students who might also might be negatively impacted by ongoing (neo-)colonialism and/or who are implicated and involved in the productive friction signalled by the "post" (i.e., an ever partial but nonetheless productive attempt to move beyond (neo-)coloniality).

21. For example, as Arapaho scholar Michael Marker (2019) signals, without wanting to diminish the efforts of individual Indigenous students, the desire to tell stories of Indigenous successes of participating within science can also mask not only the stories of those who have not successfully negotiated and navigated the space, but also can take attention away from the systems and structures which make such a journey a difficult one to begin with. As Ahenakew (2017) states, this can often be read as "a move to distract from more unsettling Indigenous demands for decolonization" (p. 85).

22. Here, inheritance comes to bear in a meaningful way: if we take Indigenous (e.g., Cajete, 1994) and quantum (e.g., Barad, 2007) metaphysics seriously, our pasts continue to present themselves even if they are but absent presences (see also Derrida, 1994/2006). This bears relevance as the debate is left unresolved (despite the occasional claim of resolution); it continues to haunt what science education is and can be in the present and into the future.

23. Decolonizing, Indigenous, and post-colonial scholarship share many similar facets. However, as McKinley (2007) states,

> *Postcolonialism* is controversial among many groups... For many [I]ndigenous researchers [and allies] the term signals that the European imperial project, and the appropriation of the 'Other' as a form of knowledge, has been assigned to an historical past... This understanding is always present in postcolonialism... [However,] postcolonialism can be used to mean "beyond;" instead of arguing lineal progression of before and after a point in history, another dimension is added with this alternative meaning... "beyond" suggests that boundaries or borders have become blurred. (p. 201, emphasis in original)

Most famously, Linda Tuhiwai Smith's (1999/2012) rejection of post-colonialism (as and through its first meaning of "after" colonization) is most cited today, even if she has since revisited and revised her earlier statement to consider the second interpretation to be deeply productive in practice:

> the idea that postcolonialism is more than what I have previously viewed with scepticism as a not very good historical moment *because it does not really exist, there is no post to colonialism*; more than a methodology one can deploy to study difference, it is, rather, an emergent, growing body of knowledge; there is a knowingness that a postcolonial research disposition can reach, can see, can seek, can come to know. (Smith, 2005, p. 552)

24. With respect to coloniality, neo-coloniality, and their relationship, Spivak (1999) states the following:

> Let us learn to discriminate the terms colonialism – in the European formation stretching from the mid-eighteenth to the mid-twentieth centuries – neocolonialism – dominant economic, political, and culturalist maneuvers emerging in our century after the uneven dissolution of the territorial empires – and postcolonialism – the contemporary global condition, since the first term is supposed to have passed or be passing into the second. (p. 172, emphasis in original)

By highlighting that the contemporary global post-colonial condition is supposed to have passed from colonialism to neo-colonialism, Spivak brings attention to the ways in which coloniality and neo-coloniality are bound by a relationship of constitutive exclusion. In other words, even though they are often framed as historically distinct (i.e., past and present), the ongoing project of territorial imperialism (colonialism) is never absent but always already present, even if it is increasingly tied to a project of economic imperialism (neo-colonialism). I signal this assumption through the use of the term *(neo-)colonial* throughout the book.

25. Battiste's (2013a, b) framework is overarching. An implicit message throughout her scholarship is that the work of decolonizing education is multi-faceted, multi-sited, divergent, and pluralistic in nature. In turn, decolonizing education resists the notion that there is *a* way of doing it – that difference and diversity in positionalities, contexts, approaches, and inclinations are strengths rather than weaknesses. This is particularly

significant (in general and within this book) as the metaphysics of modernity, as well as its Eurocentric and Cartesian modes are not uniform, but rather are differentially articulated across diverse locations.

26. For Spivak, education is a post-colonial site that discursively produces the very conditions of ethical im/possibility: education places teachers with (unlike) others while institutionally framing learning as knowing what is best for the other (see Andreotti, 2007, 2011; Spivak, 1993/2009).

27. Importantly, the post-colonial inflection on decolonizing that is de/colonizing similarly employs a two-prong process. As Rhee and Subreenduth (2006) unpack, de/colonizing approaches employ "two interrelated moves" (p. 547):

> First, projects of de/colonization need to reveal and disrupt the ways in which imperialism constructs the inferior Others within and beyond the West through complex apparatuses of oppression.... Second, [they] examine the process[es] by which societies and individuals interpret, negotiate, subvert and re-construct such knowledge/power to create performative possibilities for themselves. (p. 547)

28. In her ethnography of school-based science education, Deborah Pomeroy (1994) came to refer to the "standard account" curriculum of WMS as one of *White Male Science*.

29. Of course, "decolonizing school science *begins* at the stage of 'acceptance'" (Aikenhead, 2006c, p. 393, emphasis in original): an acceptance of IWLN and that decolonizing school science is a goal that is worthwhile and important (see also Kuokkanen, 2007).

30. However, integration of Indigenous perspectives does not always entail or require "acceptance" (see Chapter 3). Furthermore, even an intent to accept Indigenous science is not necessarily unproblematic.

31. Latour (1993) refers to this as "particular universalism": a framework in which Nature is stable and outside of Culture in which diverse cultural positionings mediate access to knowledge about Nature, but in which "one society - and it is always the Western one - defines the general framework of Nature with respect to which the others are situated" (p. 105). In other words, it is a conceived of and enacted as an epistemic privilege.

32. This is a significant location to labour as some scholars, such as Cobern and Loving (2008), problematically articulate the corollary argument that the epistemology of WMS (i.e., epistemic realism) should be considered the *best* way of knowing Nature of its high level of alignment with a Cartesian ontology (see Chapter 5).

33. While Battiste (2013a, b) does not come to state explicitly how she understands deconstruction or whom she draws upon, a persistent theme

throughout this book is generous and generative (mis)readings. Rather than criticize what some might perceive as a lack (as negation forecloses possibility), such indeterminacy can be read as a gift of potentiality and of meaning that is productively on the move that might come to respond to diverse contexts (see Kuokkanen, 2007).

34. Donald's (2012) use of deconstruction/reconstruction primarily hinges upon Indigenous-non-Indigenous relationships that are already being enacted. Here, deconstruction does not entail a destruction of the hybrid, complex, and contradictory space interfacing Indigenous and Western thought and being. Rather, it entails keeping an eye out for porous locations in order to reconfigure, rethink, and differently enact the relations that are there to create new and renewed possibilities for ethical relationality (see also Nakata, 2007a, 2007b). This metaphor is productive in differentially coming-to-understand the ways in which the post-colonial concept of de/colonizing is enacted throughout this book.

35. While still under-employed and -explored in science education, deconstruction provides the possibility of (re)openings in a multiplicity of seemingly stuck and sedimented locations by deferring and differing concepts which problematically present themselves as stable such as representation and self-identity (e.g., Gilbert, 2001; Gough & Price, 2009; Sammel, 2010).

36. However, the question *What is deconstruction?* is always fraught; it is an approach that works against the metaphysical stasis that comes with the word "is" (see Derrida, 1976). So much so, that Derrida (1976) takes up the complicity between *is* and *and* seriously: "in French, 'is' (*est*) and 'and' (*et*) 'sound the same'" (Spivak, in Derrida, 1976, p. 30) but share a much more complex relationship of signification. Because the statement of *this is that* is never fully achieved or achievable due to *that* always already being an unfaithful reproduction of *this*, an *is* statement is always to a certain degree an *and* statement. This is to say that, for Derrida, Being (i.e., to *be*) or presence is always deferred.

   More importantly, as Spivak (1993/2009) suggests, Derrida "does not develop a systematic description of this mode of operation. (There is, after all, no useful definition of deconstruction anywhere in Derrida's work)" (p. 31). Thus, any account of deconstruction must always be partial as deconstruction is always already on the move; the discontinuity that is deconstruction is in itself dis/continuous such that Derrida does not have the final word on deconstruction (see Barad, 2010; Kirby, 2011).

37. Importantly, deconstruction simultaneously is not, and should not become, theory for the sake of theory. As Derrida (1994/2006) offers about deconstruction, "what remains irreducible to any deconstruction, what remains as undeconstructible as the possibility itself of deconstruction is, perhaps, a certain experience of the emancipatory promise" (p. 74).

38. For example, at the time of writing, even auto-correct suggests that ontology is a singular affair (via grammatical suggestions; i.e., ontology rather than *an* ontology).

REFERENCES

Ahenakew, C. R. (2016). Grafting Indigenous ways of knowing onto non-Indigenous ways of being. *International Review of Qualitative Research, 9*(3), 323–340.

Ahenakew, C. R. (2017). Mapping and complicating conversations about Indigenous education. *Diaspora, Indigenous, and Minority Education, 11*(2), 80–91.

Aikenhead, G. S. (1997). Toward a First Nations cross-cultural science and technology curriculum. *Science Education, 81,* 217–238.

Aikenhead, G. S. (2001). Students' ease in crossing cultural borders into school science. *Science Education, 85,* 180–188.

Aikenhead, G. S. (2006a). *Science education for everyday life.* London, ON: Althouse Press.

Aikenhead, G. S. (2006b). Cross-cultural science teaching: Rekindling traditions for Aboriginal students. In Y. Kanu (Ed.), *Curriculum as cultural practice: Postcolonial imaginations* (pp. 223–248). Toronto, ON: University of Toronto Press.

Aikenhead, G. S. (2006c). Towards decolonizing the pan-Canadian science framework. *Canadian Journal of Science, Mathematics and Technology Education, 6*(4), 387–399.

Aikenhead, G. S., & Elliot, D. (2010). An emerging decolonizing science education in Canada. *Canadian Journal of Science, Mathematics and Technology Education, 10*(4), 321–338.

Aikenhead, G. S., & Huntley, B. (1999). Teachers' views on Aboriginal students learning Western and Aboriginal science. *Canadian Journal of Native Education, 23*(2), 159–175.

Aikenhead, G. S., & Michell, H. (2011). *Bridging cultures: Indigenous and scientific ways of knowing nature.* Toronto, ON: Pearson Canada Inc.

Aikenhead, G. S., & Ogawa, M. (2007). Indigenous knowledge and science revisited. *Cultural Studies of Science Education, 2*(3), 539–591.

Andreotti, V. (2007). An ethical engagement with the other: Spivak's ideas on education. *Critical Literacy: Theories and Practices, 1*(1), 69–79.

Andreotti, V. (2011). *Actionable postcolonial theory in education.* New York, NY: Springer.

Apffel-Marglin, F. (2011). *Subversive spiritualities: How rituals enact the world.* New York, NY: Oxford University Press.

Association of Canadian Deans of Education (ACDE). (2010). *Accord on Indigenous education.* Retrieved from June 6, 2011, from http://mediarelations.con cordia.ca/pdf/Accord%20June1%202010.pdf.

Bang, M., & Marin, A. (2015). Nature–culture constructs in science learning: Human/non-human agency and intentionality. *Journal of Research in Science Teaching, 52*(4), 530–544.

Bang, M., Warren, B., Rosebery, A. S., & Medin, D. (2012). Desettling expectations in science education. *Human Development, 55*(5–6), 302–318.

Barad, K. (2000). Reconceiving scientific literacy as agential literacy. In R. Reed & S. Traweek (Eds.), *Doing Science + Culture* (pp. 221–258). New York, NY: Routledge.

Barad, K. (2007). *Meeting the universe halfway: Quantum physics and the entanglement of matter and meaning.* Durham, NC: Duke University Press.

Barad, K. (2010). Quantum entanglements and hauntological relations of inheritance: Dis/continuities, spacetime enfoldings, and justice-to-come. *Derrida Today, 3*(2), 240–268.

Barnhardt, R., & Kawagley, A. (2005). Indigenous knowledge systems and Alaska Native ways of knowing. *Anthropology and Education Quarterly, 36*(1), 8–23.

Barnhardt, R., & Kawagley, A. (2008). Indigenous knowledge systems and education. *Yearbook of the National Society for the Study of Education, 107*(1), 223–241.

Battiste, M. (2005). You can't be the global doctor if you're the colonial disease. In P. Tripp & L. J. Muzzin (Eds.), *Teaching as activism* (pp. 121–133). Montreal, QC: Queen's University Press.

Battiste, M. (2008). Research ethics for protecting Indigenous knowledge and heritage: Institutional and researcher responsibilities. In N. K. Denzin, Y. S. Lincoln, & L. T. Smith (Eds.), *Handbook of critical and Indigenous methodologies* (pp. 497–509). London, UK: Sage.

Battiste, M. (2013a). *Deconstruction and reconstruction: Roles, responsibilities and implications of a decolonizing framework.* Retrieved November 11, 2012, from http://www.indigenouseducation.educ.ubc.ca/transformation/indigenous-perspectives/.

Battiste, M. (2013b). *Decolonizing education: Nourishing the learning spirit.* Saskatoon, SK: Purich Publishing.

Belczewski, A. (2009). Decolonizing science education and the science teacher: A white teacher's perspective. *Canadian Journal of Science, Mathematics and Technology Education, 9*(3), 191–202.

Biesta, G. (2009). Witnessing deconstruction in education: Why quasi-transcendentalism matters. *Journal of Philosophy of Education, 43*(3), 391–404.

Blackstock, C. (2011). The emergence of the breath of life theory. *Journal of Social Work Values and Ethics, 8*(1), 1–16.

Blaut, J. (1993). *The colonizer's model of the world: Geographical diffusionism and Eurocentric history.* New York, NY: Guilford Press.

Bohm, D. (1994). *On creativity.* New York, NY: Routledge.

Butler, J. (2005). *On giving an account of oneself*. New York, NY: Fordham University Press.

Cajete, G. (1994). *Look to the mountain: An ecology of indigenous education*. Durango, CO: Kivaki Press.

Cajete, G. A. (1999). *Igniting the sparkle: An Indigenous science education model*. Durango, CO: Kivaki Press.

Cajete, G. (2000). *Native science: Natural laws of interdependence*. Santa Fe, NM: Clear Light Books.

Cajete, G. (2006). Western science and the loss of natural creativity. In F. Arrows (Ed.), *Unlearning the language of conquest: Scholars expose anti-Indianism in America* (pp. 247–259). Austin, TX: University of Texas Press.

Cajete, G. (2015). *Indigenous community: Rekindling the teachings of the seventh fire*. St Paul, MN: Living Justice Press.

Canadian Council on Learning (CCL). (2007). *The cultural divide in science education for Aboriginal learners*. Retrieved from http://www.ccl-cca.ca/pdfs/LessonsInLearning/Feb-01-07-The-cultural-divide-in-science.pdf on 12/09/2009.

Carter, L. (2004). Thinking differently about cultural diversity: Using post-colonial theory to (re)read science education. *Science Education, 88*(6), 819–836.

Carter, L. (2005). A place for alternative readings: Can they be of use? Responding to comments on "Thinking differently about cultural diversity: Using postcolonial theory to (re)read science education". *Science Education, 89*(6), 913–919.

Carter, L. (2006). Postcolonial interventions within science education: Using postcolonial ideas to reconsider cultural diversity scholarship. *Educational Philosophy and Theory, 38*(5), 677–692.

Carter, L. (2010). The armchair at the borders: The 'messy' ideas of borders, border zones and epistemological diversity in multicultural science education. *Science Education, 94*, 1–20.

Chinn, P. (2007). Decolonizing methodologies and Indigenous knowledge: The role of culture, place and personal experience in professional development. *Journal of Research in Science Teaching, 44*(9), 1247–1268.

Cobern, W. W., & Loving, C. C. (2001). Defining "science" in a multicultural world: Implications for science education. *Science Education, 85*, 50–67.

Cobern, W. W., & Loving, C. C. (2008). An essay for educators: Epistemological realism really is common sense. *Science & Education, 17*, 425–447.

Dion, S. D. (2007). Disrupting molded images: Identities, responsibilities and relationships—Teachers and indigenous subject material. *Teaching Education, 18*(4), 329–342.

Derrida, J. (1976). *Of grammatology* (G. C. Spivak, Trans.). Baltimore, MD: John Hopkins University Press.

Derrida, J. (1994/2006). *Specters of Marx: The state of the debt, the work of mourning, & the new international* (P. Kamuf, Trans.). New York, NY: Routledge.

Donald, D. (2012). Indigenous Métissage: A decolonizing research sensibility. *International Journal of Qualitative Studies in Education, 25*(5), 533–555.

El-Hani, C. N., & de Ferreira Bandeira, F. P. S. (2008). Valuing Indigenous knowledge: To call it "science" will not help. *Cultural Studies of Science Education, 3*(3), 751–779.

Erickson, G. (2000). Research programmes and the student science learning literature. In R. Millar, J. Leach, & J. Osborne (Eds.), *Improving science education: The contribution of research* (pp. 271–292). Philadelphia, PA: Open University Press.

Gilbert, J. (2001). Science and its 'Other': Looking underneath 'woman' and 'science' for new directions in research on gender and science education. *Gender and Education, 13*(3), 291–305.

Gough, N., & Price, L. (2009). Rewording the world: Poststructuralism, deconstruction and the 'real' in environmental/science education research. In R. Vithal, M. Setati, & R. Dhunpath (Eds.), *Researching possibilities in mathematics, science and technology education* (pp. 55–70). New York, NY: Nova Science Publishers.

Grande, S. (2004). *Red pedagogy: Native American social and political thought.* New York, NY: Rowman & Littlefield Publishers.

Grande, S. (2008). Red pedagogy: The un-methodology. In N. K. Denzin, Y. S. Lincoln, & L. T. Smith (Eds.), *Handbook of critical and Indigenous methodologies* (pp. 225–254). London, UK: Sage.

Haraway, D. J. (1989). *Primate visions: Gender, race, and nature in the world of modern science.* New York, NY: Routledge.

Higgins, M. (2014). De/colonizing pedagogy and pedagogue: Science education through participatory and reflexive videography. *Canadian Journal of Science, Mathematics and Technology Education, 14*(2), 154–171.

Higgins, M., & Madden, B. (2017). (Not so) monumental agents: De/colonizing places of learning. *Canadian Social Studies, 49*(1), 34–38.

Higgins, M., & Madden, B. (2018). (Not idling at) the Flâneur in Indigenous Education: Towards being and becoming Community. In R. Irwin & L. Cutcher (Eds.), *The Flâneur and Education Research: A Metaphor for Knowing, Being Ethical and New Data Production* (pp. 1–28). New York: Palgrave Pivot.

Higgins, M., & Madden, B. (2019). Refiguring presences in Kichwa-Lamista territories: Storying with Indigenous place. In C. Taylor & A. Bayley (Eds.), *Posthumanism and higher education: Reimagining pedagogy, practice and research* (pp. 293–312). Basingstoke, UK: Palgrave Macmillan.

Higgins, M., Madden, B., & Korteweg, L. (2015). Witnessing (the lack of) deconstruction: White teachers' "perfect stranger" position in urban Indigenous education. *Race Ethnicity and Education, 18*(2), 251–276.

Higgins, M., Mahy, B., Aghasaleh, R., & Enderle, P. (2019). Patchworking response-ability in science and technology education. *Reconceptualizing Educational Research Methodology, 2, 3*(2), 356–382.

Higgins, M., & Kim, E. J. (2019). De/colonizing methodologies in science education: Rebraiding research theory-practice-ethics with Indigenous theories and theorists. *Cultural Studies of Science Education, 14*(1), 111–127.

Higgins, M., & Tolbert, S. (2018). A syllabus for response-able inheritance in science education. *Parallax, 24*(3), 273–294.

Jackson, A. Y., & Mazzei, L. A. (2012). *Thinking with theory in qualitative research: Viewing data across multiple perspectives.* New York, NY: Routledge.

Kawagley, A. O. (2006). *A Yupiaq worldview: A pathway to ecology and spirit.* Long Grove, IL: Waveland Press.

Kayumova, S., McGuire, C. J., & Cardello, S. (2019). From empowerment to response-ability: Rethinking socio-spatial, environmental justice, and nature-culture binaries in the context of STEM education. *Cultural Studies of Science Education, 14*(1), 205–229.

Kim, E. J. A., Asghar, A., & Jordan, S. (2017). A critical review of traditional ecological knowledge (TEK) in science education. *Canadian Journal of Science, Mathematics and Technology Education, 17*(4), 258–270.

Kirby, V. (2011). *Quantum anthropologies: Life at large.* Durham, NC: Duke University Press.

Kuokkanen, R. J. (2007). *Reshaping the university: Responsibility, Indigenous epistemes, and the logic of the gift.* Vancouver, BC: UBC Press.

Kuokkanen, R. (2010). The responsibility of the academy: A call for doing homework. *Journal of Curriculum Theorizing, 26*(3), 61–74.

Lather, P. (2007). *Getting lost: Feminist efforts toward a double(d) science.* New York, NY: State University of New York.

Latour, B. (1993). *We have never been modern.* Cambridge, MA: Harvard University Press.

Lewis, B., & Aikenhead, G. (2001). Introduction: Shifting perspectives from universalism to cross-culturalism. *Science Education, 85,* 3–5.

Little Bear, L. (1994). Preface to Routledge classics edition. In D. Bohm (Ed.), *Of creativity* (pp. vii–xiv). New York, NY: Routledge.

Little Bear, L. (2000). Jagged worldviews colliding. In M. Battiste (Ed.), *Reclaiming Indigenous voice and vision* (pp. 77–85). Vancouver, BC: University of British Columbia.

Little Bear, L. (2016, June). *Big Thinking and rethinking: Blackfoot metaphysics 'waiting in the wings'.* Keynote address at 2016 Congress of the Humanities and Social Sciences, Calgary, Alberta.

MacIvor, M. (1995). Redefining science education for Aboriginal students. In M. Battiste & J. Barman (Eds.), *First Nations education in Canada: The circle unfolds* (pp. 73–98). Vancouver, BC: University of British Columbia Press.

Madden, B., & McGregor, H. E. (2013). Ex(er)cising student voice in pedagogy for decolonizing: Exploring complexities through duoethnography. *Review of Education, Pedagogy, and Cultural Studies, 35*(5), 371–391.

Marker, M. (2006). After the Makah whale hunt: Indigenous knowledge and limits to multicultural discourse. *Urban Education, 41*(5), 482–505.

Marker, M. (2019). Indigenous STEM success stories as disquieting decolonization: thoughts on new times and old thoughts about place-ness. *Cultural Studies of Science Education, 14*(1), 199–204.

McKinley, E. (2001). Cultural diversity: Masking power with innocence. *Science Education, 85*(1), 74–76.

McKinley, E. (2005). Locating the global: Culture, language and science education for indigenous students. *International Journal of Science Education, 27*(2), 227–241.

McKinley, E. (2007). Postcolonialism, Indigenous students, and science education. In S. K. Abell & N. G. Lederman (Eds.), *Handbook of research on science education* (pp. 199–226). Mahwah, NJ: Lawrence Erlbaum.

McKinley, E., & Aikenhead, G. (2005). Comments on "Thinking differently about cultural diversity: Using postcolonial theory to (re)read science education". *Science Education, 89*(6), 901–906.

McKinley, E., & Stewart, G. (2012). Out of place: Indigenous knowledge in the science curriculum. In B. Fraser, K. Tobin, & C. J. McRobbie (Eds.), *Second international handbook of science education* (pp. 541–554). Dordrecht, NL: Springer, Netherlands.

Nadeau, R., & Désautels, J. (1984). *Epistemology and the teaching of science.* Ottawa, ON: Science Council of Canada.

Nakata, M. (2007a). *Disciplining the savages: Savaging the disciplines. Exploring inscriptions of Islanders in Western systems of thought.* Canberra, Australia: Aboriginal Studies Press.

Nakata, M. (2007b). The cultural interface. *The Australian Journal of Indigenous Education, 36*(S1), 7–14.

O'Loughlin, M. (1992). Rethinking science education: Beyond Piagetian constructivism toward a sociocultural model of teaching and learning. *Journal of Research in Science Teaching, 39*, 791–820.

Parry, G. A. (2008). *Seed graduate institute: An original model of transdisciplinary education informed by Indigenous ways of knowing and dialogue.* Unpublished doctoral dissertation, California Institute of Integral Studies, Purdue University, San Francisco, CA.

Peat, D. (2002). *Blackfoot physics: A new journey into the Native American universe.* Newbury Port, MA: Weiser Books.

Peat, F. D. (2007). *Pathways of chance*. Pari, IT: Pari Publishing.

Patel, L. (2016). *Decolonizing educational research: From ownership to answerability*. New York, NY: Routledge.

Pillow, W. (2003). Confession, catharsis, or cure? *International Journal of Qualitative Studies in Education, 16*(2), 175–196.

Pomeroy, D. (1994). Science education and cultural diversity: Mapping the field. *Studies in Science Education, 24*, 49–73.

Rhee, J. E., & Subreenduth, S. (2006). De/colonizing education: Examining transnational localities. *International Journal of Qualitative Studies in Education, 19*(5), 545–548.

Sammel, A. (2009). Turning the focus from 'other' to science education: Exploring the invisibility of whiteness. *Cultural Studies of Science Education, 4*, 649–656.

Sammel, A. (2010). Finding the crack in everything: Exploring the causal promise in science education. *Canadian Journal of Science, Mathematics and Technology Education, 6*(4), 325–337.

Shapin, S., & Schaffer, S. (1985). *Leviathan and the air-pump*. Princeton, NJ: Princeton University Press.

Smith, L. T. (1999/2012). *Decolonizing methodologies: Research and Indigenous People* (2nd ed.). London, UK: Zed Books.

Smith, L. T. (2005). Introduction. *International Journal of Qualitative Studies in Education, 19*(5), 549–552.

Smith, L. T., Maxwell, T. K., Puke, H., & Temara, P. (2016). Indigenous knowledge, methodology and mayhem: What is the role of methodology in producing Indigenous insights? A discussion from mātauranga Māori. *Knowledge Cultures, 4*(3), 131–156.

Snively, G., & Corsiglia, J. (2001). Discovering Indigenous science: Implications for science education. *Science Education, 85*, 6–34.

Spivak, G. C. (1976). Translator's preface. In J. Derrida, *Of grammatology* (G. C. Spivak, Trans.) (pp. ix–lxxxvii). Baltimore, MD: Johns Hopkins University Press.

Spivak, G. C. (1988a). Can the subaltern speak? In C. Nelson & L. Grossberg (Eds.), *Marxism and the interpretation of culture* (pp. 271–313). Urbana, IL: University of Illinois Press.

Spivak, G. C. (1988b). *In other worlds: Essays in cultural politics*. New York, NY: Routledge.

Spivak, G. C. (1993/2009). *Outside in the teaching machine*. New York, NY: Routledge.

Spivak, G. C. (1994). Responsibility. *boundary 2, 21*(3), 19–64.

Spivak, G. C. (1999). *A critique of postcolonial reason*. Cambridge, MA: Harvard University Press.

St. Pierre, E. A. (2011). Post qualitative research: The critique and the coming after. In N. K. Denzin & Y. S. Lincoln (Eds.), *The Sage Handbook of Qualitative Research* (4th ed., pp. 611–626). Thousand Oaks, CA: Sage.

Subreenduth, S. (2006). 'Why, why are we not allowed even...?': A de/colonizing narrative of complicity and resistance in post/apartheid South Africa. *International Journal of Qualitative Studies in Education, 19*(5), 617–638.

TallBear, K. (2013). *Native American DNA: Tribal belonging and the false promise of genetic science.* Minneapolis, MN: University of Minnesota Press.

Traweek, S. (1992). *Beamtimes and lifetimes: The world of high energy physicists.* Cambridge, MA: Harvard University Press.

Tuck, E. (2009). Suspending damage: A letter to communities. *Harvard Educational Review, 79*(3), 409–428.

Tuck, E., & Yang, W. (2012). Decolonization is not a metaphor. *Decolonization: Indigeneity, Education & Society 1*(1), 1–40.

van Eijck, M., & Roth, W. M. (2007). Keeping the local local: Recalibrating the status of science and traditional ecological knowledge (TEK) in education. *Science Education, 91*(6), 926–947.

Wallace, M. F. (2018). The paradox of un/making science people: Practicing ethico-political hesitations in science education. *Cultural Studies of Science Education, 13*(4), 1049–1060.

CHAPTER 2

The Homework of Response-Ability in Science Education

The purpose of this chapter[1] is to introduce response-ability as a concept and practice to (re)open science education's understanding and enactments of responsibility towards Indigenous ways-of-living-with-nature (IWLN) and traditional ecological knowledge (TEK). This is significant as even well-intentioned forms of responsibility are often and inadvertently overcoded by the (neo-)colonial logics that it sets out to refuse and resist: responsibility and the ability to respond are often not one and the same. Within this chapter, I revisit a significant personal pedagogical encounter in which this distinction made itself felt and known. Thinking with the work of Sami scholar Rauna Kuokkanen, this narrative provides a platform to explore practices of epistemic ignorance and its (co-)constitutive relation to knowledge, as well as what she refers to as "the homework of response-ability" required to (re)open the norms of responsiveness towards the possibility of heeding the call of Indigenous science from within the structure of science education. Concluding thoughts underscore the promise of deconstruction (rather than destruction)[2] as a theoretical, methodological, and ethical tool to resist the (fore)closure of responsibility towards hospitably receiving Indigenous science on its own terms.

© The Author(s) 2021
M. Higgins, *Unsettling Responsibility in Science Education*,
Palgrave Studies in Educational Futures,
https://doi.org/10.1007/978-3-030-61299-3_2

PREAMBLE: RESPONSIBILITY
AND DE/COLONIZING SCIENCE EDUCATION

There is no single, simple, exhaustive answer to the complex question of how we can know the "other" .... We must redefine the problem not in terms of knowing the "other" but in terms of learning to "see" the existence of epistemes [ways-of-knowing] that have long been rendered invisible. We need to redefine the question in terms of a convergence of epistemes. (Kuokkanen, 2007, p. 120)

As explored in the previous chapter, "there is no single, simple, exhaustive answer to the complex question of how we can know the 'other'" (Kuokkanen, 2007, p. 120) *in* (and importantly *of*) science education. This is in part because even when Indigenous science (see Cajete, 2000) is included within school science, rather than excluded, it is often ways that differ from or defer its intended meanings. Differentially, each exclusion or problematic inclusion enacts (partial) dialectic negations of Indigenous science by sublating, subsuming, or suturing over it. "We must redefine the problem not in terms of knowing the 'other' but in terms of learning to 'see' the existence of epistemes [ways-of-knowing] that have long been rendered invisible" (Kuokkanen, 2007, p. 120) as the very framing of the problem is not without consequence. For Indigenous, diasporic, and other post-colonial students, such school science regularly produces experiences of cultural *assimilation* and *acculturation* rather than *enculturation*. In other words, rather than a harmonious interfacing of cultures (i.e., enculturation), encounters of school science are more likely to house potential for dialectical negation that is either actualized (i.e., assimilation) or remains un-actualized through students' complex and complicated curricular navigation (i.e., acculturation). For these students whose daily lived experiences continue to be negatively impacted by colonial logics (e.g., Eurocentrism), this manifests as a form of epistemic violence. Here, as mentioned within the previous chapter, science education and educators have a responsibility for which there are but only alibis for turning away.

However, in thinking with Sami scholar Rauna Kuokkanen (2007), whose scholarship centres Western modern educational and institutional responsibility towards Indigenous ways-of-knowing and ways-of-being, it is not sufficient to take up this responsibility: "We need to redefine the question" (p. 120) of responsibility itself. As stated above, and simply put, we need responsible ways to be responsible: science education cannot be responsible *towards* Indigenous science if it cannot perceive it and be responsible *for* the ways in which it has and continues to render it invisible. For there to be responsibility, there must be an ability to respond: how we (re)define and approach the question of IWLN in science education matters, in both senses of the word (i.e., matters and materializes).

As an emergent scholar pursuing unsettling science education and aspiring ally, the primary orientation that guides my efforts is ethically heeding the call of Indigenous science (e.g., traditional ecological knowledge [TEK], Indigenous ways-of-living-with-Nature [IWLN]). I continue to wrestle with the question: *How is Indigenous science to-come with/in the context of science education?* As Tewa scholar Gregory Cajete (2000) explains, Indigenous science displays and has, since time immemorial, always deployed "ingenuity, creativity, resourcefulness, and ability of people to learn and to teach a harmonious way of existence with Nature" (p. 78). Accordingly, the guiding question I pose is not intended to signal a science yet-to-exist. Rather, *to-come* calls on both the ways in which Indigenous science has not yet (wholly) arrived within the context of science education, as well as a responsibility of hospitality towards that which is to-come. If science education is to hospitably receive Indigenous science, it must address the ways in which its structures—the assumptions, terms, modes of organization, practices, and beliefs—contribute to exclusion of Indigenous science, as well as inclusion that disciplines, differs from, and defers Indigenous science (to-come). Equally significant is the exploration of the following: *how can the culture of the discipline of science education be (re)opened and re(con)figured to receive Indigenous science to-come, on its own terms, and in ethical relation? Importantly, when that which is to-come (here, Indigenous science) is never (fully) knowable within and distorted by the current frames of science education, what modes, practices, and enactments of responsibility are available? Lastly, how must responsibility be unsettled (i.e., decolonized and deconstructed) in order to (re)open the space of responsiveness?*

Significantly, in *unsettling* responsibility, it is important to understand the ways in which responsibility can be at once taken-for-granted (i.e., settled) and (re)produced through (neo-)colonial logics (i.e., settling). With such a double(d) understanding in mind, Kuokkanen (2007) quickly describes the conventional and (neo-)colonial concept of responsibility as such:

> Western liberal notions of responsibility are often constructed as a social Darwinist "burden of the fittest," with the benevolent imperialist self-cast as "helping" those less fortunate (read "privileged"). In this discourse, responsibility becomes nothing more than a duty... [producing] a hierarchy in which the "helper" enjoys moral superiority, which often manifests itself as a patronizing attitudes and practices. (p. 41)

Responsibility, at least as such, cannot and does not account for the ways in which the subject of responsibility might always already be in Indigenous-Western relations and the ways in which Western subjects already have responsibility in the juridical sense.[3] Further, it doubly individualizes responsibility, making responsibility an individual affair and rendering the individual the adjudicator of responsible action (e.g., "if an individual [educator] consciously believes that she or he is not racist, that is the end of the issue for that person and the end of her or his responsibility" [Kuokkanen, 2007, p. 63]).

Elsewhere (Higgins, 2014), I have begun asking similar questions of the relationship between responsibility and the (in)ability to respond within educational research: asking questions of response-ability (see also Higgins, 2017). This earlier exploration began exposing and troubling the ways in which I attempted[4] to account for and be accountable to Indigeneity (e.g., IWLN) from within (naturalized and normalized) (neo-)colonial discourses (e.g., Eurocentrism, whiteness), even though I was actively working against this power differential (see also Higgins, Madden, & Korteweg, 2015). Stated otherwise, as the result of a (neo-)colonial curriculum that is hidden in plain sight (see Battiste, Bell, Findlay, Findlay, & Henderson, 2005), efforts to work against and beyond (neo-)colonial categories, concepts, and structures often come to reify that which is laboured against; decolonizing approaches may come to be de/colonizing. In a nutshell,

> De/colonizing underscores the complexity of the material-discursive structures, commitments, and practices of educational institutions and the Indigenizing initiatives they pursue. It suggests that decolonization need not be (and conceivably cannot be) constructed in neat opposition to colonization, and calls for consistent examination of colonial logics and productions that seep into hybrid colonizing and decolonizing contexts. (Higgins & Madden, 2017, p. 35)

To take seriously de/colonizing is to be hyper-vigilant of the ways in which colonial logics and productions seep (even) into decolonizing efforts (see also Madden & McGregor, 2013). This, as Lyn Carter (2004, 2010) and Ali Sammel (2009) point out, is much needed in spaces of science education.

Within this chapter, I "begin" this exploration herein with a focus on the relationship between response-ability and my own practice as de/colonizing science educator through a narrative of a significant personal pedagogical encounter in which the distinction between responsibility and the ability to respond made itself felt and known.

Encountering the Subtle Yet Important Difference Between Response-Ability and Responsibility in My De/Colonizing Science Education Practice

Because we need to "begin" some-where and some-time, let's "begin" in Iqaluit, Nunavut in July of 2009.[5] At the time of authoring this book, the narrative I am about to tell is one that dates about ten years. However, it is one that I continue to heed as it continues to bear relevance on how I understand myself in relation to responsibility and the in/ability to respond. During the summer of 2009, I was delivering curriculum that I developed that engaged Indigenous (here, Inuit) and non-Indigenous youth in exploring, constructing, and documenting differential cultural constructions of science (i.e., ways-of-knowing-Nature) through participant-driven videography in their home community of Iqualuit, Nunavut. As my first major research project towards decolonizing science education (see Higgins, 2014), I was poised to learn a difficult lesson about the distinction between ability and willingness that Kuokkanen (2007) presents in how she defines response-ability: "an ability to respond, to respond to the world beyond oneself, as well as a willingness to recognize its existence" (p. 39). But then again, "decolonization ... is not as straightforward a strategy as it first seems" (p. 144). My willingness, intentionality, or desire to recognize the otherness that is Indigenous science (in relation to science education) was not sufficient in and of itself. But I am getting ahead of myself here.

Through this work, I took up the important call to decolonize science education through what Mi'kmaq scholar Marie Battiste (2013) describes as the "two-prong process" of decolonizing education. It simultaneously and iteratively entails *deconstruction* of (neo-)colonial structures and strategies, and *reconstruction* that centres and takes seriously Indigenous, diasporic, and other post-colonial ways-of-knowing and ways-of-being towards reshaping the place-based processes and priorities of education and educational research. Both prongs are of significance given my positionality as a white, male, fourth-generation Euro-settler of Irish and Scottish descent who is working to honour my ever-shifting relationships as a science educator working with/in diverse First Nations, Métis, and Inuit communities.

Responding to the first prong (i.e., deconstruction), I engaged in examining and challenging the ways in which Eurocentrism—a pervasive discursive force that (re)centres Western modern(ist) culture, people, places, and histories as the normative standard against which other ways-of-knowing are judged, usually as lesser and deficient (Battiste, 2005)—works to maintain the status quo. This was done through working to disrupt the concepts and categories that tend to create, and are utilized to uphold, inequality within science education, as well as the systems under which these inequalities become possible (e.g., "what counts" as science in science education and its entangled apparatus of norms; see Chapters 3–6). This process largely involved engaging in critical self-reflexive questioning, responding to queries such as:

> How does my Western training in the world of science (i.e., in physics) differentially produce my conceptions of the nature of science, what it is, what it is perceived as, and what it can be? How do I work against the problematic foreclosure of such knowledge in order to maintain pedagogical flexibility? How do I work within and against the implicit Eurocentric notions of validity, empirical worth, and instrumentality that I have received in order to make space for Indigenous knowledges? (Higgins, 2014, p. 163)

In engaging in the second prong of decolonizing education (i.e., reconstruction), students were collaboratively involved in creatively juxtaposing Western modern science (WMS) and *Inuit Qaujimajatuqangit* (i.e., Inuit traditional knowledge) to reveal, (re)structure, and (re)direct the multiple ways that the gaze of dominance is maintained. For example, this gave youth an agentic role in resisting problematic constructions of Indigeneity with respect to ways-of-knowing-Nature (e.g., science). This participant-directed videography took various shapes, notably documentary-style interviews with diversely positioned community members (e.g., traditional knowledge holders, health practitioners, environmental scientists), alongside their own short movies that were a form of digital storytelling. Despite well-laid plans, in attempting to put to work a decolonizing sensibility and taking up responsibility (both that of pedagogy and educational research), there were nonetheless ways in which I was not able to respond. This subtle but importance differentiation between responsibility and response-ability manifested most noticeably though my self-reflexive work around the youth's engagement in digital storytelling (see Higgins, 2014). While my research was originally planned around youth engaging in documentary film, early on the youth made it clear (through scrunched brows, an Inuit way of saying no) that they did not wish to only make movies about their perception of science during what was, for them, a STEM summer camp. Cognizant of the multiple gradients of power across which dissent was being articulated, as well as the ongoing problematic research relationships between research institutions and Indigenous communities (see Battiste, 2013), I knew I would have to "let go" of the research project as designed.

While the youth agreed to participate in interviewing community members around ways-of-knowing-Nature, we negotiated that the youth would *primarily* engage in digital storytelling practices as their major project to share with parents and community members by the end of the programme. As an emerging decolonizing educator, I recognized the importance of respecting learners' choices. However, as a budding science education researcher, it was difficult to shift away from a focus I had been developing (i.e., exploring cross-cultural ways-of-knowing-Nature). Because youth were spending less time accessing those who "know" about and with Nature (e.g., traditional knowledge holders, health practitioners, environmental scientists), I had trouble conceptualizing the youth's storytelling practices as enacting ways-of-knowing-Nature. This dissonance was perhaps most heightened when some of the youth explored *Oreo eating Olympics* as a central story topic! In this sense, I was not able to take up responsibility much beyond the ways the ways in which I had conceived of prior the research; there were ways in which I was not able to respond.

Significantly, plugging Indigenous metaphysics into an educational framework organized by a Western modern metaphysics resulted in excesses that were far too often sutured over, subsumed, and/or sublated by an approach to knowing and being with nature that could not account for Indigenous ways-of-knowing-in-being's excesses.[6] Specifically, I had come to the research with the settled, Western modern scientific assumption that *science* is strictly a human, epistemological affair (see Barad, 2007; Cajete, 2000). Because of the ways this dialectic negation operates, as a result, I was blind(ed) to the ways in which some of the digital storytelling practices that youth engaged in (e.g., (re)telling of traditional story of *Mahaha*, the aptly named traditional Inuit monster who chases after children to tickle them to death with his long claws) were not simply stories *about* place but were told *with* place (i.e., having and being had by an Indigenous "sense of place"; see Cajete, 1994, 2000). In the videos, place makes itself intelligible through the beings that come to (co-)constitute the ecology of relationships that make the eastern arctic a beautiful, yet dangerous place if not respected on its own terms. Their stories "starred" an ecology of relationships with which Inuit peoples have developed ways-of-knowing-in-being premised on Nature's flux and processes, deeply guided with and through relational ethics, as well as practices of regeneration. The stories were never the students' (and the humans they worked with) alone (despite the frames brought to the viewing); the natural world always makes itself intelligible and participates in the construction of knowledge about itself, whether we acknowledge it or not (see Cajete, 1994, 2000).

Elsewhere (Higgins, 2014), I stated that the decolonizing curriculum (e.g., border crossing) and pedagogies (e.g., culture broker) available to me worked both within and against a problematic centre. As such, curriculum, pedagogy, and pedagogue were exceeded in pedagogical practice by the very coloniality the approach worked against, thus becoming de/colonizing: pedagogical slippage occurred. Here, thinking with Kuokkanen (2007), coloniality overcoded the ability to respond, making me unable to (fully) take up the responsibility of heeding the call of Indigenous science. Specifically, I could not (wholly) respond to the natural world and Indigenous-ways-of-living-with-Nature because I could not recognize its existence (beyond that which made itself intelligible within my frames). I could not responsibly heed the call of Indigenous science because I could not hear the call as such. Again, a willingness to recognize the otherness that is Indigenous science is not sufficient[7]; it is specifically for this reason that Kuokkanen (2007) suggests that response-ability entails addressing epistemic ignorance.

EPISTEMIC IGNORANCE AND/IN SCIENCE EDUCATION

Kuokkanen (2010) states, "if knowledge is a prerequisite for responsibility, ignorance presents a serious threat to responsible, response-able behaviour and thinking" (p. 64). Yet, as illuminated by my desire to recognize the existence of Indigenous science from within a Western modern(ist) episteme, working to know Indigenous-way-of-living-with-Nature from such a perspective is also a project that is fraught. As Kuokkanen (2008) suggests, the relation between knowledge and ignorance is not so linear or dichotomous; not knowing is not necessarily an absence of knowledge, but can also be the result of knowledge.[8] Kuokkanen (2008) refers to this knowledge-as-ignorance, as well as traces the discursive forces and flows through which it emerges as epistemic ignorance:

> Epistemic ignorance refers to ways in which academic theories and practices ignore, marginalize and exclude other than dominant Western European epistemic and intellectual traditions. These "other" epistemic and intellectual traditions are foreclosed in the process of producing, reproducing and disseminating knowledge to an extent that generally there is very little recognition and understanding of them. Epistemic ignorance is thus not limited to merely not-knowing or lack of understanding. It also refers to practices and discourses that actively foreclose other than dominant epistemes and refuse to seriously contemplate their existence. Epistemic ignorance is thereby a form of subtle violence. (p. 63)

As Michiel van Eijck and Wolff-Michael Roth (2007) underscore, this is certainly the case in science education regarding the relationship between WMS, TEK, and IWLN. Drawing on Michel Foucault, they explain that the logics of science education can often be characterized as a "regime of truth". Regimes of truth are marked by circular relations: each "truth" is but a differential articulation of the systems of power that produces it, whose articulation in turn (re)produces the systems of power. Such a circular relation can be read in two ways: first, as the capillary circulation of power from one conceptual node to another; and second, signaling a (quasi-)hermetic circle, a (fore)closure of knowledge.

Foreclosure, as post-colonial scholar Gayatri Spivak (1999) utilizes it, signals instantiated pre-emergence of meaning. It indicates the ways in which the language we possess also possesses us. It is when the knowledge shapes how we intake experience, preventing experiences of otherness to be anything more than what can already be known within the already existing, and rigidified, circular relations of closure. In other words, how what we know acts as barrier to engaging with what we do not; a form of closure that is a priori to meaning-making.

The foreclosure resulting from epistemic ignorance is of particular relevance considering the ways in which IWLN and TEK are often only considered science when they fit the criteria of "valid" science (which often happens to be that of WMS). Or, as Kuokkanen (2008) summarizes, the foreclosure resulting from epistemic ignorance makes it such that "Indigenous people 'cannot speak'; that is, when they speak from the framework of their own epistemic conventions, they are not heard or understood by the academy" (p. 60). As mentioned within the previous chapter, after Ngāti Kahungunu ki Wairarapa and Ngāi Tahu scholar and science educator Liz McKinley (2007), the ways in which Indigenous science can be articulated in relation to WMS maps onto four general categories: (a) where Indigenous science can be explained within WMS; (b) where Indigenous science *could be* explained through WMS, but the explanation has yet to be developed; (c) where there is a link between Indigenous science and WMS's knowledge claims, albeit through different knowledge principles and practices; (d) where WMS cannot accept aspects of Indigenous science (e.g., spirituality, animism). This cartography of relations comes to shape if, as well as when and how, Indigenous science is to be included within school science curriculum. Importantly, the degree to which "included" Indigenous science differs from its intended purposes or is deferred through non-inclusion depends highly upon the degree to which it is already articulable within the terms of WMS, as well as science education's ability to ethically respond to difference (from itself). As a result, some forms of Indigenous science "cannot speak" (Kuokkanen, 2008) and remain more "to-come" than others.

It is for this reason that Kuokkanen (2010) states that "the responsibility towards the other must not emerge from hierarchical relations" (p. 69) as these often come to reproduce the very structures of said hierarchy (e.g., here, the epistemic privileging of WMS over other ways-of-knowing-Nature such as TEK and IWLN). Not only do these hierarchal relations potentially produce foreclosure of (the possibility of) knowing otherwise for those who would wish to uphold the hierarchy, but possibly also for those who are critically within and against it (as illustrated by my story earlier). The reproduction of such hierarchies of relationships (which also remain on-the-move) is not always a conscious choice; even work founded in best intentions to challenge inequitable relations may come to reify problematic structures. Integrating Indigenous science into an educational programme that has not come to examine the ways in which it (re)produces and is (re)produced by forms of epistemic ignorance runs the risk of (re)producing similar problematics, albeit differently. For example, this can result in enacting pedagogies or curriculum that work towards "'rescuing' the 'other' or knowing what is best for the 'other'" (Kuokkanen, 2010, p. 69). It can also corral Indigenous ways-of-living-with-Nature into a (neo-)colonial space of intelligibility without accounting for or be accountable to the ways in which it differs and exceeds such framing. In turn, the work of responsibility towards the ways in which Indigenous science comes to be othered, as well as to-come, in science education requires more than a desire for the relationship to be otherwise if and when the possibility of ethical relationality is (fore)closed by epistemic ignorance.

There is work to be done: work that addresses not only what we do not know, but also how what we know prevents us from knowing what we do not. This is, following Kuokkanen (2007), the homework of response-ability.

The Homework of Response-Ability (Towards Indigenous Science) in Science Education

Doing homework is an ongoing practice that includes learning as much as possible about the area where the academic takes risks. However, familiarizing oneself with areas one knows little about still amounts to hegemonic practice if we do not engage in the "home" part of the homework.... Homework starts from where we are. (Kuokkanen, 2007, p. 117)

As stated, earlier, responsibility is often premised upon the possibility of knowing the other(ness) to which we are responding. But, as the discursive formations of science education often come to foreclose the very possibility of (wholly) heeding such a call (through varying degrees of epistemic ignorance), attempting to know about Indigenous science requires that we engage with, as Kuokkanen (2007) invites, the "'home' part of the homework" (p. 117) for it to be more than a "hegemonic practice" through which responsibility becomes response-inability through its enactment (even when the individualistic desire is otherwise).[9]

The "home" part of homework can take many meanings: home as cultural, disciplinary, geographical, historical, epistemological, ontological, among others. However, *home*work is always risky as it threatens to rupture who we (think we) are, what we (think we) know, and what we (think we) do. Addressing the ways in which the multiplicity of "homes" in homework are (fore)closed when responding to otherness to-come such as Indigenous science is a project that can be unsettling. In part, this is because it asks critical science educators to examine and sit with the ways in which (their) science education practice continues to uphold problematic practices of subsuming, sublating, and suturing over of Indigenous science. Yet, it must bear risk if we are to (re)open responsiveness and the ability to respond to the (constructed) otherness of Indigenous science which is to-come: "responsibility with an inventive rupture implies, first and foremost, the ability of interrupting the self, of moving beyond the 'I' as the ethical subject" (Kuokkanen, 2010, p. 65). Moving beyond the "I" as the ethical subject means considering the ability to respond as being more than individualistic (without excusing the "I" from responsibility): responsibility must be more than a (double(d) practice of) self-accounting for individually doing or having done the right thing. This is a key component of homework. As Kuokkanen (2010) explains, addressing Western modernity as the cultural "home" of science education entails addressing its "worldview of individualism and the notion of the Cartesian subject, [in which] dependency on others is considered a burden" (2010, p. 62), as well as the ways in which this comes to shape responsibility.[10]

For science educators, moving beyond the "I" as the ethical subject entails considering the self-in-relation as always already (co-)constituted by vectors of power such as whiteness, Eurocentrism, (neo-)coloniality, modernity, neoliberalism, amidst many others and their respective but irreducibly linked historicities and futurities-to-come, even when working against them. This is all the more important for those who, like me, occupy markers of identity that are privileged by these systems. It is the homework of attending to the ways in which the forces and flows of dominance come to produce the (fore)closure of both self and otherness (making both invisible the normalization of normativity as well as that which lay beyond). Considering the self-in-relation also entails the unheroic work of not assuming that critical pedagogy will always be empowering (Ellsworth, 1989), but examining the ways in which this approach may always already be disempowering and prevent participants from responding (from "home"). It is an attempt at a double(d) reversal of the gaze of dominance: a gaze that entails both the literal reversal of studying those who do the studying (i.e., in order to reverse the direction of the gaze), as well as the study of the ways in which those who do the studying study (i.e., in order to reverse the ways in which the gaze is produced and producible).

Yet, while such a double(d) reversal is important, it does not reduce accountability for and towards the other. As mentioned earlier, decolonizing science education must be a movement that creates openings in (neo-)colonial systems and also leverages openings towards making space for honouring Indigenous peoples, places, practices, and priorities. We must still attempt a response within this relation of responsibility, even if response-ability may never be (fully) achieved. Battiste (2005) underscores this (im)possibility using the example of Eurocentrism: "Eurocentrism is not like a prejudice from which informed peoples can elevate themselves" (p. 122). It is for this reason that there is need to reconfigure the normative processes through which we respond and enact responsibility (within the "home" that is science education):

> What is more, 'starting from here' involves a subtle but radical shift from 'knowing the other' to learning, and more specifically, learning to receive. Rather than assuming the possibility of knowing the other, we need to learn to think in a fundamentally different way. Instead of thinking that 'we must know' or even 'we are entitled to know'—positions that, by retaining the sense of ownership as well as distance, allow very little room for hospitality...—we need to draw a difference, however provisional, between knowing and learning. (Kuokkanen, 2010, p. 68)

In other words, moving beyond the "I" as the ethical subject entails recognition of the ways in which the Other is always already an irreducible and (co-)constitutive part of the self-in-relation of response-ability.[11] Rather than the individualistic project of knowing the other (which, as mentioned earlier, cannot be disassociated from forms of epistemic ignorance), we are called to learn from the other (something that requires, by definition, relationality). As Kuokkanen (2010) states, not all learning results in knowing the other: response-ability "requires not only patience but acceptance that there will always be gaps, the 'other' can never be

fully known" (p. 70). As we learn from the other (recognizing the important, yet subtle distinction between learning *about* and *from*), we can
learn to learn. Stated otherwise, as we attempt to heed the call of Indigenous science (that is not wholly intelligible as such within the epistemes
of science education), we must not only listen, but also listen to how
we listen (for the ways in which listening prevents us from hearing).[12]
It is a subtle and attentive movement that necessarily vacillates between
knowing and not knowing in order to (re)open the norms of responsiveness in order to not only heed the call of Indigenous science but
also work towards hospitably receiving this plurality of diverse Indigenous
ways-of-living-with-Nature.

CONCLUSION: RESPONSE-ABILITY AS MOVING WITHIN, AGAINST, AND BEYOND THE (FORE)CLOSURE OF EPISTEMIC IGNORANCE OR DECONSTRUCTION AS LEARNING TO LEARN

Science education *always already* has a responsibility towards TEK and
IWLN; such responsibility precedes its being.[13] However, it is not always
able to enact and uphold this task. As explored within the significant
encounter I opened with, the ways in which I became science educator
(fore)closed my ability to respond to Indigenous science because I could
not heed its call as such. For example, at the time, I could not respond to
the ways in which TEK and IWLN are always already more than strictly
a human practice that is enacted by the other-than-humans that come
to constitute place. My knowledge of what science *is* was knowledge
that acted as epistemic ignorance towards what science *could be*; specifically, what it always already is and continues to be since time immemorial
in the form of TEK and IWLN. There was and continues to unsettle
responsibility, in both senses of the word.

   In turn, thinking with this experience generated the central question
explored throughout: How is the irreducible responsibility that science
education has towards Indigenous science to be enacted when Indigenous
ways-of-knowing-in-being are made unintelligible, undesirable, and/or
invisible through science educations' very systems of thought? This
chapter coalesces around the (co-)constitutive relation between knowledge and ignorance (e.g., knowledge-as-ignorance) through Kuokkanen's
(2008) conceptualization of epistemic ignorance, such that the inability to

respond can be framed as rendering unintelligible that which lay beyond epistemic registers, and also inefficacy to account for or be accountable to the ways in which engagement is fraught through the naturalization of said frames. Accordingly, the homework of responsibility through which we labour to transform response-inability into response-ability must entail a (re)opening of the closure through which the other cannot be heard, while simultaneously working to heed the call:

> doing one's homework implies unlearning one's privilege and learning... It requires the critical examination of one's beliefs, biases, and assumption as well as an understanding of how they have developed and become naturalized in the first place. (Kuokkanen, 2007, p. 115)

It is important to note that addressing the "home" of homework in attempting to move beyond the (fore)closure of knowledge that is epistemic ignorance is not only deeply productive, but also necessary. Addressing the multiplicitous "home" in the homework of response-ability must begin from the ways in which we are shaped by "home" towards its (co-)constitutive exteriority. Kuokkanen (2007) states, as we engage within the very structures that produce epistemic ignorance, this work requires "subtlety and responsibility". Homework that too quickly attempts to evacuate the ways in which we carry "home" elsewhere runs the risk of reproducing the same problems in a new context, albeit slightly differently. Rather, it is important to continue labouring within and against "home" as we attempt the move beyond; to move too quickly to a theory-practice beyond without attending to the "home" of homework runs the risk of as "proceeding in any other way would eventually backfire and merely too tight[ly] reinforce existing structures and discourse [through]... 'irresponsibilizing destruction'" (Kuokkanen, 2007, p. xx). In other words, Kuokkanen (2007) advocates for a deconstructive and critical inhabitation of these structures rather than an attempt to move beyond through their destruction: "the process of decolonization can only emerge from *within* those structures of domination, from inside" (p. 146).[14]

Deconstruction provides a way out of the (fore)closure of knowledge by (re)opening the interiority to its co-constitutive exteriority, and revealing the ways in which the "philosophical category of the centre (named Eurocentrism)" (McKinley & Aikenhead, 2005, p. 902) operates. It is to attend to porosity between the two in order to displace, disrupt, and decentralize that which was placed with/in (and in turn with/out). As McKinley and her ally Glen Aikenhead state: "deconstruction is the decentralization and decolonization of European thought… Hence, deconstruction is a deconstruction of the concept, the authority, and the assumed primacy of the category of 'the West'" (2005, p. 902). Deconstruction provides a means of engaging with the interplay of knowing and not knowing that is inextricably linked to epistemic ignorance, as well as the possibility of placing self and other in relations that (re)open the possibility of learning to learn (as opposed to "knowing the other" within the structures afforded).[15]

Importantly, the possibility of hospitably *receiving* Indigenous science is not only an ethical call. Working to heed the call is ontologically, epistemologically, ecologically, and politically generative: science education stands to learn much from Indigenous ways-of-knowing-in-being and its practices of relational balance, (re)generation, and renewal.

NOTES

1. An earlier version of this chapter appears in Bazzul and Siry's (2019) *Critical Voices in Science Education Researsch* (pp. 223–233), and is reprinted with permission.
2. As responsibility is at once necessary yet inadequate, "deconstruction may offer a new way of challenging conventional understandings of responsibility by seeking to move beyond traditional interpretations of politics and ethics" (Kuokkanen, 2007, p. xx).
3. If not in the individual sense, then responsibility is juridical at least in the structural sense. See Chapter 6 for a lengthier discussion on the tie between legal and scientific systems.
4. It is worth latching onto the word *attempt*, momentarily, as it is significant. It marks an important disjuncture between taking up responsibility and actualizing it that was the deconstructive *snag* that was used to begin unsettling responsibility (e.g., here, intentionality is often a double(d) settled meaning that is equated with responsible action).
5. As explored within the previous chapter, the question of where to "begin" an inquiry is trouble(d) from its very articulation: with each and every *here-now* there is a multiplicity of *there-thens* which are enfolded into its being and becoming. This turn of phrase, to appear in every chapter, serves first to remind the reader that to "begin" as if nothing had been done, what Kuokkanen (2007) refers to as "know-nothing-ism," is its own form of irresponsibility as it has *always already* begun *elsewhere* and *elsewhen*. Second, it is to suggest that because we are *already* within the

inquiry, there are many productive junctures with/in which the inquiry can delve into.

Significantly, linking this to problematizing position (rather than positionality) statements, Kuokkanen (2007) offers:

> The responsibility of academics [and educators] cannot be limited to neutral descriptions of who we are, as has become common practice at least. In the more self-reflective, critical academic circles; it must also link itself to the concrete, physical locations of our enunciation. (p. 117)

The last part is significant as, for many educators in settler-colonial states (e.g., the US, Canada, Australia), physical location places one in a direct and irreducible relation with Indigeneity (either as someone ancestrally belonging to the place since time immemorial; or as a guest, whether invited or not):

> Considering how many universities are located on [I]ndigenous peoples' lands, [I]ndigenous epistemes have always existed in the physical space of the university, however invisible or ignored. Without waiting to be invited, [I]ndigenous epistemes are already "in" the academy. The problem is not how to bring [I]ndigenous knowledge to the university, since it is already there. The problem is the epistemic ignorance that prevails because of the gift of [I]ndigenous epistemes remains impossible within the academy. (p. 108)

Again, this is to say that we are *always already* within the question of Indigeneity within science education: the *somewhere* where we "begin," whether a university or other educational institution, is on Indigenous Land and places us firmly in relation. Importantly, as we are already in relation, part of the work is addressing science education's epistemic ignorance which prevents us from recognizing it as such, and in ways that are of consequence to our ways-of-knowing-in-being. "Beginning" *somewhere* matters (in both senses of the word).

6. While not wanting to reduce this to language, it is important to recognize that the language we possess also possesses us (see Spivak, 1976): language matters in terms of the phenomena we can bear witness to. As David Peat (2002) explains:

Our [scientific] language disposes us to employ concepts that are entirely inappropriate for the quantum world.... [David] Bohm rejected the idea of a reality composed of objects in interaction in favor of processes and activities in a continuous movement of unfolding and enfolding. Moreover, this reality is not confined to matter but extends to thoughts, feelings, and emotions unfolding within the brain and body. To Bohm there was no dichotomy between inner and outer, mental and physical, subjective and objective, for all are aspects of one underlying movement. The English language [as intended], however, keeps bringing us back to a world of objects. (p. 237)

David Bohm, in his search for a means of accounting for and being accountable to the process, flux, and relationality of quantum phenomena, recognized that these phenomena not only exceeded the enacted practice of science but also the language with which science thought about these phenomena. Bohm's (1980) frustrations with language centred largely around the subject/object binary enacted through much of the English language and its inability to account for contexts which exceed this framing:

The subject-verb-object structure of language, along with its world view, tends to impose itself very strongly in our speech, even in those cases in which some attention would reveal its evident inappropriateness. For example, consider the sentence "It is raining." Where is the "It" that would, according to the sentence, be the "rainer that is doing the raining?" Clearly, it is more accurate to say: "Rain is going on." Similarly, we customarily say, "One elementary particle acts on another," but... each particle is only an abstraction of a relatively invariant form of movement in the whole field of the universe. So it would be more appropriate to say, "Elementary particles are on-going movements that are mutually dependent because they merge and interpenetrate." (p. 37)

For Bohm, language is more than a representational tool that mirrored reality; it is also an enacted enfolding and an unfolding of our epistemologies and ontologies (see also Barad, 2007; Peat, 2002; Spivak, 1993/2009). Language would then come to shape what was and what was not knowable and do-able, as well as if and how it could come to be. Thus, in order to differently consider and respond to these excessive quantum phenomena, Bohm laboured to envision a hypothetical language that he called the "rheomode" (see Bohm, 1980).

Stemming from the greek word *rheo*, meaning *to flow*, the language-to-come that is the rheomode would have been a language that dealt with processes and activity, transformation and change. This would not only be of use for the quantum phenomena he observed, but also for the interconnected work to which he brought these understandings of process and flux to, such as dialogue and creativity (see Bohm, 1994, 1996). As Peat (2002) explains, the rheomode "is based primarily on verbs and grammatical structures deriving from verbs. Such a language, Bohm argued, is perfectly adapted to a reality of enfolding and unfolding" (p. 238).

However, not unlike Barad (2007), Bohm (1980) states that while the linguistic structures that serve us to quickly describe the macro-world are potentially better served by subject/object language than quantum phenomena, it too is exceeded:

> The same sort of description holds on the larger-scale level. Thus, instead of saying, "An observer looks at an object," we can more appropriately say, "Observation is going on, in an undivided movement involving those abstractions customarily called 'the human being' and 'the object he is looking at.'" (p. 37)

In taking the double(d) meaning of unsettling seriously, one of most frequent critiques of engaging with the differential configurations of the English language that scholars such as Bohm (1980) and Barad (2007) propose and enact that I continue to encounter is one of unintelligibility. "Why do we need a new language that is so complicated?" some might ask. There is importance in the un/intelligible to push us beyond what we know, how we come to know, and in turn, what we can know. This is not only to differentially become aware of the epistemological and onto-logical configurations that are always already present within language, but also to (re)open that space, to foster a space of response-ability, a space which allows us to consider otherwise excluded otherness and act in turn.

However, in closing this story (and lengthy footnote), what Bohm had conceived of as a hypothesis and a theoretical language already had lived and long-lasting analogues in place. The similar yet different vocabulary of the "new physics" that stretches the boundaries of Western science was part of the everyday (but differently articulated) vocabulary of Indigenous people:

> A few months before his death, Bohm met with a number of Algon[qu]ian speakers and was struck by the perfect bridge between

their language and worldview and his own exploration of philosophy. What to Bohm had been a more breakthrough in human though – quantum theory, relativity, his implicate order and rheomode – were part of the everyday life and speech of the Blackfoot, [Mi'kmaq], Cree, and Ojibwa[y]. (Peat, 2002, p. 238)

It is potentially the case that Bohm did not know because he *could not* know if the "lines of making sense" (Spivak, 1993/2009) were laid out such that science, particularly at the time, perceived IWLN as wholly other; but this marks the importance of unsettling settler structures in order to (re)open the space of responsiveness.

7. For Kuokkanen (2007), response-ability requires an ability and willingness to recognize a world beyond oneself. However, and relevant to this story, she cautions that recognition itself must be both an ongoing practice and one that is not isolated from complimentary practices. She states, "recognition cannot be merely an item on a list that, once checked, requires no further consideration" (p. 93), as we often see as the case in practices such as Land acknowledgements at the beginning of events. In these instances,

Recognition can, therefore, also become a proxy for avoiding any responsibility for doing the homework of finding out about things that are unfamiliar – a way of closing the doors and windows rather than granting an unconditional welcome. (p. 91)

Explicitly stated, recognition is always fraught. In taking deconstructive complicity or co-constitution seriously, recognition and misrecognition share an intimate relation:

To recognize someone is always to misrecognize others and render them and their work invisible. Thus, recognizing someone's work … always involves conforming to certain predetermined norms, that is, to a set of prevailing assumptions about what is worth recognizing. (p. 91)

This is to say that good intentions alone are not sufficient in moves of recognition (even if a willingness to recognize is necessary, an important first step): if practices of recognition cannot be answerable to its own practices (e.g., here, attempting to perceive Indigenous ways-of-knowing-in-being through a Western modern scientific worldview), recognition can remain little more than lip service.

8. Further, linking to the previous chapter's discussion of forms of know-nothing-ism, ignorance "is not merely an innocent lack of knowledge

but an intentional not-knowing that serves the economic interests of the status quo" (Kuokkanen, 2007, p. 68). This can be thought of in numerous ways: a refusal (or, softer, a deferral) to learn (despite numerous resources), not taking seriously that which is offered (as a form of epistemic superiority and hubris), or other.

9. Taking seriously the "home" in the "homework of response-ability" includes accounting for and being accountable to our participation in any of the phenomena, scientific or otherwise, that we come to bear witness to. For example, Kuokkanen (2007) invites us to differently consider the problematic relation between Indigenous and Western knowledges systems:

> Cultural discontinuity is a consequence, not a cause. It is the result of two things: the willful ignorance that is embedded in the mainstream middle-class culture; and the logic of Eurocentric rationalism, which denies the existence of intellectual conventions and perceptions of the world other than those rooted in the Enlightenment. (p. 54)

Here, there is a clear responsibility. However, there is homework to be done to be able to: first, receive such knowledge; and, second, to respond to and with it.

10. Butler (2005), along these lines rhetorically asks the following: "does the postulation of a subject who is not self-grounding, that is, whose conditions of emergence can never be fully account for, undermine the possibility of responsibility and, in particular, of giving an account of oneself?" (p. 19). Where it is the common assumption that one must "know oneself" (as a stable, containable, and individualistic subject) in order to be morally responsible, Butler (2005) suggests that ethics can only emerge because the "I" of ethics only emerges with/in relation and that a self-contained self becomes a source of "'moral narcissism' whose pleasure resides in its ability to transcend the concrete world that conditions its actions and is affected by them" (p. 105). Relation begets and precedes responsibility; knowledge of a partially knowable self is still, nonetheless knowledge upon which one can act.

11. Along similar lines, Barad (2007) suggests,

> There are no individual agents of change. Responsibility is not ours alone. And yet our responsibility is greater than it would be if it were ours alone. Responsibility entails an ongoing responsiveness to the entanglement of self and other, here and there, now and then. (2007, p. 394)

Because of this entanglement, responsibility is not simply a moral imperative but rather an ethico-onto-epistemological enactment that shapes and is shaped by our relational "being-of-the-world." Here, being-of-the-world is a response to Heidegger's being-in-the-world which necessitates an 'I' which precedes the relationship with the world, inevitably masking the conditions of emergence for the 'I' which notably only comes to be because there is a world from which it was never separated or separable.

Thus, as Barad (2010) states, responsibility precedes us because it is produced with/in the co-constitutive relationships through which the 'I' of responsibility *becomes*:

> Entanglements are relations of obligation–being bound to the other–enfolded traces of othering. Othering, the constitution of an 'Other', entails an indebtedness to the 'Other', who is irreducibly and materially bound to, threaded through, the 'self'–a diffraction/dispersion of identity. (Barad, 2010, p. 265)

The (Western modern) self does not come to be without *an* Other; we are always indebted to the inheritance of the other to whom we are with/in relation, regardless of whether this *self* can or does respond to such responsibility (see Higgins & Tolbert, 2018).

12. This is to refuse and resist "listening-as-benevolent-imperialism" (Spivak in Harasym, 1990, p. 59), or the ways in which the attempt to listen through dominant frameworks does not always result in hearing. This underscores, as brought up in the preface, the need for a slower cross-cultural or decolonizing science education practice that is rife with unsettling pauses.

13. See Chapter 6 for a lengthier exploration of the temporality put forth by this statement.

14. It is not only productively multiplicitous but also, as Derrida (1976) reminds, an inescapable and necessary condition:

> The movements of deconstruction do not destroy structures from the outside. They are not possible and effective, nor can they take active aim, except by inhabiting those structures. Inhabiting them *in a certain way*, because one always inhabits, and all the more when one does not suspect it. Operating necessarily from the inside, borrowing all the strategic and economic resources of subversion from the old structure, borrowing them structurally, that is to say without being able to isolate their elements and atoms, the enterprise of deconstruction always in a certain way falls prey to its own work. (p. 24, emphasis mine)

There is no outside of "where we are," only the differential ability to respond to the relationality and responsibility that precede and shape us through critical inhabitation *in a certain way:* active engagement with the norms that structurally and invariably shape our becoming "all the more when one does not suspect it" (Derrida, 1976, p. 24; see also Spivak, 1976, 1993/2009).

Or as scientist Mazzocchi (2006) states on the question of (re)opening possible possibilities within the context of cross-cultural scientific endeavours:

> We need to open ourselves to participating in the experience of others, and yet we should also be aware that this opening can only start from where we already are – from our point of view or the tradition to which we belong. (p. 465)

15. In making the case that there are commensurabilities between Indigenous and deconstructive thought, Kuokkanen (2007) speaks to the ways in which deconstruction offers not only a way out of the (fore)closure of epistemic ignorance, but also the closure marked by *epistemology itself*.

> ... both [I]ndigenous thought and deconstructive practice recognize that human existence is embedded in intricate webs that can never be fully grasped either fully or once all.... But perhaps it is productive to recognize that deconstructive practices have something to offer [I]ndigenous scholarship, which sometimes tends to ground itself in modernist views rooted in linear reality, and in assumptions that easy access to a neutral truth or the human consciousness is somehow possible. (p. 62)

Without wanting to romanticize deconstruction, as "many theorists of deconstruction are heavily invested in the West's ontological and philosophical traditions" (Kuokkanen, 2007, p. 62), it is productive to engage in with deconstruction in the double(d) movement of unsettling science education (here, deconstructing *epistemic ignorance* towards its co-constitutive ontology), a topic to be explored in greater depth in the chapters to come.

REFERENCES

Barad, K. (2007). *Meeting the universe halfway: Quantum physics and the entanglement of matter and meaning.* Durham, NC: Duke University Press.

Barad, K. (2010). Quantum entanglements and hauntological relations of inheritance: Dis/continuities, spacetime enfoldings, and justice-to-come. *Derrida Today, 3*(2), 240–268.

Battiste, M. (2005). You can't be the global doctor if you're the colonial disease. In P. Tripp & L. J. Muzzin (Eds.), *Teaching as activism* (pp. 121–133). Montreal, QC: Queen's University Press.

Battiste, M. (2013). *Decolonizing education: Nourishing the learning spirit.* Saskatoon, SK: Purich Publishing.

Battiste, M., Bell, L., Findlay, I., Findlay, L., & Henderson, J. (2005). Thinking place: Animating the Indigenous humanities in education. *The Australian Journal of Indigenous Education, 34,* 7–18.

Bohm, D. (1980). *Wholeness and the implicate order.* New York, NY: Routledge.

Bohm, D. (1994). *On creativity.* New York, NY: Routledge.

Bohm, D. (1996). *On dialogue.* New York, NY: Routledge.

Butler, J. (2005). *On giving an account of oneself.* New York, NY: Fordham University Press.

Cajete, G. (1994). *Look to the mountain: An ecology of indigenous education.* Durango, CO: Kivaki Press.

Cajete, G. (2000). *Native science: Natural laws of interdependence.* Santa Fe, NM: Clear Light Books.

Carter, L. (2004). Thinking differently about cultural diversity: Using postcolonial theory to (re)read science education. *Science Education, 88*(6), 819–836.

Carter, L. (2010). The armchair at the borders: The 'messy' ideas of borders, border zones and epistemological diversity in multicultural science education. *Science Education, 94,* 1–20.

Ellsworth, E. (1989). Why doesn't this feel empowering? Working through the repressive myths of critical pedagogy. *Harvard Educational Review, 59*(3), 297–325.

Harasym, S. (1990). *The post-colonial critic: Interviews, strategies, dialogues.* New York, NY: Routledge.

Higgins, M. (2014). De/colonizing pedagogy and pedagogue: Science education through participatory and reflexive videography. *Canadian Journal of Science, Mathematics and Technology Education, 14*(2), 154–171.

Higgins, M. (2017). Post-qualitative mo(ve)ments: Concluding remarks on methodological response-abilities and being wounded by thought. *Reconceptualizing Educational Research Methodology, 8*(3), 89–101.

Higgins, M., & Madden, B. (2017). (Not so) monumental agents: De/colonizing places of learning. *Canadian Social Studies, 49*(1), 34–38.

Higgins, M., Madden, B., & Korteweg, L. (2015). Witnessing (the lack of) deconstruction: White teachers' 'perfect stranger' position in urban Indigenous education. *Race Ethnicity and Education, 18*(2), 251–276.

Higgins, M., & Tolbert, S. (2018). A syllabus for response-able inheritance in science education. *Parallax, 24*(3), 273–294.

Kuokkanen, R. (2010). The responsibility of the academy: A call for doing homework. *Journal of Curriculum Theorizing, 26*(3), 61–74.

Kuokkanen, R. J. (2007). *Reshaping the university: Responsibility, Indigenous epistemes, and the logic of the gift.* Vancouver, BC: UBC Press.

Kuokkanen, R. J. (2008). What is hospitality in the academy? Epistemic ignorance and the (im)possible gift. *Review of Education, Pedagogy, and Cultural Studies, 30*(1), 60–82.

Madden, B., & McGregor, H. E. (2013). Ex(er)cising student voice in pedagogy for decolonizing: Exploring complexities through duoethnography. *Review of Education, Pedagogy, and Cultural Studies, 35*(5), 371–391.

Mazzocchi, F. (2006). Western science and traditional knowledge. *EMBO Reports, 7*(5), 463–466.

McKinley, E. (2007). Postcolonialism, Indigenous students, and science education. In S. K. Abell & N. G. Lederman (Eds.), *Handbook of research on science education* (pp. 199–226). Mahwah, NJ: Lawrence Erlbaum.

McKinley, E., & Aikenhead, G. (2005). Comments on "Thinking differently about cultural diversity: Using postcolonial theory to (re)read science education". *Science Education, 89*(6), 901–906.

Peat, D. (2002). *Blackfoot physics: A new journey into the Native American universe.* Newbury Port, MA: Weiser Books.

Sammel, A. (2009). Turning the focus from 'other' to science education: Exploring the invisibility of whiteness. *Cultural Studies of Science Education, 4,* 649–656.

Spivak, G. C. (1976). Translator's preface. In J. Derrida, *Of grammatology* (G. C. Spivak, Trans., pp. ix–lxxxvii). Baltimore, MD: Johns Hopkins University Press.

Spivak, G. C. (1993/2009). *Outside in the teaching machine.* New York, NY: Routledge.

Spivak, G. C. (1999). *A critique of postcolonial reason.* Cambridge, MA: Harvard University Press.

van Eijck, M., & Roth, W. M. (2007). Keeping the local local: Recalibrating the status of science and traditional ecological knowledge (TEK) in education. *Science Education, 91*(6), 926–947.

Critical Possibilities and Possible Critiques Through Deconstructive Play in/of the Multicultural Science Education Debate

Serious Play: Inflecting the Multicultural Science Education Debate Through and for (Socratic) Dialogue

The purpose of this chapter is to differentially revisit the *multicultural science education debate*, which is a central curricular location in science education that acts as both a potential entry point and problematic gate-keeping device for Indigenous science to-come, by inflecting it with a potentially less oppositional mode of meaning-making. In short, by inspecting and inflecting the culture of this debate, this chapter engages in the *cultural* homework of response-ability. Within this debate, it is generally agreed upon by science educators that there is a clear moral imperative to respect students from diverse cultural backgrounds within the multicultural science education classroom.[1] However, what constitutes respect and how it is enacted continues to be hotly debated; in turn, this presents itself as a rich location to unsettle science education. A significant contributing factor is how conceptions of respect are deeply intertwined with, including influenced and impacted by, considerations of "what counts" as science. This has produced two largely incommensurable positions around the inclusion of Indigenous ways-of-living-with-Nature (e.g., ethnoscience, Indigenous knowledge systems, Indigenous science): those who contest its status as scientific knowledge and those who champion it. However, as the process of debate enacted is commonly one of opposition, there is little room for meaning made across positions. Above and beyond addressing the sources of knowledge that continue to uphold this serious debate, this chapter plays with/in the

© The Author(s) 2021
M. Higgins, *Unsettling Responsibility in Science Education*,
Palgrave Studies in Educational Futures,
https://doi.org/10.1007/978-3-030-61299-3_3

debate processes as a means of opening these foreclosed spaces in science education as both form and content lead to the excluding, differing, and deferring of Indigenous science to-come.

As with most other chapters in this book, this one begins with a positional vignette meant to simultaneously introduce the topic, situate myself within the inquiry, and provide furtive glances at the concepts, questions, and curiosities to-come.

PRELUDE TO (A) SERIOUS PLAY

Because we need to "begin" some-where and some-time, let's "begin" in Thunder Bay, Ontario in February of 2009.[2] During a graduate student conference (and as a graduate student), I was asked to take down a poster I was presenting minutes after I put it up. The campus-wide graduate research poster session had barely begun, as many graduate students hurried to hang their posters up. During this time, a member of the university's Faculty of Science took note of my poster as he was walking through the exhibit, seemingly on route elsewhere. However, he stopped upon seeing my poster, his face reddening as his pace accelerated. "I'm going to request that you take this poster down", he tersely demanded. Unsure as to why the request was being made, and unable to make sense of the physical cues he was exhibiting, I nervously asked, "Why?". The point of contention, he said, was the title of the poster, "Shared horizons: A dialogue between Indigenous and Western science", as well as its content below. In short, the poster highlighted my own work in cross-cultural science education in which I endeavoured to juxtapose and braid Western modern science (WMS)[3] and local enactments of traditional ecological knowledge (TEK),[4] namely Inuit (i.e., Indigenous) science in the circumpolar region (see Higgins, 2011, 2014). Feeling my pulse quicken, I asked him to elaborate with noticeable defensive and deliberate emphasis: "What is *your* issue with this?" The short of his response was that the friction was stemming from the cultural prepositions (i.e., Indigenous and Western) preceding the word science. In other words, as he told me, "there's no such thing as Indigenous science, or Western science for that matter". Agitatedly, he added, "There is only *science*". Reading the issue as one of not recognizing Indigenous knowledge systems as valid and productive ways of knowing nature, I too was visibly frustrated. In an attempt to recover ground, I situated the cross-cultural work within a longstanding and ongoing conversation in science education. To this, he retorted that this type of engagement was not happening "in science". With the tension escalating, both of us growing increasingly irritable, and neither hearing nor being heard, he repeated his demand to take down the poster. I outright refused. Our exchange ended as the faculty member went to seek out a member of the graduate student conference's organizing committee to enforce his request.

As a science educator who has worked over ten years in Indigenous communities, making space for diverse ways-of-knowing-nature was and continues to be a commitment that is both personal and political. I could not simply take down my poster. While it cannot be stated with certainty, it is likely that the science faculty member also had his own commitments that beckoned him to firmly take the position expressed. Nonetheless, there is a part of me that wishes that this exchange could have played out differently. In hindsight, and giving the faculty member benefit of the doubt, the poster could have acted as a productive conversational pivot for both involved. What if I perceived his position of science as singular and universal as something other than a potentially disrespectful "the ends justify the means" approach to morality in teaching science? What if he perceived my position of advocating for scientific pluralism as something other than "anything goes" epistemic relativism? While I would like to think that we shared a common desire for science and science education that is rigorous and empirical engages nature yet not indoctrinating through cultural imposition, it appeared as though we had implicitly agreed to disagree before said conversation could begin. If the demand for me to take down my poster had not shut down the possibility of conversation, my reaction, rebuttal, and refusal certainly did. The act of attempting conversation further entrenched our respective positions, giving the impression that a combative and antagonistic relationship was the only type we could inhabit.

While this is but one experience from my perspective, the character-istics that mark it are not isolated. Michiel van Eijck and Wolff-Michael Roth (2007) state, "one can be surprised about the fierce debate that currently shakes the foundations of science education" (pp. 927–928) and, I would add, the plurality of locations in which it is and continues to be occurring (e.g., science teacher education). Often referred to as the multicultural science education debate,[5] it is not strictly isolated to a singular place but extends to plural locations that are continuously and differentially shaped by Western colonial relationships. These include

locations: (a) where settlers remain and have become numerically dominant (e.g., New Zealand, Australia, Canada, United States, Peru, Taiwan);
(b) where colonial settlers have never reached majority and/or that have
undergone formal decolonizing as defined by the United Nations (e.g.,
India and many African nations); and (c) in which displaced diasporic
communities live, whose forced migration from the lands in which their
cultural identity developed is the result of colonialism past and present
(e.g., descendants of chattel slaves in former British and French colonies,
Hmong immigrants [formerly from Thailand] in China and the US)
(Aikenhead & Ogawa, 2007; McKinley, 2007).

At stake are notions of "what counts" as science within the context
of the multicultural classroom,[6] and how their entanglement, impact,
and influence constitute respect and how it is enacted towards culturally
diverse students. Seemingly most polarizing, and of central significance in
this inquiry, is the status of TEK as scientific knowledge (or *equally valid*
to scientific knowledge).

I would argue that the ongoing debate is not the result of poorly
formulated scholarly arguments. Rather, it is a symptom of the oppositional and antagonistic modes through which the multicultural science
education debate operates (see McKinley & Stewart, 2012), such as was
the case in the exchange between myself and the Faculty of Science
member in the introductory vignette. For the multicultural science
education debate to move towards shared meanings and understandings, it must become and remain an open process rather than a sedimented product. Open channels of communication allow for productive
engagement across and between positions, rather than a protective and
prohibitive form of disengagement, such as that demonstrated within
the introductory vignette. As Elizabeth St. Pierre (1997) reminds us,
the goal of educational research should be to both "produce different
knowledge and knowledge differently" (p. 175). As such, this chapter
seriously engages both within and against the norms that shape the
culture of the multicultural science education debate itself by differentially (re)presenting the multicultural science education debate literature.
However, the purpose here is not to produce new knowledge through the
advancement of either universalist or cross-culturalist positions as is traditionally the case. Instead, I undertake, and provide tools for readers to
engage the task of producing knowledge differently through a differential
critical engagement with the knowledge production process that occurs
within this debate in order to move towards the possibility of shared
meanings.[7]

Programme for (a) "Serious Play" To-Come

Differential knowledge (re)production takes the form of (a) "serious play" as means of possibly working within and between the often "scripted" (i.e., well-established and entrenched) positions of universalist and cross-culturalist in order to (re)open these foreclosed[8] spaces of meaning-making. In short, the "serious play" introduced here but expanded upon later signals an entangled conceptual apparatus comprising Socratic dialogue, Bohmian dialogue, and Derridean "play". Together, these inform the textual strategies used as well as the practices of reading advocated for. There are sharp distinctions between Socratic dialogue (i.e., Platonic form of representation of live, face-to-face discussions between two "scripted" positions), and Bohmian dialogue (i.e., the stream of meaning through which something is made *in* common), as well as *a* play (i.e., theatrical drama), and Derridean play (i.e., the always already present possibility and process of (re)signification; see Derrida, 1976). However, through (a) "serious play", this very term takes a double(d) meaning in that it is both a theatrical drama (i.e., *a* play) engaging with serious topics as well as a serious commitment and engagement with the (re)signification (i.e., Derridean play) of the concepts and terms within. Similarly, while the format of Socratic dialogue traditionally represents discussions between two parties, it also acts as an invitation to the reader who is open to being in a Bohmian dialogue with the text to create movement of meaning between the two positions (re)presented.

The goal of this chapter is to encourage and invite "serious play", which can be read in two distinct ways. On one hand, it can be understood as a call for science educators to seriously (re)engage in the drama of foreclosed conversations around notions such as "what counts" as science and other hotly debated issues within multicultural science education. On the other hand, it can be interpreted as an invitation for science educators to engage with the always already possible play of (re)signification of these notions and the associated positions. Together, serious play calls for the dialogical movement of meaning that occurs through and during the possible play of (re)signification, be it between people, concepts, or (scripted) positions, in order to produce a different set of possible possibilities[9] emerging from discussions such as the one at the beginning of the paper as well as the one that is mimicked within the Socratic dialogue to come.

There are four "acts" to this serious play. The first act sets the stage for the dialogue. I further expand upon the conceptually entangled notion of "serious play" through suggesting Bohmian dialogue as a means of working towards non-adversarialism, expanding upon the Derridean "play" of (re)signification, and proposing Socratic dialogue as a means of representing and producing this play. In the second "act," reading notes are provided. I describe the universalist and cross-culturalist positions participating within the Socratic dialogue and beyond, and also make suggestions for readers looking to engage in the serious play of Bohmian dialogue. In the third "act", the Socratic dialogue on multicultural science education is "played out". Herein, both universalist and cross-culturalist characters enunciate points of contention and agreement within multicultural science education (e.g., "what counts" as science) while providing a space for readers to potentially engage in differential meaning-making around these issues. The fourth and final "act" that follows the Socratic dialogue is an exploration of how knowledge is both reproduced and potentially differentially produced within the multicultural science education debate. This section also engages with recent literature that endeavours to open up multicultural science education through (re)signification of locations that threaten to foreclose the possibility of further play (i.e., those which remain unresolved and unresolvable within the debate).

ACT I: SETTING THE STAGE FOR (A) "SERIOUS PLAY"

From the Dialectic of Discussion to Bohmian Dialogue: An Ethic for Seriously Playing Together

The positioning and approach enacted between the Faculty of Science member and myself within the introductory vignette could be stated to be what Janice Moulton (1983) calls the "adversary method". She characterizes it by its aim "to show that the other party is wrong, challenging them on any possible point, regardless of where the other person agrees" (p. 156). Similarly, David Bohm (1996) might qualify the above engagement between the faculty member and myself as "discussion". Discussion, having the same suffix as percussion and concussion, evokes imagery of verbal jousting in which speakers must beat, bang, and thump one another's arguments in order for one meaning to emerge victorious through the dialectic negation of the other. In order to achieve this, the discussant must make common their view by subsuming, sublating, or suturing over those with whom they are discussing. This "making common" is often aggressive and adversarial, the shared assumptions required to engage in conversation on common ground are rarely reached. In turn, not only is it "not a good way to convince someone who doesn't agree with you" (Moulton, 1983, p. 156) but it also breaks down the very possibility of communication (see also Latour, 2004a; Kirby, 2011). As was the case between the faculty of science member and myself, our respective refusal to consider the other's point of view (re)entrenched our respective positions. As Bohm (1996) asks, "how can you share if you are sure you have the truth and the other...[similarly] has the truth, and the truthes don't agree?" (p. 43). However, this did not mean we needed to concede our respective viewpoints. While there is always moments in which one must inhabit imposed norms within communicative spaces as a necessity, such inhabitation often requires the one conforming to allow parts of themselves to be dialectically negated. This type of subsuming, sublating, and suturing over often results in either *a* communication (i.e., a singular imposed meaning) or a *null* communication (in which silences are rife with meaning; see Mazzei, 2007) but not necessarily communication as an open process of back-and-forth.

For these reasons, there is a call from scholars such as Bohm and Moulton for dialogue rather than the dialectic of discussion, a call to listen rather than strictly talk. While both dialectic and dialogue begin from an encounter in which two differing views on a similar or same topic encounter one another, their ethic of resolution differs. Dialogue's Greek roots entail through (*dia*, as opposed to *di* which would simply signal two)[10] the meaning of the word (*logos*). Rather than a dialectic contest between dichotomized views, dialogue acts as a stream of meaning, a process of communication in which those engaged are not concerned with defeating propositions or in which meaning is to be made common through imposition but rather a process through which meaning is being made *in* common. Because it is a non-adversarial model, a process through which shared meanings are made together, it does not require that the meanings interfaced together to be negated for something new to emerge. This creates space for the possibility for meaning-making positions in-between that are often lost and foreclosed in dialectic and discussion.[11]

In and through dialogue, Bohm (1996) calls for "suspended action". The suspension is a call to listen that is framed as both a listening to others and to oneself in that the act "listening" to how we listen to others can tell us much about ourselves, including the values that frame what and how we "hear". Between the science faculty member and myself, there were diverse epistemic as well as affective cues that we could have sensed of ourselves and of the other (e.g., physical indicators of tension). They may have signalled the ways that we were being played by our respective personal assumptions. For example, if I could take my quickening pulse as an indicator of meaning on the move to partially glimpse at my then held assumption of epistemic universalism being diametrically opposed to respectful multiculturalism, perhaps I could have differently participated in the conversation. The purpose of "suspended action" is then to come to awareness, albeit partial, of how values are inflected, deferred, and deflected through our selves. It allows us to re-think the self-in-relation to the norms that shape how, who, and what we can be (see also Butler, 2005; Foucault, 1997; Mazzei, 2007; Peat, 2007). The action following a suspension period in which we consider gentle ways in which the situation can be re(con)figured, is a non-adversarial process through which shared meanings can potentially be made together.

The Serious Play of (Re)Signification

Within the introductory vignette, one of the potential issues at hand was that the very terms and conditions that shaped engagement, such as "what counts as science", were sedimented, stratified, and thus unable to move towards the shared meanings called for in Bohmian dialogue. In order for conversations around and about multicultural science education to be modes of dialogue rather than discussions, it required that the terms of engagement not be foreclosed before they are brought up. In order words, there is a need for the possibility of play. While the play that I am advocating for here is not the same as that of a child who might re-imagine a branch as wand, it is a useful metaphor with which to think as play is deeply tied to the act and possibility of (re)signification. (Re)signification is a conceptual process of un-binding and interchangeability of what something "is not" (e.g., a wand) and "is" (e.g., a stick). It is an interplay of absence and presence with respect to what is signified while retaining the signifier (Derrida, 1976).[12] In turn, it allows for movement within the stream of meaning-making through the disruption and destabilization of stratified and sedimented meaning. Such decentering creates a space of meaning-making that allows for the production of positions in-between "present" and "absent". Play is not a complete unravelling and an undoing, but rather a critical and complicit use of that which is played with, so that new meanings can take hold (Lather, 2007; Spivak, 1993/2009). Accordingly, when that which is played with is no longer the material of children's games, "play can be serious business" (Mazzei, 2007, p. 22).

Socratic Dialogue as (a) Serious Play

For revitalizing critical conversations that mobilize meanings of what quality multicultural science education "is" and "is not", it is useful to think through a medium that works within and against the adversarial and scripted discussions operating within the field: Socratic dialogue. Plato's Socratic dialogues are often read as a representation of a live, face-to-face discussion between two scripted positions. These positions are usually in unresolved, and potentially unresolvable, opposition to one another in a manner that is not so dissimilar from the introductory vignette. However, as Richard Smith (2011) reminds us, while it could be said that Socratic dialogues are but another form of representation, they are even more so a textual invitation to think across and between the positions presented within. As Socratic dialogue is never fully captured by one definition or the other, holding these two differing readings of Socratic dialogue in tension make it an effective choice for thinking about multicultural science education. Given the ways in which the positions discussing multicultural science education are scripted, Socratic dialogue offers itself as an effective medium through which to represent a possible discussion, hence allowing for working within the script (e.g., TEK's status of "counting as science" defended or denied). Furthermore, Socratic dialogue's implicit invitation to dialogue across differently positioned concepts and commitments allows for working against the script (i.e., towards the possibility of shared meanings).

In addition to this textual invitation, Socratic dialogue provides the disruption and destabilizing through serious play required for the stream of meaning associated with dialogue. As Smith (2011) elaborates: (a) Socratic dialogues are strictly imitations, (b) imitations are always a form of play, and, accordingly, (c) Socratic dialogues are never more than (a) play. Mimesis or imitation creates "an opportunity to adopt 'alternative persona' as a mechanism for addressing difficulties" (Turnbull & Mullins, 2007, p. 94) and playing with/in difficult positions.[13] Although the universalist and cross-culturalist characters (re)presented within the upcoming Socratic dialogue are but imitations, seriously and respectfully engaging with them and their interplay means not turning them into caricatures of their positions. While Socratic dialogue is long known to engage in serious topics, it is a mode in which seriousness and playfulness are always in tension and inverting so that "they do not form a fixed binary" (Smith, 2011, p. 230) as they are always already enacting the play of (re)signification. In addition, as it is (a) play with serious topics, "most attempts in the dialogues to reach a definition—of courage, friendship and so on—fail resoundingly" (Smith, 2011, p. 223). This "failure" to reach fixed and (fore)closed conclusions or outcomes has much to do with play's productive prevention of foreclosure through keeping meaning on the move. Accordingly, this generates a space for meaning-making that is never fully prescriptive (i.e., meaning is not already made for the reader but rather always open to a certain degree of interpretation), and always open to further play (i.e., meaning is open to the possibility of (re)signification as the context under which it is signified differs). For this reason, Socratic dialogue acts as an invitation to the reader to engage in their own playfulness with the ideas found within, to play with serious notions that desire to be (re)produced differently.

ACT 2: THE PROGRAMME FOR (A) "SERIOUS PLAY": A PRIMER FOR PLAYING ALONG

Who Is Playing (or Played)?

The Socratic dialogue intentionally plays within the period when this debate reached its peak. As such, this dialogue primarily focuses on and draws from what van Eijck and Roth (2007) refer to as "one of the landmarks" (p. 927) of multicultural science education (see also McKinley & Stewart, 2012),[14] as well as some of the pieces the symposium authors were responding to. Accordingly, the "key players" and their respective scholarship primarily informing and inspiring the universalist account are the work of William Cobern and Cathleen Loving (2001), Michael Matthews (1994), Harvey Siegel (1997, 2001), and Sherry Southerland (2000). Similarly, the cross-culturalist position which is "played out" draws on the scholarship of Bradford Lewis and Glen Aikenhead (2001), Gloria Snively and John Corsiglia (2001), as well as William Stanley and Nancy Brickhouse (1994, 2001).

Within the field of science education, universalists are those who uphold the belief that WMS is the best, most valid, and/or powerful way of knowing the natural world. Furthermore, universalists see WMS as a knowledge-practice that is quasi-objective, quasi-neutral, and/or quasi-transcendental vis-à-vis culture. On the subject, Matthews (1994) states that universalists:

> ... regard science as an intellectual activity whose truth-finding goal is not, in principle, affected by national, class, racial or other differences: science transcends human differences... This universalist view recognizes that while aspects of culture do influence science, nevertheless cultural considerations do not determine the truth claims of science. (p. 182)

WMS's epistemic superiority with regard to knowing nature is substantiated by WMS's ability to explain, predict, empiricize, and stabilize natural phenomena, and apply this scientific knowledge to produce technologies that are unique (e.g., airplanes, modern medicine). Because WMS can at once be framed as culturally specific and universal, universalists within the multicultural science classroom continue to deliver science education curriculum which is about enhancing students' scientific literacy by developing an appreciation, and understanding, and applications of WMS in everyday life. However, recognizing that universalism and multiculturalism are not a mutually exclusive and dichotomous, universalists endeavour to instructionally provide culturally diverse students whose opportunities to learn WMS in ways that are respectful of that diversity.

Within this Socratic dialogue, the *universalist* character's account and engagement is primarily inspired and informed by the aforementioned scholars who take this stance with respect to multicultural science education. Also, in keeping with the goal of "serious play", these accounts are inflected by recent scholarship that revisits the multicultural science education debate as a means of working towards shared meanings from a universalist position (e.g., Cobern & Loving, 2008; El-Hani & de Ferreira Bandeira, 2008; El-Hani & Mortimer, 2007; Zeyer, 2009).

Cross-culturalists are those who take up the pluralization and proliferation of scientific perspectives, and the localization of scientific knowledge without characterizing knowledges as relativistic (see McKinley, 2007). Cross-culturalists accept and promote the idea that TEK can and should be considered "equal" to that of WMS. This does not entail treating them *equal* (i.e., the same) but as *equally valid*. Cross-culturalists work from the assumption that because WMS stems from the dominant, hegemonic culture, many students who are not of that culture will experience science education as a form of dominance and hegemony. This experience is, in part, rooted in the valuation of WMS through decentering and devaluing of their own cultural knowledge. As a result, cross-culturalists uniquely endeavour to reform curriculum in addition to modes of instruction. With respect to science education curriculum, there is a commitment to create space in which TEK is included and simultaneously decenters WMS (see Aikenhead & Ogawa, 2007; Aikenhead & Michell, 2011; LeGrange & Aikenhead, 2016). It is important to note that what is argued for here is not a teaching of *all* non-Western-modern bodies of knowledge, which could result in relativism, but rather an inclusion of diverse and longstanding empirical ways-of-knowing the natural world. With respect to instruction within the multicultural classroom, the focus shifts from having students accept culturally diverse scientific notions towards developing understanding and appreciation. This often entails that cross-culturalist teachers assist their students in respectfully and meaningfully navigating cultural spaces that are not their own without the requirement that students abandon whichever way-of-knowing-nature that they are further developing.

In this Socratic dialogue, the *cross-culturalist* account and engagement is primarily inspired and informed by the aforementioned scholars who take this stance with respect to multicultural science education. To engage in the play of (re)signification, the cross-culturalist account is also inflected by revisitations of the multicultural science education debate that works towards shared meanings from a cross-culturalist position (e.g., Alsop & Fawcett, 2010; McKinley & Stewart, 2012; van Eijck & Roth, 2007).

It is important to note that while there are often internal similarities within both of these positions in terms of how they are both defined and enacted, there is also diversity. While the characters within the dialogue are *a* universalist and *a* cross-culturalist who are composite characters of the scholars who inspired this dialogical inquiry, there is not *a* singular way of being either a universalist or a cross-culturalist. While we should perhaps change the terms of the conversation to account for *universalisms* and *cross-culturalisms*, it is generally the case that both universalists and cross-culturalists will internally agree on the notion that WMS *is* or *is not* the most or only valid way of knowing the natural world respectively. However, the strength of such a "what counts as science" claim and other interrelated claims varies greatly within both positions, as well as the degree to which one embodies such beliefs within their research and teaching practices. It could even be argued that one need not consciously or explicitly hold the belief that WMS *is* or *is not* the most or only valid approach to be a universalist or a cross-culturalist respectively but rather that any and every science educator's curriculum and pedagogy can implicitly demonstrate and convey such commitments (see Carter, 2004; McKinley, 2000; Sammel, 2009).

These normative and counter-normative values are always already circulating and inflecting the signification of what it means to be science educator. As such, it is worth highlighting here that the strict use of two characters is not an attempt to present them as *the* universalist and *the* cross-culturalist. Furthermore, even if that were the goal, the ongoing play of (re)signification makes it such that the position, even if there is an attempt to fully account for the range of diversity, is always exceeded, ruptured, and overturned. In other words, even in a more comprehensive and expansive definition of both universalist and cross-culturalist positions, there is always already someone who identifies as either but only partially fits within. Rather than attempting to engage in the impossibility of presenting *an* essence, this dualistic choice is made as a means of, first, working within the constraints that are posited by Socratic dialogue as a methodology. Secondly, it is an attempt to reproduce some of the particularities that occur through the adversary method. In particular, the assumption that:

> the only, or at any rate, the best, way of evaluating work ... is to subject it to the strongest or most extreme opposition. And it is assumed that the best way of presenting work ... is to address it to *an imagined opponent* and muster all the evidence one can to support it. (Moulton, 1983, p. 153, emphasis in original)

This *imagined opponent* is often the most radical proponent of that which is considered the adversary. While scholars often see their own encampment as rich and diverse, through the adversary method, the opposing side often becomes a parody of itself. As a result, the imagined adversary who is addressed is often positioned in a manner that few, if any, scholars occupy. Universalists are not-so-gently cast as wholly privileging epistemic claims (*through* the norms of Western modernity) over moral or ethical ones, and vice versa for cross-culturalists. This creates perceived positions which are either strawmen or deeply untenable. If more modest positions are not more frequently discussed it is because scholars "who cannot be recast into an adversarial mold are likely to be ignored" (Moulton, 1983, p. 155). In some ways, like Socratic dialogue, the multicultural science education debate has become a scene where drama (i.e., the serious matters of "what counts" as science) and comedy (i.e., the parody-like positioning of adversaries) alternate and seep into one another.

Rules for (a) Serious Play

Serious play is not something that occurs on its own.[15] Rather, the act of "playing" (i.e., (re)signification) is a process that requires both players (e.g., "readers") and played (e.g., texts). Furthermore, "playing" occurs through the movement in-between the two. Consequently, as a reader you are encouraged to engage with the Socratic dialogue in a manner that is dialogical (i.e., so that there is movement in-between your views and the views presented within the text and something is made *in common* through reading), rather than dialectic (i.e., so that either your views or the views presented within the text are *made* common through reading). Here, the practice of suspended action is significant as a lived practice of dialogue and a (partial) coming-to-awareness of what we think, as well as how what we think is produced and producible.[16]

Achieving dialogue through suspended action, be it between people, texts, or ideas, is not such a simple task as we often hold, and are held by, assumptions that make it difficult for us to know differently. On this, Bohm (1996) states that the assumptions we always already hold are entangled within the meaning-making process. They not only shape and are shaped by what we know (as representations), but also how we come to know (as the world presents itself to us):

> ... representation is not only present in thought or in imagination, but it *fuses with the actual perception or experience*. In other words, the representation fuses with the "presentation," so that what is "presented" (as perception) is already in large part a re-presentation. (p. 64, emphasis in original)

In other words, as difference presents itself in the world, the assumptions that frame our thought subsume what is perceived within a representation, hence producing sameness. If we seriously desire being open to difference, Bohm (1996) states that there is a need to "go into all the pressures that are behind our *assumptions*... into the process of thought *behind* the assumptions, not just the assumptions themselves" (p. 9, emphasis in original). However, as Bohm (1996) states, "thought is not proprioceptive" (p. 29). In other words, thought is incapable of perceiving itself. It is impossible to simultaneously think about something and think about how we are thinking about something. Furthermore, the privileging of one is almost always at the expense of the other (see also Barad, 2007; Lather, 2007). Despite the impossibility of fully thinking about how we think, this does not mean that we are "off the hook".

Because thought is incapable of perceiving itself, Bohm (1996) suggest paying attention to the affective and embodied movements occurring alongside thought. This, he suggests, might offer a means of differentially thinking about how we think while working towards the suspension of our assumptions. As Moulton states (1983), such "reasoning has largely been ignored by [scholars] because it is different from the reasoning used to address an adversary and it is too complex and interrelated to be evaluated by counterexamples" (pp. 160–161). As such, considering that which is usually excluded from consideration in meaning-making processes can become a useful tool in producing knowledge differently.

These affective and embodied movements of thought are not only much more (self-)perceptible than thought but also hold a direct connection to thought:

Movements are taking place inside you – physical feelings – the heart beat, the blood pressure, the way you breathe, the way your body feels tense; and also the kinds of thoughts that go along with these feelings. You can observe these things, be aware of them, and their connection. (Bohm, 1996, p. 84)

While a consideration of sensorial or affective cues on their own might not provide direct insight into our assumptions and could also be considered an act of navel-gazing, Bohm (1996) states that they are nonetheless indicators that the process of knowing is occurring. Furthermore, not only are they connected to thought but also part of thought. For Bohm (1996), thought is "part of a material process" which happens with/in "the brain, the nervous system, [and] the whole body", such that thought is not and should not be considered as disembodied and separate from affect as they are "all one system" (p. 94). Accordingly, sensorial or affective cues should be thought of as constitutive parts, and signals of the process of thought. Whether this process is one of knowledge production (i.e., new meaning being made of experiences or perceptions) or reproduction (i.e., new experiences or perceptions being fused over by previously held meanings) of knowledge, emotional and embodied indicators provide a rich entry point for attempting to think about how you think.

Therefore, as you read the Socratic below dialogue below, there is an explicit invitation to attempt to suspend the moments in which you would want to interject and pay attention to your feelings, both emotional (e.g., frustration) and physical (e.g., furrowing of the brow) as they become productive sites of inquiry into how we think about how we think (e.g., the systems and norms through which thought are inflected). If we take the idea that playing requires both players and that which is played, not only does a text require a reader for (re)signification, but it may be productive to think about these affective and sensorial movements as one of the ways in which the text is playing the reader.

Act 3: "Two Science Educators Walk into a Bar": A Socratic Dialogue on Multicultural Science Education

Persons of the Dialogue: UNIVERSALIST (U); CROSS-CULTURALIST (CC).

Scene: Two multicultural science educators, one a *universalist*, the other a *cross-culturalist*, meet to discuss teaching practices over a pint of their favorite form of fermentation.

> *U*: Hello [CC]. I know that in the past, we haven't always seen eye-to-eye on what multicultural science education might entail; I thought that today we could attempt to begin our dialogue from a point on which we both agree. I brought with me a definition of multicultural science education upon which we should both be able to concur:
>
>> Multicultural science education is a construct, a process, and an educational reform movement with the goal of providing equitable opportunities for culturally diverse student populations to learn quality science in schools, colleges, and universities. (Atwater & Riley, 1993, p. 664)

CC: Hello [U]. Great idea! This is certainly a view of multicultural science education that I can get behind. I take it this is also the case for you, having picked the definition.

U: You are correct.

CC: Although we may both agree on the definition, I fear that we may read, and implement the same definition differently.

U: How so?

CC: I believe we differ in how we might interpret what is meant by "quality science".

U: In that, like Southerland (2000), I view "quality science" as "inquiry characterized by reliability on evidence and reason with the goal of understanding an objective, external, physical world" (p. 290)?

CC: Yes, and that I, like Masakata Ogawa (1995), characterize science as "a *rational* perceiving of reality" (p. 588, emphasis in original).[17] That is not to say that our respective definitions are incommensurate. However, partnering words such as reliability, evidence, and external with reason gives me the impression that there is a particular form of reason that you may be privileging through this process. Nonetheless, I am more concerned with how your definition of science is enacted within and through your teaching practices, particularly around "what counts" as science. If I may inquire, which systems of knowledge describing the physical world could be, should be, and are taught as "science" within your science classroom?

U: For the most part, my science education practices revolve around canonical understandings of science. In other words, WMS is primarily what is taught in my classroom. While I do not shy away from, nor am I opposed to, teaching other scientific viewpoints and meaning-making structures for understanding the natural world, WMS is the most effective and useful system for working towards knowing nature.

CC: Is it fair here to state that the way in which you address multiculturalism in your classroom is largely instructional rather than, or in addition to being, curricular?

U: Yes, that would be a fair statement.

CC: Before we address instructional questions, I would like to ask you a few curricular ones.

U: Go ahead.

CC: Within science education curricula, the belief of WMS holding a position of scientific superiority is one that is often and simply taken-for-granted. However, this is not an accusation I am making of you. I assume that your centering of WMS is an informed choice. Could you justify this claim you make for WMS?

U: First, WMS demonstrates the properties that are at the heart of "quality science": scientific knowledge and theories that are genuinely testable, predictive, and explanatory. Secondly, this is not to say that other ways of knowing the natural world are ineffective or not of value but rather that WMS best matches the criteria I have outlined previously.

CC: While I recognize the desire to make utilitarian curricular choices within educational spaces, it is also important to think about what is being maximized through these choices as well as how and why these choices are being made. My primary concern here is that how you define science and implement educationally might be at odds with "providing equitable opportunities for culturally diverse students" (Atwater & Riley, 1993, p. 664). In other words, taking such a stance is potentially disrespectful to students whose culture might not align with that of WMS,[18] be it in the context of science education or elsewhere.

U: I'm not sure I follow as to how my curricular conception and instructional delivery of science education might be inequitable or disrespectful, but I am willing to hear you out. Go ahead. Make your case.

CC: Since science is dependent upon an intersection of particular experimental systems, communities of researchers, organizational modes, as well as historical circumstances, all of which contextualize the generation of scientific knowledge, would you not agree with me that science is a situated practice?

U: Yes, I would have to agree.

CC: Similarly, since education is linked with culture, gender, history, sociopolitical and socio-economic context, amongst other factors, would you agree that education is also a situated practice?

U: Yes, go on.

CC: Should it not be argued then that science education, like its two constitutive parts, science and education, also be a situated practice?

U: Agreed.

CC: Then, should science education, as a situated practice, not also draw upon and reflect the contexts within and through which it emerges? This might entail many different things when considering a variety of contextual elements such as the constitution of the student body. For the sake of this conversation, I would like to clarify that I am referring specifically to the inclusion and reflection of TEK, be it called ethnoscience, Indigenous science, Indigenous knowledge systems or something else altogether, to enhance students' learning within the science classroom.

U: I would cautiously agree with you that science education should draw on and reflect the educational context. As I mentioned earlier, I am not opposed to teaching bodies of cultural knowledge alongside scientific beliefs, because teaching TEK, IWLN, ethnoscience, and other forms of culturally-specific scientific understandings can enhance students' learning by illustrating what WMS can do that which other approaches to the natural world cannot. In other words, including non-canonical understandings of science can be a way of showing that not all thoughts are equal in all contexts. The inclusion of ways of knowing nature beyond WMS in the science classroom is not something that is agreed upon by all universalists. However, I do not view the exclusion of TEK or other approaches to knowing nature as an essential practice within science education. As an aside, I would like to inquire into the criteria you are utilizing for the systems of knowledge whose inclusion you are arguing for.

CC: I can certainly elaborate upon this. As you will recall, we have discussed and agreed upon earlier that science needs to be conceived as socio-culturally situated. However, this is not to say that it is only situated in this manner. As Karen Barad (2007) states, it is important "to remember that there are cultural *and* natural causes for [scientific] knowledge claims" (p. 40, emphasis in original). Accordingly, the ways-of-knowing-nature that I am arguing for are not only culturally situated, but also rooted in natural empiricism. Not only is TEK empirical in its engagement with one's natural environment, but it is also a longstanding, intergenerational human engagement whose knowledge traditions often span thousands of years. As such, I am arguing for a pluralism of ways-of-knowing the natural world in which both culture and nature are significant and significantly engaged factors in the development of the knowledge claims being made.

U: While I do not wish to make a "slippery slope" argument, would Creationism fit here?

CC: It certainly is not my intent to exclude Western ways-of-knowing the natural world that are not WMS as there are multiple forms of TEK stemming from the West. However, as I understand it, Creationists' knowledge claims about the natural world are primarily, and possibly strictly, culturally-based (i.e., around biblical scripture). As it does not meet the criteria of natural and cultural causes for knowledge claims, Creationism would not, and should not, be included with the other bodies of knowledge that "count as science". The pluralism for which I am arguing need not and should not be an "anything-goes" form of relativism.

U: Thank you for the clarification.

CC: It is my pleasure. Furthermore, I think you would agree with me here that the science classroom should not be a place of dogmatic education.

U: I do agree here.

CC: Then, if you will allow, I would like to ask further questions around your earlier statement around how, in your science education practices, you might include other bodies of scientific knowledge and compare them to WMS. In particular, I am curious as to how you might respond to the oft-enunciated claim that a WMS-centered science curriculum has been and continues to be indoctrinating when there are pluralistic perspectives with the classroom.

U: Certainly, go ahead.

CC: Do you believe that it is possible both to respect youth whose culture is reflected in TEK and represent that TEK in terms of contrast with, indeed in terms of deficit to, WMS? Furthermore, if WMS is maintained as the primary curricular content in light of plurality, how do you avoid the pitfalls of this becoming a form of dogmatism?

U: If it is done with caution, sensitivity, and care, I believe that this comparative process can be engaged in and taught respectfully. Also, one can teach science without demanding that students change their beliefs, or denying them altogether. I recognize that historically there have been many cases in which interfacing WMS with other ways of understanding the natural world have been problematic in that they have forcefully produced totalizing discourses, cultural silencing, and marginalization. As Cobern and Loving (2001) state,

> The problem is not that science dominates at what it does best: the production of highly efficacious naturalistic understanding of natural phenomena. The problem is that too often science is used to dominate the public square as if all other discourses were of lesser value. (p. 62)

I am not advocating for education that is disrespectful or impositional. The issue at hand is not the comparative work, nor the centering of WMS, but rather the educational mode through which it was done. One can engage in teaching WMS without it devolving into scientism through its unquestioned and unquestionable privileging at the expense of other forms of knowledge.

CC: Would you care to elaborate on this? I do not understand how you can simultaneously respect an individual while not affording validity to their culturally-based scientific knowledges.

U: While these comparisons may not hold certain cultural knowledges as equivalent to WMS, this is not to deny them validity. Rather, it acts as an explication of how these knowledges do not meet the scientific standard. This also entails that there are situations and contexts in which

WMS is *not* the most appropriate form of knowledge. Furthermore, and accordingly, such comparisons do not discredit the individual learner.

CC: Therefore, as I understand it, the terms we are currently discussing are not those of "what counts" as science, but rather those of what counts best, as well as when, how, and where.

U: Most certainly.

CC: However, if WMS is the only way of understanding the natural world that is recognized, perceived, or privileged as "quality science" or as best meeting its criteria, would students of non-Western cultures be required to learn WMS at the expense of their culture? Furthermore, if this is not to discredit the individual learner, could this be considered a form of epistemic violence?

U: As I said, this would have to be done with caution, sensitivity and care; I am not advocating for a form of science education that is inherently disrespectful. It seems that we may have diverging notions of what respect may entail in the multicultural classroom. Let us take a step back and discuss this term of engagement.

CC: Certainly.

U: Within a multicultural classroom, there is a clear and moral imperative to treat members of *all* cultures justly and with respect, correct?

CC: Of course.

U: Therefore, it is imperative that *all* students, as well as their cultures and cultural beliefs be treated with respect as well.

CC: I agree with you up to here. Go on.

U: However, it is important to note that respecting students, as well as their cultures does not require treating their scientific beliefs as scientific knowledge. There is nothing wrong with identifying a set of beliefs as scientifically deficient, so long as it is done in a culturally sensitive manner, of course. The goal of science education is to make students scientifically literate, not indoctrinate them through scientism.

CC: You are correct in highlighting the importance of science education not simply being about teaching and upholding *any* belief about the natural world. As mentioned earlier, science education should certainly not become an "anything goes" form of relativism. However, this is not to say that *all* cultural knowledge about the natural world should simply be treated as belief. Even if the criteria for "what counts" as knowledge of the natural world is different across various cultural knowledge systems, this does not mean that TEK needs to loses its status as knowledge once it enters the science classroom. It is also worth noting that while all culturally developed systems for understanding nature are different, there are often similarities as well. For example, Gregory Cajete (1999) makes such a comparison between Indigenous science and WMS in stating: "Indigenous science is a broad category that includes everything from metaphysics to philosophy to various practical technologies practiced by Indigenous past and present" (p. 83). Similarly to WMS, Indigenous science "has models which are highly contextual to tribal experience, representational and focused on higher order thinking and understanding" (p. 85).

U: Perhaps I should rephrase my earlier statement. I do not wish to deny diverse culturally-specific knowledge about the natural world the status of knowledge. Furthermore, the degrees of similarity are in fact worth noting as many forms of ethnoscience have rich histories of empirical naturalistic observation. However, is there not something lost in calling TEK "science?"

CC: Not always, but it is certainly a possibility that is always present. When placing multiple and diverse competing claims next to one another, there always runs the risk that they become homogenous in a way that erases their diversity and distinctiveness.

U: This is not only the case for TEK, but also WMS.

CC: I certainly agree.

U: If so, then why is calling TEK science or having it "count" as science of such importance?

CC: So long as "science" continues to be used in some educational locations as a term that excludes rather than includes, there is a need to continue working within that space even if it is complicated, complex, and sometimes contradictory.

U: I agree with you here that there continues to be science educators who deny the inclusion of any other way of knowing nature beyond WMS, but that is not what I am advocating for here.

CC: There is, however, one complexity or complication that I would like to further discuss with you when both TEK and WMS are included within a curriculum. Here, we both agree that such juxtapositions potentially affect both TEK and WMS. However, as I see it, these processes of interaction are not always even and equal in what they produce.

U: How so?

CC: Without going into a longstanding history of uneven interactions between TEK and WMS, it is nonetheless worth considering the positioning of both bodies of knowledge within the science classroom. As you mentioned earlier, you believe that it is possible to treat non-Western-modern ways of knowing nature with respect while simultaneously treating them as scientifically deficient in comparison to WMS. To me, treating TEK as less valid than WMS sends a message that is not so dissimilar from treating it as invalid. While "what counts as science" has changed to be more inclusive, this type of valuation seems to indicate that the norms through which it comes to count remain similar, if not the same. However, perhaps I do not understand your position correctly. Would you care to elaborate upon how you come to validate scientific knowledge within the science classroom?

U: Certainly. It is worth repeating here that I do not wish to deny TEK the status of knowledge about the natural world. However, this does not mean that it wholly stands up to WMS or the criteria of "quality science". If we return to our earlier and agreed upon definition of science education as providing students opportunities to learn "quality science", I think it is worth noting that quality is not something that varying or competing forms of knowing the natural world have or do not, but rather that some approaches are better than others. As Siegel (2001) asserts, "knowledge prized by WMS goes beyond the observational, in that it seeks ... theories which are testable, and which predict and explain naturalistic observations" (p. 809). These are not only qualities attributed to the knowledge but also the knowledge building process. WMS seeks "the discovery, articulation and, explanation of regularities in nature in terms of *laws* which are invariably universal and invariant across time and place. This is not an assumption of universalists, but rather a methodological directive of WMS" (p. 805, emphasis in original). Do note that I am not attributing perfection to WMS as there are many contexts in which it is not the most appropriate form of knowledge such as the arts and the social sciences. However, I can confidently claim that culturally-specific understandings of nature such as TEK, as Siegel (2001) states, are *"less adequate as scientific understandings* than those of WMS" (p. 809, emphasis in original). While culturally-specific understandings of nature share naturalistic observation as a property with WMS, they do not achieve or produce the same quality of explanatory, testable, and predictive theories around, or towards further, observation. As such, I believe that it would be a disservice to students to not teach them primarily about the system that has reached the deepest understanding of the natural world, WMS.

CC: I agree with you that not teaching students WMS would be to do them a disservice. If you will recall, I am not arguing for the exclusion of WMS but rather for an inclusion of other ways of knowing the natural world that also takes them seriously. While I cannot fault your logic for the claim of scientific superiority that you are making for WMS when testability, predictability, and explanatory potential are the criteria, I do take issue with the criteria and how they are shaped. However, we can always come back to this as I am further preoccupied by the underlying message that you believe that such valuation can be done respectfully. Where I am unable to follow your argument is in your statement that denying a student's cultural knowledge the status of scientific equivalency can be done without disrespecting the student: morality and episteme go hand in hand.

U: While I agree with you that ethics and knowing go hand in hand, I do not agree with the conclusion you reach through this association. Rather, I would argue that not engaging in such comparative work would be disrespectful to students. I am making this claim on the basis that within the relationship between ethics and knowledge, there is a moral imperative to treat learners as fully rational persons or, at very least, in the process of becoming fully rational. Being fully rational means being capable of distinguishing between justified beliefs and beliefs that are either unjustified or less justifiable. Hence, providing students opportunities to exercise such rationality is to respect their capacity as learners.

CC: Granted, not treating students as capable of learning and providing opportunities for rich learning is highly problematic and disrespectful. However, being respectful of students and their ability to learn need not be at the expense of their cultural knowledges.

U: I agree here. As I mentioned earlier, providing opportunities to exercise reason need not be culturally insensitive, nor is this what I am arguing for. To learn as to why certain accounts, such as WMS, are worthy of belief is not the same as imposing knowledge or requiring the student to renounce their cultural knowledge.

CC: While we agree on not dichotomizing this issue, I believe that perhaps we are attributing different weights to the ethical valuation of respecting a student's capacity to learn and respecting the culturally-specific scientific knowledge they bring with them into the classroom. I am not certain to which extent changing the explicit instructional message changes the implicit curricular one. As I do not see us moving forward on this at this point in time, let us return to our earlier discussion about science, education, and by extension, science education as situated learning processes. If we consider science, and hence science education, to be a situated practice, could we not include culture to the ways in which WMS' knowledge is constructed?

U: As I have stated earlier, I do not deny the situatedness of scientific knowledge processes, nor do I deny that culture shapes scientific knowledge. As Siegel (2001) states, there is often a misconception that universalists are perceived as framing science as acultural and, accordingly, deniers of social constructivism. Rather, universalists happily agree that scientific theories are human constructions in the sense that they are conceived, formulated, articulated, and revised by human scientists... [They] agree that culture influences the truth claims – i.e., what is claimed at a time by scientists to be true – of science. (p. 807)

CC: Then let us reframe an aspect of this conversation. I am going to ask you to reconsider the status that WMS is granted above other ways of knowing the natural world. If we consider this to be a cultural claim, would you agree that the privileging of WMS would then be considered culturally insensitive for those who do not belong to this culture?

U: If this were strictly a cultural claim, I would have to agree, but this is certainly not the case. The privileging of WMS over other forms of understanding the natural world is not rooted in cultural presumptions or values, but rather on the criteria of "quality science": testable, predictive, and explanatory scientific theories.

CC: Surely these criteria concerning what is considered "quality science" must also have their own cultural origins as well, mustn't they?

U: Yes, they would.[19] However, the fact that an idea has particular cultural origins need not mean that it is true or relevant only to or within that culture.

CC: My concern here is that criteria for what counts as "quality science" are the same as those that define WMS. Furthermore, they are defined through the same set of cultural values. This circularity would lead me to believe, with WMS as the meter stick, that no other science could be considered "quality science" by virtue of it not being WMS. Would these criteria then not be self-serving, and therefore culturally presumptive as well as assimilationist when they become the basis for a "respectful" multicultural science education?

U: ... but the criteria for good science are "universal" in that they are universally applicable. By this I do not mean that WMS transcends culture. Rather, if we are to take an instrumentalist approach, it is the most reliable, reproducible, and productive knowledge about the natural world. As such, it is applicable above and beyond the cultural circumstances in which it originated. For the sake of argument, even if it wasn't, teaching WMS need not be a form of cultural imperialism. It can be taught in a manner that is culturally respectful and sensitive, there can be science education without scientism.

CC: Both of these clauses are debatable. While I believe that I have already addressed the first when I brought up the circularity of the argument, I am more concerned with the second. It has been argued that regardless of how culturally relevant, culturally responsive or culturally respectful science education is, if the end goal is to teach youth of WMS and bring them into its culture, it continues to be an assimilationist, marginalizing practice. Like Ali Sammel (2009),

> I wonder if there is a science curricula that does not indoctrinate., ... given the pervasiveness of assimilationism in Western science education... The system is pushing for assimilation of students into Western science ontology.... The structural inequities are given little attention while the mantra *of we all have a chance to succeed in science* reigns supreme. (p. 653, emphasis in original)

It may very well be that if, as science educators, we wish to reach our earlier agreed upon goal of "providing equitable opportunities for culturally diverse students", we may need to do so partially at the expense of what has traditionally and singularly been considered as "quality science".

U: On this, we will have to agree to disagree. Like Charbel Niño El-Hani and Eduardo Fleury Mortimer (2007), I find myself asking:

> How can we avoid demanding that our students change their beliefs by learning science, and, yet, intend that they apply in their lives what they learn in the science classroom? Surely, we could give up the second intention. But then why should we bother about teaching science at all? (p. 673)

If we need to take the "science" out of science education, then I will have no part of it. I think we are done here for today.

ACT 4: PLAYING OUT THE (RE)PRODUCTION OF KNOWLEDGE

When (Re)Signifying Is Signifying *Again* Rather Than *Anew*

Like *all* Socratic dialogues, the above dialogue on multicultural science education produces some shared understandings but fails to reach a conclusion beyond an agreement to disagree. This disagreement is both vis-à-vis "what counts as science" as well as how one might respectfully engage multicultural science education in terms of curriculum and instruction. It could be said that this result is, in part, a function of the medium through which a mimetic production of these debates within the science education literature is filtered and inflected. Another factor is that the predominant message being presented by the literature is that the debate was and continues to be left unresolved and unresolvable.[20] However, regardless of whether it *is* or *is not* resolvable, the Socratic dialogue implicitly acts as an invitation to engage and participate in the always already occurring play of (re)signification.

Above and beyond the dialogue between universalist and cross-culturalist characters that occurs within the play, there is also a dialogue and/or a dialectic that occurs between the reader and the text. Should you have made meaning *in* common (i.e., dialogically) rather than *made common* (i.e., dialectically), you may have located creative junctures in-between the two positions that rupture and open up both sedimented scripts. However, if as a reader, you interpret the perspectives as irreconcilable, this does not mean that you have not engaged in the play of (re)signification. Rather, it may mean that the act of signification that you engaged is one that (re)plays that which is already there. In other words, through your engagement in the play of (re)signification, you have either signified *again* rather than *anew* the meanings that were already there. Recall that unmooring meaning is neither an easy nor individual affair; the "you" who (re)signifies is not only you—the language we possess also possesses us—we inherit and are haunted by longstanding and lingering meanings within science education. The production of meaning is always already produced between the text and yourself, and as such, both you and the text are partially implicated in whether the meaning made was complicit and/or critical as a function of the relational flow of meaning (Derrida, 1976; Lather, 2007; Mazzei, 2007; Spivak, 1976).[21]

Should you find yourself in the former position (i.e., complicit) rather than the latter (i.e., critical), it is worth considering a few key differences between the Adversary method and Socratic dialogue. While both methods rely on refuting arguments by shaking up beliefs that are held, the way they go about it and the purpose is different. Moulton (1983) elaborates that whereas the purpose of the Adversary method is to achieve epistemic victory, the goal of engagement in Socratic dialogue is to position the other party such that "they can begin philosophical inquiries with a more open mind" (p. 156). Furthermore, the success of the Socratic method depends on persuading the other person whereas the success of the Adversary method depends "on showing [the other person's] views to be wrong to others" (p. 156). Accordingly, while the Adversary method might be an effective means of proving someone wrong, it certainly is not an effective means of getting someone to agree with you.

Bohm (1996) elaborates upon this by stating that such a mode of address is a double(d) form of violence. First of all, this puts the recipient on the defensive:

> The opinions that we are defending ... are not merely opinions, they are not merely assumptions; they assumptions with which we are identified – which are therefore defending, because it is as if we defending ourselves. (p. 39)

Secondly, this self-defense psychically prevents not only the reception of that which is being levied against you but also productive dialogical thought that would allow you to find common ground:

> You cannot defend something without first *thinking the defense*. There are those thoughts which might question the thing you want to defend, and you've got to push them aside. That may readily involve self-deception – you will simply push aside a lot of things you would rather not accept by saying they are wrong, by distorting the issue, and so on. Thought defends its base assumptions against evidence that they may be wrong. (p. 12, emphasis in original)

Through this process, sensory, and affective indicators often precede reactive thought in the form of anger, frustration, a gut feeling, a clenched jaw, a change in heart rate, or other. However, as Bohm (1996) states, you don't always make the connection between your sensory-affective reaction, what occurred, and what you are thinking—it is incredibly common for those schooled with, in, or though Western modernity to fragment the thought from the thinker (and the mind from the body with/in the thinker) (see also Cajete, 1994, 2000). Accordingly, you use the feeling or the sensation to justify the thought, or inability thereof. In short, the Adversary method simply does not encourage others to consider that which is causing this violence, but rather a (p)re-entrenchment, a position from which defense might be possible. In turn, this makes for poor scholarly engagement if engagement is defined as something more than an individualistic endeavour.

Instead, Socratic dialogue "looks for premises that the other person will accept and that will show that the original belief was false" (Moulton, 1983, p. 156). It is certainly worth noting here that what each and every reader might need to be convinced will be different, as well as what they might be convinced of. Similarly, but holding patterns of difference,

Bohmian dialogue searches for shared meanings. This does so by asking the question "is it absolutely necessary?" (Bohm, 1996, p. 26): is it absolutely required that the other be wrong or that there be an absolute truth? For Bohm (1996), "dialogue may not be concerned directly with the truth—it may arrive at truth, but it is concerned with *meaning*" (p. 43, emphasis in original).[22] Such meaning is the type of shared meaning that emerges when assumptions and opinions are collectively opened and the play of (re)signification can occur without the need to negate previously held beliefs. Within recent literature, there have been some scholars who have been attempting to produce intermediary positions within the within the multicultural science education debate that extend (potential) points of agreement without negating firmly held claims. These dialogical possibilities yield potential for Indigenous science to-come (and are pulled through as productive insights for (re)opening the structures of science education in upcoming chapters).

What Continues to (Not) Be at Play? Possibilities for Further Dialogue Through the Play of (Re)Signification

As it stands, the debate continues within multicultural science education literature largely around conflicting notions of "quality science" (i.e., "what counts as science", how it is valued and evaluated) as well as questions of "quality education" (i.e., what is learned, how it is learned, questions of what constitutes respectful teaching and learning). However, a shared commitment across positions to a notion of science education that respects diverse knowers, as well as a commonly held view that it can be productive to include TEK and other non-WMS knowledges in the classroom exists, despite divergences on how these are conceptualized and enacted.

One such location is the growing consensus that students "should become scientifically literate, but not indoctrinated by scientism" (Zeyer, 2009, p. 1100). In other words, what in the past was primarily a cautionary note stemming from cross-culturalists (e.g., Lewis & Aikenhead, 2001; Sammel, 2009; Snively & Corsiglia, 2001; Stanley & Brickhouse, 2001) is now general recognition of the possibility that science education can be and has been impositional in its nature. However, while Southerland (2000) states that one should not conflate of WMS's claim to universalism with the impositional nature of scientism,[23] other universalists are increasingly taking the position that it is nonetheless a possibility, and one that *all* science educators (regardless of their positioning within the multicultural science education debate) should be actively working against (see also Cobern & Loving, 2001, 2008; El-Hani & de Ferreira Bandeira, 2008; El-Hani & Mortimer, 2007; Zeyer, 2009).

Despite these gains for diversity and equity, there continues to be a multiplicity of sedimented and stuck categories and concepts that represent a serious challenge to dialogical interfacing of cross-culturalist and universalist positions. Accordingly, some science education scholars have welcomed the always already extended yet implicit invitation to engage in the play of (re)signification in order to labour these meaning-full disjunctures. They have opportunistically used shared yet never fully converging meanings (e.g., the goal of science =/= scientism) to labour that which continues to (not) be at play (e.g., Cobern & Loving, 2008; van Eick & Roth, 2007; El-Hani & Mortimer, 2007; McKinley & Stewart, 2012). In other words, they look to that which is conceptually absent yet present (e.g., taken-for-granted implicit and tacit meanings) as a point of differential entry instead of (or in addition to) engaging in the dialectic negation of those with whom they would disagree. For example, and explored in further depth in Chapter 5, Cobern and Loving (2008) illuminate the absent presence of ontology within the multicultural science education debate in order to justify and defend a universalist position (in turn, denouncing cross-culturalism). With these absent presences, they inflect already present terms to mean something anew and askew.

Of particular interest is van Eijck and Roth's (2007) effort "to contribute to overcoming the multiculturalism-universalism incompatibility in science education" (p. 927) through the deconstruction of the theory/practice binary (i.e., theory as always already present within practice and vice versa). Specifically, they invite both universalists and cross-culturalists to (mis)read knowledge as knowledge-practice. Drawing from cultural-historical activity theory and Bruno Latour's work of reading Louis Pasteur's discovery (of pasteurization) through Actor-Network Theory (Latour 1988), van Eijck and Roth (2007) extend a framework in which: (a) scientific knowledge derived through WMS is not only a product but also a dynamic and heterogeneous process; (b) never independent of human activity such that "there is no possibility for someone to 'have' knowledge and simultaneously 'fail to enact' it" (van Eijck & Roth, 2007, p. 934); and (c) scientific knowledge is always in context such that it is both naturally *and* culturally situated (see also Alsop &

Fawcett, 2010; Barad, 2000, 2007; van Eijck & Roth, 2009). They also signal the processual nature of scientific knowledge and the ways in it plays out in the production and application of the knowledge.

As presented in the play, both universalists and cross-culturalists agree that constructing reliable knowledge about the natural world, whether it be through WMS or TEK, is a process that is at once culturally mediated *and* employs naturalistic empiricism (Aikenhead & Ogawa, 2007; Cobern & Loving, 2001, 2008; Siegel, 1997, 2001; Snively & Corsiglia, 2001). However, the processual quality of meaning-making with nature is often an absent presence that upholds arguments but does not fully come to bear on the culture of the multicultural science education debate. Importantly, (re)signifying knowledge as knowledge-practice does not require either party to renounce their position; rather, it potentially opens up fruitful locations for possible dialogue.

How might considering scientific knowledge as knowledge-practice assist us in collectively working towards the shared goal of working against scientism in science education? This is significant to the overall scope of this book as scientism often comes-to-be a central mechanism (alongside exclusivity and Eurocentrism; see McKinley & Stewart, 2012) through which Indigenous science is excluded, differing, and deferred. The following are three contingent insights for potentially (re)opening the spaces of science education which will come to inform the work of the chapters to come, notably that scientific knowledge-practices are: (a) always situated (Chapter 4), (b) culturally hybrid (Chapter 7), and (c) ontologically situated (Chapters 5 and 6).

Scientific Knowledge-Practices as Always Situated. First, as van Eijck and Roth (2007) discuss, one salient example of such an opening produced by considering scientific knowledge as knowledge-practices is around the contested status of WMS as universal. Recall that for universalists, the knowledge produced by WMS is true in *all* contexts, with the cautionary note that it might not be the most appropriate form of knowledge when addressing particular topics and milieus (e.g., arts, social sciences, humanities, the super-natural; Cobern & Loving, 2001, 2008; El-Hani & de Ferreira Bandeira, 2008; El-Hani & Mortimer, 2007; Zeyer, 2009). For cross-culturalists, what is at stake are the ways in which WMS is often granted epistemic privilege as a result of undiscerning translation into diverse contexts that supersedes longstanding, traditional, and complex knowledge-practices that are shaped in relation with the contexts in which they emerged (see Aikenhead & Ogawa, 2007; Aikenhead &

Michell, 2011; Snively & Corsiglia, 2001; Stanley & Brickhouse, 2001). Van Eijck and Roth (2007) diverge from these either/or positions by focusing on the *when* and *how* of universal knowledge-practice, rather than the *is* or *is not* of universal knowledge (see also Harding, 2008). In short, they focus not only on the practice of developing knowledge that can be applicable in all contexts but also what it means to apply this knowledge in a wide range of contexts:

> a twofold price must be paid for scientific knowledge to transcend local contexts and to become *apparently* static, singular, and homogeneous. First, for scientific knowledge to be produced and to be transcendent, the local contexts to be investigated must be reduced to scientific praxis. Second, for the scientific knowledge to be applied in another local context, this target context must be transformed again to scientific praxis. (van Eijck & Roth, 2007, p. 938, emphasis in original)

In other words, they differentially articulate the cross-cultural statement that scientific knowledge-practices are decontextualized as well as the universalist one that scientific knowledge transcends context by firmly placing it within *a* knowledge-practice context. This context, whether real or imagined, is the scientific laboratory:

> the reduction of local contexts to scientific praxis is inherent to the static, homogeneous, singular, and transcendent nature of scientific knowledge and that transcendent scientific knowledge is useless unless local contexts are changed to function like scientific laboratories. (van Eijck & Roth, 2007, p. 935)

While such conclusion may not necessarily be agreed upon by both sides *as truth*, it is nonetheless posits the possibility of shared meanings. For universalists, this does not require the abandonment of the notions that WMS is universal (i.e., applicable across multiple contexts) or "the best at what it does" (i.e., producing claims about nature). Treating the knowledge produced from empirical observation and Western modern scientific modes *as if* the context in which it was produced or applicable in were a scientific laboratory does not deny it of reliability or repeatability. Instead, it situates the conditions through which such a knowledge-practice is produced and applicable as well as the "price to be paid" for it to be so. For cross-culturalists, the situating of WMS as a knowledge-practice whose real or imagined context of production and applicability is the laboratory produces a space in which a localized TEK provides a welcome supplement to WMS that is more than the sum of what WMS is not. Because TEK is produced and producible through a knowledge tradition that accounts for and is accountable to the ecological contexts in which they are developed, TEK's knowledge-practices around nature are differentially produced with many of the exclusions that WMS must make to treat a context as a laboratory in mind. The "price to be paid" here is the difficulty in translating across diverse contexts. For the multicultural debate in science education here, this potentially produces a position in which the knowledge-practices of WMS and TEK might be able to share a differential yet overlapping space in which one can complement the other without the need to renounce one's position through the situating of one's knowledge-practice.[24] These shared meanings might become ripe for a (re)opening of science education for Indigenous science to-come by not requiring an oppositional mode through which one side must emerge victorious.

Scientific Knowledge-Practices as Culturally Hybrid. The second possibility of working towards shared meanings with/in the multicultural science education by considering scientific knowledge as knowledge-practices debate rests in culturally hybrid knowledge-practices. Van Eijck and Roth (2009), continuing their earlier work, extend science's internal dynamism and heterogeneity outwards by considering the complex and conflicting ways in which WMS and TEK *interact* and *interface* at the level of the personal learner in pluralistic science education contexts. This is significant as discussions within multicultural science education debate are often framed by a Western/non-Western binary that disallows the play of (re)signification between the two. It is worth attending in brief to the primary problematics engendered by the production of a fallacious Western/non-Western binary including the exclusion of conversations around the hybridity of science and the complexities of these cultural interfaces (see also Harding, 2008). First, the assumption that cultural traits are either separate or that they can be separated that is produced by such a binary renders invisible the many elements of WMS that stem from non-Western cultural sciences and vice versa (e.g., the "father" of modern geometrical optics was Ibn al-Haytham, a middle-eastern a man). Secondly, as these contributions are not always recognized or brought up, the historical complexities of scientific knowledge production that occur at these cultural interfaces (e.g., appropriation, assimilation) are also, accordingly, left out (e.g., the Bayer pharmaceutical company's "discovery" and subsequent synthesis of traditional willow-bark-based medicine used by Indigenous peoples that is now known as Aspirin; see Snively & Corsiglia, 2001). Lastly, the Western/non-Western binary makes it difficult for science educators to account for the complexities that occur at the currently lived and differently situated cultural interfaces that both they and their students occupy (see Belczewski, 2009; van Eijck & Roth, 2009; Higgins, 2014). However, as this binary often operates within the bounds of multicultural science education, the Socratic dialogue is written in a way that reflects it.[25] As Indigenous science's perpetual status of yet-to-come is invariably shaped by culturally hybrid relations in which WMS and TEK are unevenly and unequally interfaced, considering this interface as a knowledge-practice entails that it is not something that simply *is* but rather something that *is done*. As a *doing*, this invites the possibility that it might be *done otherwise*, (re)opening this oft foreclosed space of science education towards the possibility of Indigenous science to-come.

Scientific Knowledge-Practices as Ontologically Situated. The third possibility for collectively working against scientism through situating scientific knowledge-practice lies in a consideration of ontology. Cobern and Loving (2008), extending van Eijck and Roth's (2007) notion that knowledge-practice is always culturally situated, direct attention to the ways in which epistemologies are always already ontologically situated. This is significant as the nature/culture binary, deployed within the context of the multicultural science education debate, disallows the play of (re)signification between the two (i.e., culture -> nature-culture; see also Barad, 2007; Latour, 1993). As Cobern and Loving (2008) suggest, the primary and almost exclusive focus on epistemologies within the debate has detracted from considerations of how epistemology aligns with ontology.[26] While they use this (re)signified natural-cultural interplay to make a stronger case for universalism (e.g., "Epistemic Realism Really is Common Sense" [p. 425]), there is nonetheless room for the possibility of shared meanings. Taking seriously the notion that ontology is not a singular affair (Barad, 2007), this can be achieved not by refuting their claim but rather by situating it within *a* context: epistemic realism (i.e., the epistemology of WMS) really does align best with *an* ontology of Cartesianism (i.e., the ontology of WMS). This creates a space in which WMS achieves "distinction not privilege" (Cobern & Loving, 2008, p. 444) all the while not requiring universalists' claims of onto-epistemic alignment to be refuted. This also produces complimentary space for differently positioned, yet also onto-epistemologically aligned, forms of TEK (see Cajete, 1994, 1999, 2000).[27] Such plurality without negation is significant for Indigenous science to-come as the requirement for there to be *a* truth (read: singular) often places TEK at odds with WMS: a losing proposition when the norms of "what counts" as science best describes WMS (as "what counts" is modelled after WMS; see van Eijck & Roth, 2007).

Epilogue to (a) "Serious Play": A Call for Further Serious Play Through Dialogue

The way in which the adversary method insidiously operates within the field of multicultural science education continues to produce a space which, by paying attention to extreme positions because they are extreme,

> presents a distorted picture about what sorts of positions are worthy of attention, giving undo attention and publicity to positions merely because they are those of a hypothetical adversary's and possibly ignoring positions which make more valuable or interesting claims. (Moulton, 1983, p. 158)

The resulting entrenchment of positions leads to percussive and concussive discussions that foreclose the possibility of dialoguing across difference as well as the production of positions and positional propositions in-between those presented. Or, in my case within the example at the beginning of this text, the dramatic yet almost comedic mutual (dis)engagement protected our respective terms of engagement, prohibiting the possibility of productive conversation before it even began. Such engagements often result in communication that is protective and prohibitive rather than productive (Spivak, 1976).

While the Socratic dialogue within this chapter is meant to be but an imitation of life or a form of mimesis, this mimicry of a discussion between the educators holding universalist and cross-culturalist positions acts as an invitation to the possible play of (re)signification in order to make space for dialogue and the possibility of Indigenous science to-come. In turn, questions of respect within the multicultural science education classroom need to remain open to respond to the differentially entangled values, cultures, and histories of not only students present within the classroom, but also that of science, as well as peoples to whom a curricular geo-political commitment to is made (i.e., learning about and through local Indigenous peoples' ways-of-knowing-nature, regardless of classroom composition). Similarly, in order to keep the possibility of (re)signification of "what counts" as science open rather than foreclosed, there needs to be serious play with the assumption that non-Western bodies of scientific knowledge are not held as concurrent systems not because of their value, but how they are valued. As dialogue is about process and not product, "it's not all that important whether you agree or not... the point is that we would establish, on another level, a kind of bond" (Bohm, 1996, p. 37). It is important to not foreclose the possibility of shared meanings (rather than "universal" truths) that are made together, and within science education, this making *in* common need not necessarily exclude students, both Indigenous and non-Indigenous (see van Eijck & Roth, 2009; Higgins, 2011). This is of utmost importance as not reaching shared meanings within the multicultural science education debate is of consequence for Indigenous science to-come as it upholds a status quo which (re)produces WMS as "what counts" best as science.[28]

While the multicultural science education debate is characterized by an agreement to disagree, this is not to say that the dialogical movement of meaning has not occurred. At the very least, there is an invitation to enact science education in a way that is less "caught in the competition of whose sword is sharper" (Spivak, 1994, p. 35). On this, it is important to recognize that the nature of the debate has changed by "accepting that all systems of knowledge about nature are embedded in the context of a cultural group; that all systems are, therefore culture-laden; and that (Western) science is the system of knowledge about nature that is predominant in Western culture" (Lewis & Aikenhead, 2001, p. 3) and that even the strongest of universalists agree that multicultural science education is the direction that we should be taking in order for all students to benefit from science education. Furthermore, the three insights generated through (re)considering scientific knowledge as knowledge-practices (i.e., as always situated, as culturally hybrid, and as ontologically situated) will come to inform the work in the upcoming chapters as means of (re)opening science education towards Indigenous science to-come: creating possible possibilities for unsettling science education. However, there continues to be a need for the (re)opening of foreclosed meaning-making spaces between and across scripted positions: what counts as science within science education continues to be a necessary and significant conceptual scene for dialogical engagement, as well as a productive rather than prohibitive point of entry for (further) serious play.

Notes

1. Respecting Indigenous, diasporic, and other post-colonial students within the science education classroom is one of the key motivators within the culture of the multicultural science education debate for the inclusion of TEK. However, in classroom practices, the central focus is often on achievement; more specifically an "achievement gap" (without coming to understand the norms under which uneven scholastic achievement comes to be produced and producible; see McKinley, 2007; McKinley & Stewart, 2012). However, as I have addressed elsewhere (Higgins, 2011, 2014), the incorporation of TEK and IWLN within the science classroom is beneficial for *all* students.

2. The now mantra of beginning some-where and some-time is not only a troubling of the notion of "beginnings" in the question of Indigenous science in science education, but also a reminder that because we are always already within the question, we can begin in a multiplicity of spatial and temporal locations.

    This even includes chapters within the larger narrative of science education that have been deemed closed, such as the multicultural science education debate. The call of placing this moment in the past may well be a move to render this debate one that is passed over (see Derrida, 1994/2006). Rather, because it has not yet, and perhaps never will be, resolved, its logics continue to haunt the ways in which we, as science educators, approach the question of Indigeneity within science education. For example, and most significantly, I encounter many of the arguments that are made within the literature in science teacher education courses when engaging students with how we might respond to and meaningfully include Indigenous science within school science practice. *Unsettling*, in its double(d) meaning, requires that we attend to such sedimented and stuck locations that continue to bear on the ways in which settler colonial logics are perpetuated within science education.

    Further and related, even if the temporality of this "beginning" is one that (re)opens a past passed over, it also points towards an enfolded and slow practice of *serious listening* in the here-now (to this lingering critique of the past). For Kuokkanen (2007), "listening seriously means that the listener is able to go beyond the sort of benevolent imperialism that makes arrogant assumptions about the speaker and her background" (p. 82). Listening *too quickly*, as unpacked in the previous chapter, can become a form of "'listening-as-benevolent-imperialism'— that is, listening through the dominant discourses and epistemes and therefore not hearing what is actually said" (p. 75); this has the unintended consequence of deferring a problematic past and (re)producing it in the present. Rather, a *slow* listening requires that the listener inhabits the lengthy duration of the critique. As Kuokkanen (2007) suggests, we must not move too quickly towards resolution as this "can remove the hegemonic listener from the present and in this way either remove that listener from the present critique or place the hegemonic subject outside of that critique" (p. 142). For example, the binary logics of Western modernity often equate inaction and irresponsibility; blinding the porous slippage that occurs in the ways action itself can (re)produce the very problem being responded to in a de/colonizing moment and movement (see Higgins, 2014).

3. It is generally agreed upon by science educators that perceptions of WMS are often partially (mis-)informed by particular stereotypical and monolothic images of scientists, as well as the notion that there is *a* way to do scientists think (e.g., *the* scientific method; see Aikenhead, 2006). Nonetheless, there are still often similarities enacted across Western modern *sciences* with respect to beliefs about science held and enacted by many (but likely not all) scientists (see Aikenhead & Ogawa, 2007; Aikenhead & Michell, 2011). These include, but are not limited to:

- Nature is governed by laws that are quantifiable, generalizable, and predictable—hence—nature is knowable (or can and will be known in instances in which knowledge technology does not allow for the study of phenomena).
- Nature can be reproduced (for realists) or represented (for socio-constructionists) through scientific knowledge;
- The production of scientific knowledge is embedded within social contexts. However, "the *rigour of [WMS]* decreases subjectivity as much as possible" (Aikenhead & Michell, 2011, p. 43, emphasis in original) to make knowledge claims (quasi-)objective;
- Because scientific knowledge reproduces or represents nature (the latter with as little social and subjective noise as possible), scientific data speaks for itself;
- Constructing knowledge about the natural world (i.e., scientific knowledge) is the purview of humans and is represented or reproduced upon a static and (recti-)linear space and time.

4. Traditional ecological knowledge (TEK) is a term used by biologists and ecologists that became prevalent in the 1980s that usually signals "experience acquired over thousands of years of direct human contact with the environment" (Snively & Corsiglia, 2001, p. 11). TEK is often synonymous with terms such as ethnoscience, Indigenous or Native science, as well as Indigenous knowledge and Indigenous knowledge systems; of which, it has become the most prevalent in usage (McKinley & Stewart, 2012). These other terms are often preferred because TEK, like WMS, is a concept that is often (mis-)understood as a result of prevalent, pervasive, and problematic understandings (see previous endnote). These include

"tradition" being defined in opposition to civilization and contemporaneity, "ecological" being reducible to ecology as defined by modern biology, and "knowledge" as discrete and separate not being an adequate referent for the relational knowledge processes from which TEK stem (see Kim, Ashgar, & Jordan, 2017). Such mis- and missed representations are often complicated by under-representation (Aikenhead & Ogawa, 2007; Aikenhead & Michell, 2011; Snively & Corsglia, 2001).

Similarly to WMS, there are many forms of TEK that are as diverse as the longstanding Indigenous peoples' traditions from which these knowledges stem. TEK is not the binary opposite of WMS, however this is not to say that there are not significant differences between the two (Aikenhead & Ogawa, 2007; Aikenhead & Michell, 2011). However, there are often some beliefs about nature that are sometimes but not always shared:

- Reality is not dualistic (i.e., Cartesian) but rather monist. This entails that nature is not separate from culture, and that the physical and metaphysical are always already entangled. In turn, this entails that parts of Nature (i.e., the frequent purview of science) only make sense within and can never be separated from the whole of nature or reality. This whole can be referred to as an ecology of relationships or a "sense of place" (see Cajete, 1994, 2000).
- Knowledge of nature emerges through natural-cultural relationships with knowledge keepers who are either human (e.g., Elders) or other-than-human (e.g., plants).
- Relationships entail responsibility. Accordingly, since everything is interrelated, the ecology of relationships is sustained by responsibility.
- Nature, which includes space, time, and matter, is in constant flux and in a state of indeterminacy, whose indeterminacy and flux is dis/continuous (e.g., non-linear time). In turn, while nature is empirically observable, it is never fully knowable; ways-of-knowing-nature must remain open and dynamic to ongoing patterns of difference.
- Models developed empirically across generations do not function as representations *of* nature but rather as flexible and adaptive pedagogical tools for coming-to-know nature and be relationally accountable to one's relationships.

5. McKinley and Stewart (2012) suggest that the topic of IWLN in school science is one "that has, in the past, been subsumed under wider concepts, such as multiculturalism, equity, and the like" (p. 541). However, as postcolonial scholars of science education, they do not dismiss the locations in which inclusion happens as solely problematic. Rather, they critically inhabit these spaces as a site of both possibility and problematics through

what could be read as taking a deconstructive stance. Taking a deconstructive stance is to "persistently to critique a structure that one cannot not (wish to) inhabit" (Spivak, 1993/2009, p. 284) by critically inhabiting the practices (linguistic or otherwise) at hand that shape us while at once troubling them.

As Spivak (1976) suggests, the language we possess also possesses us; thus, too simply moving beyond is "to run the risk of forgetting the problem or believing it to be solved" (p. xv) by reproducing it elsewhere, albeit differently. Thus, while recognizing the language of multiculturalism in science as a problematic entry point to consider the inclusion of IWLN, it is nonetheless the predominant entry point into conversations of, as well as gatekeeping device for, Indigenous knowledges in the science classroom. In turn, I also use and trouble this language.

6. The occurrence and recognition of classrooms *as* multicultural is recent. As McKinley and Stewart (2012) state:

> Solid decades of economic growth, and increased sensitivity to human rights, post-World War II, supported a steady improvement in education outcomes for [I]ndigenous students. As globalisation proceeded, teachers in Western countries faced classrooms of increasing cultural diversity, and anti-ethnocentrism was one aspect of the response, with teachers challenged to overcome their own deficit thinking. (p. 546)

7. This distinction is significant. For example, Cobern and Loving (2008) engage a similar representational mode (i.e., a conversation) with the explicit "intention... to present the critical arguments in common sense terms" (p. 438) and the implicit one of demonstrating that the terms of their opposition (e.g., cross-culturalism through relativism) are untenable. However, this opposition is one that is largely imagined: cross-culturalists often repeat that they are advocating for pluralism rather than relativism and that they too do not wish the latter (see McKinley, 2007; McKinley & Stewart, 2012). Furthermore, as discussed further within this chapter, discussions whose aim are to emerge victorious, rather than to listen to one another, are pyrrhic indeed. These victories not only ring hollow following the defeat of an imagined opponent, but such combativeness does not generate new and shared meanings but works towards making meaning common by suturing over *other* meanings. As such, these engagements always already fail in their ability to produce common grounds from which to address the complex questions of multicultural science education: they (re)produce ideological divisions which take us

away from the possibility of learning from one another and working towards shared goals (see Bohm, 1996; Latour, 2004a, 2004b).

8. Reminder (from Chapter 2), I use of "foreclosure" to signal instantiated pre-emergence of meaning. In other words, foreclosure signals the ways in which the language we possess sometimes appears as already fully formed, stable, and signifying an ontologically stable location; foreclosure as the semiotic locations which resist (re)opening due to their naturalization as closed before the fact (Spivak, 1999).

9. *Possible possibilities* is an expression used throughout the book meant to signal that not all possibilities are possible, nor equally possible. However, as possibility and impossibility forever vacillate, possible possibilities are always on the move (Barad, 2007). However, and importantly, not all possibilities are desired nor desirable.

10. While theories of dialogue as practice (e.g., Bohm, 1996) do not explicitly limit the number of participants involved within a dialogue to two, it implicitly groups them in and treats them as reversible communicative pairs (i.e., transmitter and receiver) in a manner similar to Socratic dialogue.

    Further, note that, as Spivak (1994) states: "dialogue is, in fact, the accepted proper name of responsibility as exchange-of-responses, implicitly understood as the flow of propositions or constatations rather than responses from both sides" (p. 45). Dialogue *as* responsibility presents itself as a mode to be able to respond, or of response-ability, to Indigenous science to-come (see Little Bear, 1994; Parrry, 2008).

11. Alternately, the process and product of dialectic and dialogue can be unpacked as mathematical operators. Consider two competing premises, positions, or postulates: *A* and *B*. When brought into proximal relation, the type of ethic of resolution (i.e., the operator *x*) shapes what is produced (i.e., the resultant side of the mathematical equation). Ideally, for dialectic, this interaction resembles such:

$$A \times B = C$$

Here, *C* is a third position or statement that would be a "best of both worlds" that is developed in an equitable mode of collaboration. However, due to always uneven relations of power, this ideal is almost never achieved or achievable. In instances in which the unevenness is more pronounced, the dialectic often bears closer resemblance the following:

$$A \times B = A$$

This is what is referred to as dialectic negation: the absorption and/or annulment of the other term (here, *B*).

Some respond to this by articulating an ethic of incommensurability in which:

$$A \times B = A + B$$

This entails that the two positions do not, cannot, and/or should not enter a proximal relation of co-production.

Recognizing that relations of power are always already uneven and unequal from the get go, dialogue strives to reach shared meanings without requiring the annulment of either meanings:

$$A \times B = A + AB + A$$

This entails that both propositions or positions (i.e., A and B) stand while also producing a shared meaning as the multiplicative cross-product of the two (AB).

12. As Spivak (1976) suggests, such deconstructive play might offer "a way out of the closure of knowledge" (p. lxxvii). For further examples of deconstructive play in science education, see Carter (2005) and McKinley and Aikenhead (2005).

13. While I lean towards a cross-culturalist position, it is productive for me to 'occupy' a position that leans towards universalism, even if it may be difficult for me, in order to think in ways that are productive rather than protective and prohibitive (see Spivak, 1976).

14. This landmark is, notably, a symposium in the journal *Science Education* (volume 85, issue 1). Despite its age, it remains relevant and significant. As McKinley and Stewart (2012) suggest, that since the publication of this symposium, "we have not progressed far with our arguments" (p. 547).

15. As dialogical approaches are infrequent within science education, this following section is intended to provide insights into how one might engage in this process. Furthermore, as Bohm (1994) posits from his work with artists, science has much to learn about questions of aesthetics, affect, and emotion and their respective role in process of knowledge production (and reproduction):

> Long before the scientist is aware of the details of a new idea, he [or she, or they] may 'feel' it stirring in him [or her] in ways that are difficult or impossible to verbalize. These feelings are like very deep and sensitive probes reaching into the unknown, while the intellect ultimately makes possible a more detailed perception of what these probes have come into contact with. Here, then, is a very fundamental relationship between science and arts, the latter must

evidently must work in a similar way, except that the whole process culminates in a sensually perceptible work of art, rather than in an abstract theoretical insight into nature's structural process. (p. 46)

16. (Re)opening science educator as location is an important project in the (re)opening of science education towards Indigenous science to-come. This is because educators are important performative agents who *do* and *undo* science education's circulating constructs, categories, concepts (see Higgins, 2014).

17. As Aikenhead and Ogawa (2007) elaborate and specify:

> The word *rational* does not signify a universalist rationality, but a rationality founded within the cultural context of use... *Perceiving* means both the process of constructing what is perceived to be reality through the participation of a group of people, and their resultant mental constructions of reality. (pp. 543–544, emphasis in original)

This is significant: there needs to be attention to the ways in which we differ and defer attempts at making meaning across difference by slipping back into conventional understandings of science education as we bring related but not identical concepts into the picture (see Chapter 5; McKinley, 2000; Sammel, 2009).

18. As Glen Aikenhead and Dean Elliot (2010) point out, the culture of school science potentially produces experiences of *assimilation* or *acculturation* rather than *enculturation* for the vast majority of students (~90%; see also McKinley, 2007). In other words, most students run the risk of experiencing science education as a form of dialectic negation rather than a dialogical encounter.

19. See Chapter 6 for more on the operationalization of the culture of Western modern science.

20. While it is not wholly agreed upon that the debate is unresolvable (e.g., Hansson, 2018), unresolvability remains nonetheless a productive feature of the multicultural science education debate: it always remains, to a certain degree, open to being otherwise (e.g., open to Indigenous science to-come). However, as knowledge is a performative enactment (e.g., Butler, 1993, 2005, 2010), this (albeit partial) openness is contingent upon continued engagement.

For Hansson (2018), the argument is made that the debate could be resolved if we could come to appreciate that "science builds on the age-old traditions of joint fact-finding that we can find in these [other] cultures" (p. 522). While not wholly in disagreement, I also understand this as coming-to-perceive *a* location in which the homework of response-ability should be engaged in. As explored more fulsomely in the chapters to come (Chapter 6 particularly), there is a nexus of reasons in which the disjuncture between Western modern and traditional knowledges is sedimented and stuck: there are multiple systems and structures in place which work to operationalize this disjuncture in favour of the Western modern Self at the expense of its (constructed) Other (e.g., a teleology of progress). Without also addressing these systems and structures, there runs the risk of this becoming a de/colonizing move: (re)producing a *settled* status quo, albeit differently.

21. Meaning-making of, or rather with texts is always a relational affair. As Spivak (1976) provocatively posits, this goes beyond deciphering *the* (i.e., singular) meaning intended by the author as it never was or will be possible for a text (or for a reader or an author) to reach the status of identity (i.e., self-sameness):

> two readings of the "same" book show an identity that can only be defined as a difference. The book is not repeatable in its "identity": each reading of the book produces a simulacrum of an "original" that is itself the mark of the shifting and unstable subject..., using and being used by a language that is also shifting and unstable. (p. xii)

In other words, readings are always temporary, transient, and contingent because there was never is *a* stable text to being with. Accordingly, readings are never singular either but rather multiplicity. This multiplicity emerges from the multiple relationships in flux that come to shape each and every reading.

22. The means of knowledge production never come to fully coincide with its ends. In other words, the pursuit of truth never (fully) produces truth (Derrida, 1976). Nonetheless, while truth is always already deconstructing, deconstruction "is not the exposure of an error, it is a vigilance about the fact that we are always obliged to produce the truth" (Spivak in Harasym, 1990, p. 46).

23. However, as later explored in Chapters 5 and 6, universalism and scientism are not so easily disentangled. While the two should not be conflated as they are not one and the same, logics of not-the-same come to mask the workings of power without an exploration of the ways in which they express themselves through co-constitutive difference.

24. The insight of scientific knowledge-practice as always already situated will come to inform the work of the Chapter 4. As scientific knowledge claims do not *mirror* reality but rather are *prismatic* inflections

of knowledge through a real or imagined laboratory, the next chapter (re)considers optical metaphors as a location for rethinking criticality in science education.

25. Scientific knowledge-practice as always already being hybrid informs Chapter 7. As the ways in which hybridity is always uneven and unequal through a dialectic relationship in which Indigenous ways-of-knowing-in-being are dialectically subsumed, sublated, or sutured over, the work of Chapter 7 considers the ways in which responsibility is also shaped within this interface. In turn, this interface must be laboured as it shapes the ability to respond (see Kuokkanen, 2007).

26. Note: Cobern and Loving's (2008) criteria of aligning with ontology problematically treats ontology as a singular affair.

27. The insight of scientific knowledge-practice as always already being ontologically situated becomes a central notion in Chapter 5 (and extended in Chapter 6), in which Cobern and Loving's (2008) call for ontological-epistemological alignment is both used and troubled to consider Cartesianism as but one ontology among many.

28. The ways in which science, education, and science education have respectively operated through forms of cognitive imperialism which have had, and continue to have, real and often negative effects on Indigenous and other non-Western peoples (McKinley, 2007).

REFERENCES

Aikenhead, G. S. (2006). *Science education for everyday life*. London, ON: Althouse Press.

Aikenhead, G. S., & Elliot, D. (2010). An emerging decolonizing science education in Canada. *Canadian Journal of Science, Mathematics and Technology Education, 10*(4), 321–338.

Aikenhead, G. S., & Michell, H. (2011). *Bridging cultures: Indigenous and scientific ways of knowing nature*. Toronto, ON: Pearson Canada.

Aikenhead, G. S., & Ogawa, M. (2007). Indigenous knowledge and science revisited. *Cultural Studies of Science Education, 2*(3), 539–591.

Alsop, S., & Fawcett, L. (2010). After this nothing happened. *Cultural Studies of Science Education, 5*(4), 1027–1045.

Atwater, M. M., & Riley, J. P. (1993). Multicultural science education: Perspectives, definitions, and research agenda. *Science Education, 77*(6), 661–668.

Barad, K. (2000). Reconceiving scientific literacy as agential literacy. In R. Reed & S. Traweek (Eds.), *Doing science + culture* (pp. 221–258). New York, NY: Routledge.

Barad, K. (2007). *Meeting the universe halfway: Quantum physics and the entanglement of matter and meaning*. Durham, NC: Duke University Press.

Belczewski, A. (2009). Decolonizing science education and the science teacher: A white teacher's perspective. *Canadian Journal of Science Education, 9*(3), 191–202.

Bohm, D. (1994). *On creativity.* New York, NY: Routledge.

Bohm, D. (1996). *On dialogue.* New York, NY: Routledge.

Butler, J. (1993). *Bodies that matter: On the discursive limits of "sex".* New York, NY: Routledge.

Butler, J. (2005). *On giving an account of oneself.* New York, NY: Fordham University Press.

Butler, J. (2010). *Frames of war: When is life grievable?* London, UK: Verso.

Cajete, G. (1994). *Look to the mountain: An ecology of indigenous education.* Durango, CO: Kivaki Press.

Cajete, G. (1999). *Igniting the sparkle: An Indigenous science education model.* Durango, CO: Kivaki Press.

Cajete, G. (2000). *Native science: Natural laws of interdependence.* Santa Fe, NM: Clear Light Books.

Carter, L. (2004). Thinking differently about cultural diversity: Using post-colonial theory to (re)read science education. *Science Education, 88*(6), 819–836.

Carter, L. (2005). A place for alternative readings: Can they be of use? Responding to comments on "Thinking differently about cultural diversity: Using postcolonial theory to (re)read science education". *Science Education, 89*(6), 913–919.

Cobern, W. W., & Loving, C. C. (2001). Defining "science" in a multicultural world: Implications for science education. *Science Education, 85,* 50–67.

Cobern, W. W., & Loving, C. C. (2008). An essay for educators: Epistemological realism really is common sense. *Science & Education, 17,* 425–447.

Derrida, J. (1976). *Of grammatology* (G. C. Spivak, Trans.). Baltimore, MD: John Hopkins University Press.

Derrida, J. (1994/2006). *Specters of Marx: The state of the debt, the work of mourning, & the new international* (P. Kamuf, Trans.). New York, NY: Routledge.

El-Hani, C. N., & de Ferreira Bandeira, F. P. S. (2008). Valuing Indigenous knowledge: To call it "science" will not help. *Cultural Studies of Science Education, 3*(3), 751–779.

El-Hani, C. N., & Mortimer, E. F. (2007). Multicultural education, pragmatism, and the goals of science teaching. *Cultural Studies of Science Education, 2*(3), 657–702.

Foucault, M. (1997). *The politics of truth.* New York, NY: Semiotext(e).

Hansson, S. O. (2018). How to reconcile the multiculturalist and universalist approaches to science education. *Cultural Studies of Science Education, 13*(2), 517–523.

Harasym, S. (1990). *The post-colonial critic: Interviews, strategies, dialogues.* New York, NY: Routledge.

Harding, S. (2008). *Sciences from below: Feminisms, postcolonialities, and modernities.* Durham, NC: Duke University Press.

Higgins, M. (2011). Finding points of resonance: Nunavut students' perceptions of science. *In Education, 17*(3), 17–37.

Higgins, M. (2014). De/colonizing pedagogy and pedagogue: Science education through participatory and reflexive videography. *Canadian Journal of Science, Mathematics and Technology Education, 14*(2), 154–171.

Kim, E. J. A., Asghar, A., & Jordan, S. (2017). A critical review of traditional ecological knowledge (TEK) in science education. *Canadian Journal of Science, Mathematics and Technology Education, 17*(4), 258–270.

Kirby, V. (2011). *Quantum anthropologies: Life at large.* Durham, NC: Duke University Press.

Kuokkanen, R. J. (2007). *Reshaping the university: Responsibility, Indigenous epistemes, and the logic of the gift.* Vancouver, BC: UBC Press.

Lather, P. (2007). *Getting lost: Feminist efforts toward a double(d) science.* New York, NY: State University of New York.

Latour, B. (1988). *The pasteurization of France* (A. Sheridan & J. Law, Trans.). Cambridge, MA: Harvard University Press.

Latour, B. (1993). *We have never been modern.* Cambridge, MA: Harvard University Press.

Latour, B. (2004a). Why has critique run out of steam? From matters of fact to matters of concern. *Critical Inquiry, 30*(2), 225–248.

Latour, B. (2004b). *Politics of nature: How to bring the sciences into democracy.* Cambridge, MA: Harvard University Press.

Le Grange, L., & Aikenhead, G. (2016). Rethinking the 'Western tradition': A response to Enslin and Horsthemke. *Educational Philosophy and Theory,* 1–7. Advance online publication. https://doi.org/10.1080/00131857.2016.116 7656.

Lewis, B., & Aikenhead, G. (2001). Introduction: Shifting perspectives from universalism to cross-culturalism. *Science Education, 85,* 3–5.

Little Bear, L. (1994). Preface to Routledge classics edition. In D. Bohm (Ed.), *Of creativity* (pp. vii–xiv). New York, NY: Routledge.

Matthews, M. R. (1994). *Science teaching: The role of history and philosophy of science.* New York, NY: Routledge.

Mazzei, L. (2007). *Inhabited silence in qualitative research: Putting poststructural theory to work.* New York, NY: Peter Lang.

McKinley, E. (2000). Cultural diversity: Masking power with innocence. *Science Education, 85*(1), 74–76.

McKinley, E. (2007). Postcolonialism, Indigenous students, and science education. In S. K. Abell & N. G. Lederman (Eds.), *Handbook of research on science education* (pp. 199–226). Mahwah, NJ: Lawrence Erlbaum.

McKinley, E., & Aikenhead, G. (2005). Comments on "Thinking differently about cultural diversity: Using postcolonial theory to (re)read science education". *Science Education, 89*(6), 901–906.

McKinley, E., & Stewart, G. (2012). Out of place: Indigenous knowledge in the science curriculum. In B. Fraser, K. Tobin, & C. J. McRobbie (Eds.), *Second international handbook of science education* (pp. 541–554). Dordrecht, NL: Springer, Netherlands.

Moulton, J. (1983). A paradigm of philosophy: The adversary method. In S. Harding & M. B. Hintikka (Eds.), *Discovering reality: Feminist perspectives on epistemology, metaphysics, methodology, and philosophy of science* (pp. 149–164). Dordrecht, NL: Reidel.

Ogawa, M. (1995). Science education in a multiscience perspective. *Science Education, 79*, 583–593.

Parry, G. A. (2008). *Seed graduate institute: An original model of transdisciplinary education informed by Indigenous ways of knowing and dialogue.* Unpublished doctoral dissertation, California Institute of Integral Studies, Purdue University, San Francisco, CA.

Peat, F. D. (2007). *Pathways of chance.* Pari, IT: Pari Publishing.

Sammel, A. (2009). Turning the focus from 'other' to science education: Exploring the invisibility of whiteness. *Cultural Studies of Science Education, 4*, 649–656.

Siegel, H. (1997). Science education: Multicultural and universal. *Interchange, 28*, 97–108.

Siegel, H. (2001). Multiculturalism, universalism, and science education: In search of common ground. *Science Education, 86*, 803–820.

Smith, R. (2011). The play of Socratic dialogue. *Journal of Philosophy of Education, 45*(2), 221–233.

Snively, G., & Corsiglia, J. (2001). Discovering Indigenous science: Implications for science education. *Science Education, 85*, 6–34.

Southerland, S. A. (2000). Epistemic universalism and the shortcomings of curricular multicultural science education. *Science & Education, 9*, 289–307.

Spivak, G. C. (1976). Translator's preface. In J. Derrida (Ed.), *Of grammatology* (G. C. Spivak, Trans., pp. ix–lxxxvii). Baltimore, MD: Johns Hopkins University Press.

Spivak, G. C. (1993/2009). *Outside in the teaching machine.* New York, NY: Routledge.

Spivak, G. C. (1994). Responsibility. *Boundary 2, 21*(3), 19–64.

Spivak, G. C. (1999). *A critique of postcolonial reason.* Cambridge, MA: Harvard University Press.

Stanley, W. B., & Brickhouse, N. W. (1994). Multiculturalism, universalism, and science education. *Science Education, 78,* 387–398.

Stanley, W. B., & Brickhouse, N. W. (2001). Teaching sciences: The multicultural question revisited. *Science Education, 85,* 35–49.

St. Pierre, E. A. (1997). Methodology in the fold and the irruption of transgressive data. *International Journal of Qualitative Studies in Education, 10*(2), 175–189.

Turnbull, W., & Mullins, P. (2007). Socratic dialogue as personal reflection. *Reflective Practice: International and Multidisciplinary Perspectives, 8*(1), 93–108.

van Eijck, M., & Roth, W. M. (2007). Keeping the local local: Recalibrating the status of science and traditional ecological knowledge (TEK) in education. *Science Education, 91*(6), 926–947.

van Eijck, M., & Roth, W. M. (2009). Authentic science experiences as a vehicle to change students' orientations toward science and scientific career choices: Learning from the path followed by Brad. *Cultural Studies of Science Education, 4*(3), 611–638.

Zeyer, A. (2009). Public reason and teaching science in a multicultural world: A comment on Cobern and Loving: An essay for educators...in the light of John Rawls' Political Philosophy. *Science & Education, 18*(8), 1095–1100.

Mirrors, Prisms, and Diffraction Gratings: Placing the Optics of the Critical Gaze in Science Education Under Erasure (After the Critique of Critique)

> There has been in the modern Western world (dating, more or less, empirically from the 15th to the 16th centuries) a certain way of thinking, speaking and acting, a certain relationship to what one knows, to what one does, a relationship to society, to culture and also a relationship to others that we could call, let's say, *the* critical attitude. (Foucault, 1997, p. 24, emphasis mine)

The purpose of this chapter[1] is to explore what Foucault refers to as "the" critical attitude and its relationship to science education. Excavating the concepts that linger and lurk when critique is presented as atheoretical in science education, the following chapter engages with the *theoretical* homework of response-ability. Drawing from Foucault's (1997) insight that *the* critical attitude is but *a* critical attitude, I explore possibilities for and of critique that stem from and respond to the crisis and critique of critique (see Barad, 2012a; Kirby, 2011; Latour, 1993, 2004a). The possibility of critique as plural is significant as *the* mode of critique within the multicultural science education debate from the previous chapter (re)produce Indigenous science as yet-to-come. Specifically, the adversary mode therein not only excludes, differs, and defers Indigenous science to-come, but also upholds the metaphysics of modernity through its enactment (via distance, dichotomy as mutual exclusivity;

© The Author(s) 2021
M. Higgins, *Unsettling Responsibility in Science Education*,
Palgrave Studies in Educational Futures,
https://doi.org/10.1007/978-3-030-61299-3_4

discussed within this chapter). In turn, I posit that (an) unsettling criticality is not only one which critiques settler colonial logics and practices but also the taken-for-granted ways-of-critiquing which can undergird these very efforts.

Building on the insight that scientific knowledge-practice is always already situated from the previous chapter (e.g., WMS's knowledge as contextualized within a real or imagined laboratory), I consider the ways in which criticality in science education is always mediated by conceptual apparatuses. In particular, I metaphorically employ three optical apparatus—the mirror, the prism, and the diffraction grating[2]—to analyse and inform how the critical gaze might be re(con)figured within science education.

As critique is always in relation (Foucault, 1997), I begin by positioning my own critical relation in and to science education. Secondly, I propose critical and complicit (mis)reading as the deconstructive methodological approach in the potentiality of (re)signifying science education otherwise. Third, a thumbnail account[3] of the crisis of the critical stance (Latour, 1993) is presented with attention to mirror metaphor it makes operational and the outcome for critical engagement within science education. Fourth, I explore prismatic dispersal as a first optical alternative (Butler, 2001; Deleuze, 1988; Foucault, 1997) as well as the types of critiques that can made be with/in this optics. Lastly, extending upon the prism, I investigate diffraction as metaphor that builds upon prismatic dispersal (Barad, 2007, 2012a; see also Cajete, 1994, 2000; Latour, 1993). To animate this discussion, these conceptual and metaphoric critical apparatuses are employed to ask questions anew about the multicultural science education debate (from the previous chapter).

THE SUBJECT OF CRITIQUE: MY RELATION TO CRITIQUE IN/OF SCIENCE EDUCATION

Because we need to "begin" some-where and some-time, let's begin in St. Catherine's, Ontario in May of 2014.[4] During the 42nd annual *Canadian Society for the Study of Education* (CSSE) conference, I am giving a talk titled *Post-Cartesian possibilities for schools as places of learning: Putting to work an intra-active pedagogy* (see Higgins, 2016). During this presentation, given on a curriculum studies panel, new possible possibilities for critique and critical engagement began to bubble; not as the result of my own work, but rather an insightful comment about the ways in which I was being critical. Significantly, I was being invited to consider the ways in which "the" critical attitude is but *a* critical attitude. In this presentation, I discussed the taken-for-grantedness and social constructedness of conceptual norms of science education produced with/in Eurocentrism through a focus on its entangled epistemologies (e.g., epistemic realism) and ontologies (e.g., Cartesianism). This was very much in line with the ways in which I was approaching criticality at the time: as a formal and informal science educator for over 10 years at the time, I had become (and continue to be) critical of many of the ways-of-knowing as well as the ways-of-being that govern the practices within diverse spaces of science education (e.g., Eurocentrism, whiteness, masculinism). It would not be long before I familiarized myself with the multicultural science education debate in which questions of "what counts" as science are asked, by extension what counts as valid course content within school-based science curriculum. Largely at stake within this debate is the inclusion or exclusion of traditional ecological knowledge (TEK) and Indigenous ways-of-living-with-nature (IWLN) alongside Western modern science (WMS), as well as the norms through which they are included, excluded, and juxtaposed. This debate is largely between science educators who champion the inclusion of TEK and IWLN as an equally valid ways of knowing nature (i.e., cross-culturalists; e.g., Snively & Corsiglia, 2001) and those who do not consider these placed-based ways-of-knowing-nature as *equally valid* to the "universal" standard of WMS (i.e., universalists; for example, Cobern & Loving, 2001). Note that *equally valid* here does not signify that TEK and IWLN achieve equivalence or sameness with WMS, but rather that they offer something that is of similar importance (e.g., the former presents frames for ethical and sustainable practices of living with nature while the latter offers quantifiability, reproduceability, and predictability through laboratory-based experimentation; see Aikenhead & Michell, 2011).

Guided by the questions: *Who is included?*, *Who is excluded?*, and *What norms shape how participation is and can be enacted?*, I had begun examining the central constructs that often determine "what counts" as science within science classrooms, with a focus on Indigenous and non-Indigenous youth's relationship to "what counts" (see Higgins, 2011). Entangled with/in this criteria for inclusion is the question of whose knowledge counts as scientific knowledge, and under which conditions it comes to be as such (Snively & Corsiglia, 2001; Stanley & Brickhouse, 1994, 2001). Digging deeper, I shifted the critical gaze to explore the ways in which Eurocentrism is entangled with/in the culture of "school science" (Higgins, 2014; McKinley, 2001, 2007; Sammel, 2009). While these critical explorations and enactments were deeply fruitful for me in working towards decolonizing goals, they also revealed the ways in which Eurocentrism circulates in capillary[5] manners into both decolonizing processes (e.g., border crossing as pedagogy; see Aikenhead, 1997, 2001, 2006a, 2006b) and decolonizing subjects (e.g., decolonizing pedagogue; Belczewski, 2009; Higgins, 2014; see also McKinley, 2001; Sammel, 2009).[6] Alternately stated, despite the ways that available decolonizing curriculum and subject positions of teacher-as-researcher worked within and against a problematic centre, they were exceeded in pedagogical practice by the very (neo-)coloniality the approach challenged, thus becoming de/colonizing (see also Carter, 2004, 2010; Sammel, 2009). While there continues to be diverse, productive possible pedagogical possibilities that stem from de/colonizing approaches, not all opportunities are equally productive (see Spivak, 1988, 1999). Accordingly, like Sammel (2009), I was beginning to "wonder if there is a science curricula that does not indoctrinate, ... if there is really an authentic 'decolonizing science practice'" (p. 653).[7] However, critique, of that sort, could not and would not account for the multiplicity of ways in which (neo-)coloniality was always already on the move beyond and between the concepts and categories laid out for it (Spivak, 1999): it would never fit the labels of *is* and *is not* applied by critique-as-usual.

This was precisely the insight that was presented to me following the presentation. During the question period, Dr. Kent den Heyer, a co-panellist then and colleague now, suggested that it appeared as though the ways in which I come-to-see and critique Eurocentrism was by treating it as a metaphoric "waste basket," and that accordingly, "Descartes would be rolling in his grave" (K. den Heyer, personal communication, May, 2014). In other words, I was treating these constructs and systems as "prematurely naturalized objectified facts" (Latour, 2004a, p. 227). Rather than treating them in a manner that obscured their relationships as or within a complex ecology, they were all-too-simple, one-size-fits-all, pre-determined matters onto which I had passed negative judgement; they were disposable and to be disposed of. This, in turn, might make my argument easily disposable and to be disposed of: "While it may seem easy to critique Eurocentric thinking and structures, how might one 'displace' current thinking?" (McKinley & Aikenhead, 2005, p. 903). If this was the practice of perception that I was employing, I began to wonder if there were other ways of deploying a critical gaze that might be more productive.

THE OPTICS OF CRITIQUE: WHY THE OPTICAL CONFIGURATIONS WE (METAPHORICALLY) DEPLOY MATTER

The eyes have been used to signify a perverse capacity—honed to perfection in the history of science tied to militarism, capitalism, colonialism, and male supremacy—to distance the knowing subject from everybody and everything in the interests of unfettered power. (Haraway, 2001, p. 677)

Of vision, Haraway (2001) reminds us that despite the perception that we live in an age in which the plurality of technologies of sight that enhance our primate eyes are near limitless (e.g., satellite surveillance, magnetic resonance imaging, closed-circuit television, spectrometers, x-ray, radio telescopes), we never come to achieve:

a God's eye view of the universe, the universal viewpoint, the escape from perspective, with all the rights and privileges accorded therein. Vision that goes right to the heart of the matter, unmediated sight, knowledge without end, without responsibility. (Barad, 2007, p. 233)

The "god-trick" of "seeing everything from nowhere" (Haraway, 2001, p. 678) is but an "impossible dream of plenitude" (Spivak, 1976, p. xix), an end goal never reached or reachable by its means. As "direct, devouring, generative, and unrestricted vision" (Haraway, 2001, p. 678) is but an ideology never achieved in technological practice, vision is not only always a situated and partial practice but one that must be accountable to its situatedness and partiality.

As Haraway (2001) posits, this is significant as there is a persistent metaphorical reliance on vision within critique (see also Barad, 2007; Haraway, 1997; Kirby, 2011). Even when critique metaphorically deploys vision (simply) as perspective, critique is always already a "politics of positioning" (Haraway, 2001, p. 681) in which positionings are plural, fragmented, unsteady, shifting, and on the move (Barad, 2012a; Butler, 2001; Foucault, 1997; Kirby, 2011; Latour, 1993, 2004a). However, just as "vision requires instruments of vision" (Haraway, 2001, p. 681), critique requires instruments of critique. Even normative and taken-for-granted conceptions of vision (e.g., perspective, sight) are situated and partial (e.g., location, direction), as well as supported by instruments of vision (e.g., biological technologies such as the particular eyes engaged in sight, epistemic frameworks; see Butler, 2010). Similarly, even what Foucault (1997) refers to as "*the* critical spirit" (emphasis mine) employs particular instruments of critique. While critique and the optical configurations that they metaphorically employ and deploy (whether implicitly[8] or explicitly) never fully come to coincide, it is nonetheless important and productive to consider how diverse optical configurations continue to produce particular "politics of positioning" (Haraway, 2001, p. 681) within critique. Thinking with and through optical configurations can be a productive way of bringing attention to the ways in which these complex and contradictory positionings are differentially produced, framed, and (un)acknowledged. Furthermore, to think of them as configurations is an important step in working towards what Barad (2007) refers to as re(con)figuring s: the ongoing processes in which configurations are dynamic, temporary, and always already being produced differentially and anew within their current agential relations.

For example, in my own critical engagement above, it could be stated that I employed an optical configuration in which "reflection [is] a pervasive trope for knowing" (Barad, 2007, p. 72).[9] That is, I arranged constructs (e.g., "what counts as science" as a quasi-neutral cultural construct) to appear as but a false reflections of reality, while simultaneously presenting them as productions of reflections from other mirrored surfaces that I held to be true (e.g., Eurocentrism). All the while, I partially masked the mirroring process that I was engaging in, as well as the ways in which I manoeuvered from one mirror to another. Accordingly, I was metaphorically positioning and utilizing the optical apparatus that shapes my critical "*ways* of seeing" (Haraway, 2001, p. 679, emphasis in original) such that it cast my "bad" objects into darkness and, by contrast, my "good" objects into light, making me blind to both, as well as the process itself. Through producing particular points of vantage for another reader, as well as myself, I used the "poor trick that allows critique to go on" (Latour, 2004a, p. 241): mirrors upon mirrors (Barad, 2007, 2012a).

It is not that the constructs and systems I addressed were unproblematic, or that my earlier arguments held no merit or validity. Rather, these arguments could easily be (un)done through processes of critique very similar to the ones I employed. Through slight of hand, a crafty critic could easily reconfigure the optical geometry of my argument. Obscure a little here, illuminate a little there, and a convincing reversal that exposes that which I shadowed, while shading that which I presented could be produced. Such smoke and mirrors would not resist the dialectic reversal of the very same move I was making,[10] but my "prematurely naturalized objectified facts" would also do little to sway or engage with those whose *matters of fact* were oppositional to mine (see Bohm, 1996; Moulton, 1983; see also Kuokkanen, 2007; Latour, 2004a; McKinley & Aikenhead, 2005). This, of course, is without putting into question the very possibility of distance from one's own *matters of fact* (Barad, 2007; Butler, 2001, 2005; Foucault, 1997; Kirby, 2011; Latour, 2004a; Smith, 1999/2012). It is for this reason that Latour states that "there is no sure ground even for criticism" (p. 227), and especially not for critique *like that*.

For this reason, it is deeply productive to engage around Foucault's classic question of "*What is critique?*" (Foucault, 1997, p. 24, emphasis in original). What if the norms surrounding *the* critical attitude, when critiqued, revealed it to be but *a* critical spirit that, as Latour (2004a) states, has run out of steam? What might it mean to be critical *otherwise*, to engage in *another* critical mode that is productive rather than protective (see Derrida, 1976; Spivak, 1976)? Furthermore, if "vision requires instruments of vision" (Haraway, 2001, p. 681), how might a differential consideration and understanding of the optical apparatuses that we employ metaphorically inform and produce critically gazing (and the critical gazer) otherwise? How might an understanding of the physical phenomena of optics (e.g., properties of light within geometrical and physical optical configurations) entangled within critique re(con)figure the possible possibilities for critical engagement? If "optics is a politics of positioning" (Haraway, 2001, p. 681), how might we re-arrange the subjects and objects of vision? How might we do so without falling into the trap of going from partial and situated vision to an unsituated "seeing everything from nowhere" (Haraway, 2001, p. 678)? I was eager to explore what this might mean for decolonizing science education. In order to engage with these questions, I explore critical and complicit (mis)reading as *a* deconstructive approach that neither rejects the structure (as critique is still ever necessary) nor accepts it (as critique as it stands has run out of steam).[11]

CRITICAL AND COMPLICIT (MIS)READINGS OF THE OPTICS
OF CRITIQUE: SCIENCE EDUCATION UNDER ERASURE

To be at once critical and complicit methodologically is to engage in the difficult task of a double(d) reading whose "interest is in complicit practices and excessive difference" (Lather, 2007, p. 105). To unpack how I utilize the expression *critical and complicit* herein, it is also productive to do a double(d) reading of complicity and the promiscuously entangled ways in which they are articulated and enacted.

First, complicity signals the critical inhabitation that is required in a project of working within and against science education. Such a critical inhabitation resists both the critical rejection of the educational structure as well as the complicity that protects rather than productively engages with the problematics within it (Spivak, 1976). It is a research approach that recognizes that constructs, categories and contexts are always already rife with both problematics and possibilities at once. It is to "persistently to critique a structure that one cannot not (wish to) inhabit" (Spivak, 1993/2009, p. 284) by taking a deconstructive stance and placing this inadequate yet necessary structure *under erasure*. For Derrida (1976), to put something under erasure "is to write a word, cross it out, and then print both word and deletion. (Since the word is inaccurate, it is crossed out. since it is necessary, it remains legible)" (Spivak, 1976, p. xiv). Methodologically, this entails the deconstructive using and troubling of concepts, categories and constructs while recognizing that they are always already both containing and constraining while problematically and productively exceeded (Derrida, 1976; Kuokkanen, 2007; Lather, 2007; St. Pierre, 2011). This excess gestures towards the second meaning of complicity.

Secondly, complicity signals towards the collusive relations that concepts and categories hold to their constitutive exteriority. In short, textuality is always already complicitous to its otherness (e.g., "the complicity between yes and no" [Spivak in Derrida, 1976, p. 319]). Thus, being critical and complicit is not only to critique this complicity, but also a process of paying attention to the slippages in which complicities surface. However, it is not only a question of witnessing these deconstructive openings.

Thus, to place under erasure requires creative tinkering with/in those moments when the inhabited structure is self-transgressing through a two-part process. First, it entails locating a productive moment in which it reveals its undecidability. Spivak (1976) describes the process of bearing witness to undecidability as such:

> If in the process of deciphering a text in the traditional way we come across a word that seems to harbor an unresolvable contradiction and by virtue of being one word is made sometimes to work in one way and sometimes in another and thus is made to point away from the absence of a unified meaning we shall catch at that word. (p. lxxv)

In short, this entails paying attention to the ways in which concepts and categories whose meanings vacillate (see Chapter 3 for the *play of (re)signification*) between a meaning and its constitutive otherness. This constitutive otherness can be read as a relation between binary and oppositional terms (e.g., life/death, familiar/strange) as well as similar yet different terms (e.g., affect and effect, amoral and immoral). Secondly, it involves the prying open of this methodological fissures or locations that "harbor an unresovable contradiction" (Spivak, 1976, p. lxxv). In short, this entails using that which exceeds it (i.e., the constitutive otherness, whether oppositional or similar yet different) as a lever by substituting it into the methodological "text". This in turn reverses the hierarchy between intended and unintended meaning, creating the possibility for new meanings to potentially be inscribed over the trace of that which was (partially) erased (Derrida, 1976; Spivak, 1976; St. Pierre, 2011).

Like any and every structure, there are many self-trangressive moments in which the structure both encompasses and eschews itself that provide important locations to work within and against. However, the space I put under erasure within this text is the differing and deferring space between what critique with/in science education *is* and *is (not)*. The formulation of *is (not)* is intentional and is utilized to signal the need for "working with the resources of the old language, the language we already possess, and which possesses us" (Spivak, 1976, p. xv) while engaging in a disruptive "repetition [that] leads to a simulacrum, not to the 'same'" (Spivak, 1976, p. lxv). In other words, it is a commitment to working within and against critique with/in science education by differentially using the concepts, constructs, and categories available by (mis)reading[12] them through substituting similar but different iterations of the optics of critique that usually or typically frame critique with/in science education.

Critically and complicitly inhabiting this space offers rich possibilities for disrupting, displacing, and differentially enacting critique with/in science education for two distinct yet interconnected reasons. First, within the context of critique with/in science education, critique, and more precisely *the optics of critique* presents itself as a signifier whose signified is unstable and undecidable. As explored earlier within this chapter, the optics of critique are often defined and deployed in a cursory and rapid manner (i.e., vision as semiotically pervasive but often under-defined).[13] While optics are gestured to and enacted within many critiques with/in science education, it is not always clear as to which optics are being referred to. Rather, the meaning always differs and is deferred (see Derrida, 1976). As such, such a location presents itself as a site ripe for productive (mis)readings and substitutions of differential and unintended understanding of these critical optics.

Second, the disjuncture between what the optics of critique with/in science education *is* and *is (not)* offers itself as a long, and ever-lengthening lever to pry this space open while maintaining a critical inhabitation of this educational space. What the optics of critique within science education often *is* (i.e., the mirror) does not fully reflect the critical shifts, breaks, and developments with respect to ways-of-being-critical both theoretically and in its practical applications. However, prying open with what the optics of critique *is (not)* allows for a working within and against critique with/in science education that does not jettison the central metaphor of optics nor the impetus for critique that constitute it but rather considers similar yet different understandings thereof.

For Derrida, "the signifier and the signified are interchangeable" (Spivak, 1976, p. lxv). Within science education, the signifier that is the *optics of critique* is already in a state of undecidable signification. Thus, the task at hand herein is the rupturing of this space using the lever of signified that is what the optics of critique *is (not)*.

MIRROR UPON MIRRORS: MATTERS OF FACT, MATTERS OF FICTION, AND SCIENCE EDUCATION

While the Enlightenment profited largely from the disposition of a very powerful descriptive tool, that of matters of fact, which were excellent for *debunking* quite a lot of beliefs, powers, and illusions, it found itself totally disarmed once matters of fact, in turn, were eaten up by the same debunking impetus. (Latour, 2004a, p. 232).

Critique has been, for the most part, a process of disrupting and displacing particularly problematic "beliefs, powers, and illusions" (Latour, 2004a, p. 232) with *matters of fact*. While this mode has had "prodigious efficacy" in the past, modernity's "critical capacities are waning" (Latour, 1993, p. 35). This largely, but not exclusively, has to do with the notion that critique is all-too-often a process that is restricted to fault-finding and passing negative judgement (Barad, 2012a; Bohm, 1996; Butler, 2001; Foucault, 1997; Kirby, 2011; Latour, 1993, 2004a; Moulton, 1983). While this has been effective while operating within particular disciplinary lines, when it comes to critical engagement at the intersection of nature, culture, and politics such as is the case within cross-cultural science education, the usual critical modes of naturalization, socialization, and deconstruction begin to break down (Latour, 1993, 2004a, 2004b).

If we take seriously Latour's (1993) eponymous thesis statement that "We [the West/Global North] Have Never Been Modern", then the division between and mutual exclusivity of nature, culture, and politics that shape understandings and enactments of modernity never fully became actualized. In other words, even if modernity is treated as totalizing, it has never fully totalized. This was in part because natural-cultural hybrids have been an absent presence that "moderns" have been (un)consciously relying upon since the proclaiming of nature and culture as being separate and separable. Accordingly, it is not only the more obvious points of convergence such as issues of science, technology, and society that are to be included within this entanglement, but rather that everything is *always already* within nature, culture, and politics. Thus, the entanglement of nature, culture, and politics often refuse to be explained away by such critical modes as they are always already exceeded by them (see also Barad, 2007; Latour, 2004a, 2004b; Kirby, 2011). As Latour poses, "is it our fault if the intersections of nature, culture, and politics are *simultaneously real, like nature, narrated, like discourse, and collective, like society?*" (p. 6, emphasis in original).

The critical gazes offered by naturalization, socialization, and deconstruction[14] are never simple or passive operationalizations of metaphoric vision (Barad, 2007; Haraway, 2001). They usually feed on the weaknesses of the other two modes in (re)presenting a truth about the world (or in the case of deconstruction, an absence of stable natural and cultural truth; see Derrida, 1976). In offering a privileged vantage point from which to state truths, naturalization states that nature *is* this by bracketing out culture, socialization states that culture *is* that by bracketing out nature, and deconstruction states that society and nature are (*not*)[15] by denying the epistemological and ontological stability required for socialization and naturalization. These statements are, respectively, made at the expense of their other(ed) statements and without taking seriously the epistemic resources presented through other modes (Latour, 1993). Herein lies the major critique of critique that I will unpack here,[16] the notion that the objects under and utilized to pursue critical inquiry are attributed and granted either firm or flimsy positions but never viewed as complex entanglements that encompass both positions, across multiple critical gazes.

Through the mirror metaphor that is made operational in critique, the objects of inquiry are almost "never complicated enough" (Latour, 2004a, p. 234). They are rarely allowed to exist as objects that are the products of rich and ongoing complex natural and cultural histories, as well as produced by, and producing various participating agents (see also Barad, 2007, 2010; Cajete, 1994, 2000; Kirby, 2011). Rather, Latour (1993) suggests that the majority (i.e., roughly 90%) of the contemporary critical scene in the social sciences positions its objects of inquiry, whether they are conceptual or concrete, in one of two positions: *fait (i.e., fact)* or *fée (i.e., fairy)*.[17] In other words, they are presented as good reflections and bad reflections of reality, with the critic themselves acting as a mirror of the observed phenomena by giving a "clear" and "accurate" representation. Latour (2004a) quickly unpacks these two positions by explaining how critics too often deploy them. First, the critic presents to "naïve believers" that the object they are using and the way they are using it are but a *fairy*, a fantasy or fetish created through the simple projection of their wishes and desires onto the object. In other words, the first critical gesture is in re-presenting a held belief or value as but a *fairy*. Second, the fetishistic projection is "explained" through use of other objects of inquiry. These other objects, presented and levied as indisputable *fact*, defy the very possibility of the projected *fairy* from being an agential choice as they are given a fully causal treatment. Accordingly, the "naïve believer", is twice slighted, once for investing belief in a *fairy*, and twice for not being able to perceive the *fact* that shaped them to do so in the first place.

Furthermore, critique through the mirror metaphor serves to mask the practices of positioning through which the *fact* and *fairy* labels are applied through what Barad (2007) refers to as the "illusion of givenness" that is mirrored correspondence. In short, Barad (2007) explains that the illusion of givenness that is produced through a mirroring of one's social or natural reality begins to break down when we consider the ways in which the mirror itself is not the thing it mirrors:

> As with Magritte's famous painting *Ceci n'est pas une pipe*, the point is not that it really isn't a pipe but only a representation of a pipe, but rather that representations do not simply refer in ways that we have come to expect, that in fact the entire question of referentiality seems to have lost its self-evident nature and givenness has lost its transparency, and we can no longer see our way through the game of smoke [and] of mirrors that representationalism has become. Like a good magician, representationalism would have us focus on what seems to be evidently given, hiding the very practices that produce the illusion of givenness. (Barad, 2007, p. 360, emphasis in original)

It is not that particular arguments can and cannot be "reflected" as *fact* and *fairy* but rather that the illusion of givenness of the mirror metaphor works to hide the ways in which such an optical apparatus was set up to produce such a reflection by presenting the carefully produced reflection as the referent.

To give an example of what is meant by the application of the mirror metaphor within critique, we will consider a dominant and contested belief that is often held within science education. As articulated in Chapter 3, how we come to understand and know nature (i.e., science) and how we use it (i.e., technology) are commonly viewed as *almost* culturally neutral processes as the result of a primarily naturalistic ontology.[18] This configuration downplays, and often negates, the impact that culture plays in the construction and implementation of modern scientific and technological knowledge (e.g., Matthews, 1994; Siegel, 1997, 2001). A critical and, more specifically, decolonizing response necessarily must make the argument that science education is cultural in order to address the ways in which the culture of science education is damaging to Indigenous knowledges and students, as well as other students who have past and ongoing predominantly negative relationships to the (neo-)colonial culture of "school science" (Barnhardt & Kawagley, 2005, 2008; McKinley, 2007; see also Harding, 2008). Utilizing the above normative critical mode (i.e., critique) often entails, first, treating this belief as a *fairy* and presenting an oppositional claim: techno-scientific practices are *primarily* cultural and secondarily naturalistic. Second, this *fairy* is explained through *facts*: theories of Eurocentrism produce the individuals enacting this belief (e.g., Sammel, 2009). Such a critique obfuscates the possibility of diverse techno-scientific practices having diverse degrees of natural and cultural production. Perhaps more importantly, it produces a foreclosure in the very change that it advocates for: if individuals enacting this belief are already bound by systems of coloniality, how might they be otherwise? Furthermore, by presenting *fact* and *fairy*, the science education critic and the critique is either addressing those who already agree or inviting those who disagree to treat them with the same brush by reversing the *fact* and *fairy* positions to unravel the argument, as explored in further detail in the following chapter. That is, the counter-belief of techno-scientific knowledge as primarily cultural is presented as *fairy* that can be "explained" by the *fact* of an agenda of cultural politics.[19] As Latour (2004b) states, dominant conceptions of cultural politics and nature are exclusionary by their definition, one cannot enact substantive claims about nature (i.e., scientific knowledge) from such conceptions of cultural and political positions. Thus, within a construction of science and technology as culturally neutral, such a *fact* would disqualify the counter-belief as "counting" as or in the construction of techno-scientific knowledge, thus re-inscribing science and technology as negligibly cultural (e.g., Matthews, 1994; Siegel, 1997, 2001).

Accordingly, if the culture of debate around questions of epistemic pluralism and questions of cross-culturalism within science education seems to be at a standstill, locked in ongoing dialectical reversal (Aikenhead & Ogawa, 2007; Alsop & Fawcett, 2010; Cobern & Loving, 2008; van Eijck & Roth, 2007; see Chapter 3), it is perhaps because the normative critical spirit, as Latour (2004a) playfully mentions, positions the critic and critique within an optics of appearing to be right and those with whom they disagree as seemingly wrong. This occurs in part because "there is never any *crossover between the two lists of objects* in the fact position and the fairy position" (p. 241, emphasis in original). On the one hand, this entails that the objects placed in the *fact* position are never explored as if they were in the *fairy* position. By treating them as strictly causal, their "origin, fabrication, [and] mode of development" (p. 238) are left unexamined. On the other hand, objects in the *fairy* position are not given the *fact* treatment. By treating them as strictly the result of a fetishistic projection, the ways in which they could continue to be causal agents, after the anti-fetishistic move is made, are masked. However, once this repertoire of critical approaches is shown to be contradictory, the "poor trick that allows critique to go on" (p. 241) begins to break down. Engaging in questioning that reverses and disrupts the fact/fairy binary positions, the critical question of culture within cross-cultural science education begins to open up. What would it mean to consider the practice of "culturally quasi-neutral" science as a *fictional matter of fact*? What if an exploration of its construction revealed this practice to have its own cultural and political "origin, fabrication, [and] mode of development" (Latour, 2004a, p. 238)? For example, what if the practice of "cultural quasi-neutrality" was but a differential enactment of the highly political sixteenth-century cultural practice of the "modest witness" in the laboratory?[20] Would the practice of "cultural quasi-neutrality" not then a significant cultural practice in and of itself? Would that not also mean that this practice would disqualify it from counting in the production of techno-scientific knowledge if the criterion of negligible or trace cultural impact is infringed?

And what if the practice of "cultural quasi-neutrality" is treated as a *factual matter of fiction*? What if a critical identification of this problematic within a culture of science education (i.e., treating it as a fetish) did not easily disallow for its rejection or did not allow for a movement beyond?[21] If critiques that "explain" science education as a primarily cultural endeavour implicitly treat its culture as a *matter of fiction* rather than account for the ways in which the "cultural quasi-neutrality" is stubborn and sticky due to its own overarching systemic diffusion and self-erasure (see Barad, 2007), would the proposed solution fall into some of the same traps as the problem (see Bohm, 1996)? Does (re)presenting something as untrue halt its (re)production?

It is for this reason that Latour (2004a) states that critique, of this particular and normative kind, has run out of steam. This, of course, does not negate the ongoing importance of and need for a critical spirit around issues of inclusion, exclusion, and the norms that shape participation in science education. There are far too many students for whom science education remains a form of epistemic violence that threatens their ways of knowing and being with/in nature (Barnhardt & Kawagley, 2005, 2008; McKinley, 2007; Sammel, 2009). Because "the practice of critique is not reducible to arriving at judgments (and expressing them)" (Butler, 2001, p. 1) through this mode of mirror-upon-mirror, there are other metaphorical optical arrangements that are available and provide differential potential and promise.

In the following sections, I present the prism and the diffraction grating as alternative optical apparatus metaphors through which critique in science education can be productively (mis)read. Such substitutions act not only as a means of challenging the mirror metaphor's implicit operations (e.g., critique as presenting what one "sees" without coming to present how one sees what they see), but also towards providing alternatives that do not outright reject the structure of critique.

FOUCAULT'S PRISMATIC CRITIQUE: PROXIMAL AND DISPERSIVE CRITICAL RELATIONALITY

> Foucault often invokes a form of the discursive, or a form of the non-discursive; but these forms neither enclose nor interiorize anything; they are 'forms of exteriority' through which either statements or visible things *are dispersed.* (Deleuze, 1988, p. 43)

Foucault (1997), in his talk titled *What is Critique?* implicitly rejects the mirror metaphor by problematizing its condition, offering instead an optical configuration through which "statements or visible things *are dispersed*" (Deleuze, 1988, p. 43). In particular, Foucault (1997) critiques the possibility of distance between the subject and object of inquiry (see also Barad, 2007, 2010; Smith, 1999/2012). As Haraway (2001) reminds us, "the eyes have been used to signify a perverse capacity ... to distance the knowing subject from everybody and everything in the interests of unfettered power" (p. 677). Significantly, this distancing is a double(d) processes that also entails separation. The distancing required for critique through an optics of mirroring makes it such that the critic cannot be at once the subject and the object of one's own critique (Barad, 2007; Bohm, 1996; Butler, 2005); it cannot and does not account for the ways in which the critique and the critic are also formed with, in, and in response to that which is under critique. This is not only important for the status of critique in general, but also critique for decolonizing purposes as well. Linda Tuhiwai Smith (1999/2012) reminds us:

One of the concepts through which Western [Modern(ist)] ideas about the individual and community, about time and space, knowledge and research, imperialism and colonialism can be drawn together is the concept of distance.... Distance again separated the individuals in power from the subjects that they governed. It was all so impersonal, rational and extremely effective. In research, the concept of distance is most important as it implies a neutrality and objectivity on behalf of the researcher. Distance is measurable. What it has come to stand for is objectivity, which is not measurable to quite the same extent. (p. 58)[22]

Extending the earlier argument (i.e., critique as alternating flimsy and firm positions; Latour, 2004a), it can be stated that it is often the case that neither the critique nor the critic themselves are treated in the same anti-fetishistic way that the negatively judged object under inquiry is treated (i.e., the object placed in the *fairy* position). This is despite the respective importance of both the critical apparatus and the critic within the production of the optical arrangement, as well as phenomena under critique. While not discussed at length within this chapter, the oft-cited solution of placing the critic under the gaze to account for the critical production of what is seen and how it is seen often reproduces the same optical arrangement (i.e., mirror-upon-mirror), albeit differently.[23]

Foucault offers us the prism as a metaphoric optical technology for informing the critical gaze otherwise (see Deleuze, 1988). Rather than operationalize critique through distance and separation, as is the case with the mirror-upon-mirror arrangement, Foucaultian prismatic critique relies on subjects and objects being in porous and proximal relations. Accordingly, Foucault invites us to consider the ways in which neither the critique nor the critic are self-enclosed or interiorized, even when brought back into the critical analysis. Because of this relational proximity and porosity, subjects and objects disperse and are dispersed through the critical process. Not only does this differentially shape subjects and objects involved within the process, but also the process itself. Before addressing how prismatic critique plays out in critical cross-cultural science education, it is important to explore, outline, and situate Foucault's (1997) conception of what critique is and can be to ground the metaphorical optical phenomena of prismatic dispersal.

As Butler (2001) states, for Foucault, "critique is always critique *of* some instituted practice, discourse, episteme, institution... it loses its character the moment in which it is abstracted from its operation and made to stand alone as a purely generalizable practice" (p. 1). Critique is always a critique of *something*, *somewhere*, and by *someone*: the norms under critique come to shape the very critique itself. This is not to say that a critical mode developed within a particular practice is a form of critical relativism that wholly rejects translation (Latour, 1993, 2004a). Rather, it is important to come to understand the qualities and conditions of that mode of critique with/in the context in which it was developed if one endeavours to remain faithful to the intent and possibilities of critique when engaging in always already occurring process of transposing it into an elsewhere and elsewhen.[24]

To frame prismatic critique, it is important to note that Foucault was a scholar critical of the Enlightenment. In particular, Foucault's (1997) exploration centres the fundamental critical question characteristic of Western Europe in the fifteenth and sixteenth century of "*how to govern*" (p. 27, emphasis in original), and its counter-question of "how not to be governed" (p. 28) from which it cannot be disassociated. However, Foucault did not only seek to critically engage with these questions, but also engage critically with the critical process itself by seeking "to understand the kind of question that critique institutes, offering tentative ways of circumscribing its activities" (Butler, 2001, p. 2).

In his exploration of what critique is and can be around questions of governmentality, Foucault (1997) identifies three historical anchoring points. First, "critique is biblical, historically" (p. 30). As the art of governance was tied to religion, critique during that period often entailed questioning the truths that sacred texts (i.e., "the Scriptures" [p. 30]) offered, and turning them on their head to disrupt the ways in which power is maintained through these texts. Second, critique is anchored in not wanting to be governed. This resistance to governance is to address rules and laws that are unjust by putting forth irreversible and unavoidable rights to which systems of government will have to submit. Third, expanding upon not wanting to be governed, critique entails not accepting the conflation between authority and truth. This does not entail a full rejection of the truths offered by authority figures, "but rather only accepting it only if one considers valid the reasons for doing so" (Foucault, 1997, p. 31). Accordingly, as Butler (2001) states, Foucault "is not posing the possibility of radical anarchy, and that the question is not how to become radically ungovernable" (p. 6). Rather Foucault (1997) asks how "not to be governed *like that*, by that, in the name of those principles, with such and such an objective in mind and by means of such procedures, not like that, not for that, not by them" (p. 28). Thus, we are to be "both partner and adversary" (p. 28) to the very thing we are critiquing.

The shift from not being governed to not being governed *like that* is significant for two entangled reasons. First, the former is a dangerous proposition. To fully reject governability and to distance oneself from it is to risk "letting someone else say 'obey'" (Foucault, 1997, p. 35) by unavoidably stepping into other regimes of governance (see also Spivak, 1976, 1993/2009). Secondly, Foucault (1997) presents the former as an impossibility. As Butler (2005) elaborates on Foucault's account,

> There is no "I" that can fully stand apart from the social conditions of its emergence, no "I" that is not implicated in a set of conditioning moral norms, which, being norms, have a social character that exceeds a purely personal or idiosyncratic meaning. (p. 7)

Thus, "to be governed is not only to have a form imposed upon one's existence, but to be given the terms within which existence will and will not be possible" (Butler, 2001, p. 8). It is not only an impossibility to stand outside of the social norms that shape one's being, but it is also an undesirability (e.g., exile, banishment; see Peat, 2002). To be fully outside the norms by which one comes to be would entail becoming wholly unintelligible as a subject and to go without the means of one's "cultural survival" (Butler, 1990; see also Butler, 2005).

If we are always already in a proximal relation to the things that we are critical of and with, critique through a clear cut and distanced subject-object relation begins to break down, as does the mirror metaphor it makes operational. As Deleuze (1988) reminds us of Foucault, "these forms neither enclose nor interiorize anything; they are 'forms of exteriority' through which either statements or visible things *are dispersed*" (p. 43, emphasis in original). The optical metaphor enacted through *dispersal* can be productively explored and unpacked through Foucault's (1977) exploration of Bentham's *Panopticon* in *Discipline and Punish*.[25]

In short, the Panopticon is an architectural structure that works to produce the "automatic functioning of power" (Foucault, 1977, p. 201) within the disciplinary space of the prison by reconfiguring the relationship between the subject (i.e., the jailor) and object (i.e., the prisoner) of power. For readers unfamiliar with Bentham's Panopticon, it can be described as such:

> At the periphery, an annular building; at the center, a tower; this tower is pierced with wide windows that open onto the inner side of the ring; the peripheric building is divided into cells, each of which extends the whole width of the building; they have two windows, one on the inside corresponding to the windows of the tower; one on the outside allows the light to cross the cell from one end to the other (Foucault, 1977, p. 200)

Unlike the dungeon, whose purpose is also to contain and discipline by making the prisoner (socially) invisible, the Panopticon bathes the prisoners within a regime of visibility through which the jailor located within the tower can potentially see *any* of the prisoners, but not vice versa. Because the jailor does not possess the ability to see everything all of the time (e.g., limited field of sight, not always being present within the tower), this uni-directional sight is important manages "to arrange things [such] that the surveillance is permanent in its effects, even if discontinuous in its action" (Foucault, 1977, p. 201). In other words, because the prisoners may not be always watched but could be watched at any point in time without their knowledge, "the inmates should be caught up in a power situation of which they are themselves to be the bearers" (Foucault, 1977, p. 201).

Here, it is no longer appropriate to state that the governmentality that Foucault is so critical of (and in relation to) is located strictly within the jailor who is traditionally conceived of as the contact point of power when thinking about the dungeon. Furthermore, while it can be said that the prisoners in this context practice self-discipline by inscribing "in [themselves] the power relation in which [they] simultaneously plays both roles [i.e., jailor and prisoner]" (Foucault, 1977, pp. 202–203), this is not to say that they have internalized and contain the discourse of power. Rather, the architecture acts as an apparatus which places them all in proximal relation through which disciplinary power and knowledge circulate in a capillary manner, dispersing through and simultaneously (re)producing them as subjects and objects of knowledge. As Butler (1990) states of Foucault, "systems of power *produce* the subjects they subsequently come to represent" (p. 2, emphasis in original). These structures cause subjects to be "formed, defined, and reproduced in accordance with the norms of those structures" (p. 2). In his critique of governmentality through disciplinary forms of punishment, Foucault turns the "sacred text" of power as individualistic and repressive (i.e., a Marxist conception of power) on its head: disciplinary power circulates *through* nodes (i.e., subjects and objects) and is productive (e.g., produces and organizes subjects as well as objects).[26] Subjects do not unequally *have* access to power; rather they are unevenly *had* by power.

The relation to critique and the point to be made here is not that critics and their critiques are always already under a regime of visibility through which disciplinary power flows in exactly the same ways as in the example of an incarcerated prisoner. Rather, "the Panopticon...must be understood as a generalizable model of functioning; a way of defining power [and knowledge] relations in terms of the everyday life of men [sic]" (Foucault, 1977, p. 205). With respect to critique, *panopticonism* is useful to think about the ways in which both the subject enacting critique (i.e., the critic) and the objects of critique are within a proximal relation through they are differentially produced and organized. It is useful to think about the Panopticon as metaphor for disciplinary power and the ways it circulates not only as a *governmental* mode of *punishment*, but more broadly as the ways in which disciplinary knowledge (e.g., science, education) disperse through, produce, and organize subjects and objects of critique. Disciplinary knowledge is a productive double(d) meaning as it need not only be read as the knowledge content of *a* discipline. It can also be read as the ways in which knowledge is disciplined. Disciplinary subjects (i.e., both the curricular content and those conveying the content) are produced (i.e., within norms) through the operationalization as well as the possible application of panoptic disciplinary power.

As presented by Butler (2001), Foucault argues "critique will be dependent upon its objects, but its objects will in turn define the very meaning of critique" (p. 3). This is doubly important. First, this necessarily entails giving the objects of critique a more robust treatment than the *matters of fact* and *matters of fiction* that Latour (2004a) cautions against as too simple a framing will result in too simple a critique. Second, the very objects of critique also produce particular possibilities and positions for the "subjects" of critique, through subjectification of the critics themselves. In other words, the critic does not come to critique with a stable subject position/ality prior to the act of critique (i.e., being) but rather the very norms which they are critiquing shape the position(s) which they can take in relation to those norms (i.e., becoming) (see Butler, 2001). The very norms that organize what is a *matter of fact* and what is a *matter of fiction* produce (and are produced by) the critic and the epistemological context (i.e., *what* and *how* the critique can know) within which they are operating.

With Foucault's prism as an optical prosthetic technology, the task of critique becomes not one of establishing or employing a "pregiven epistemological context" (Butler, 2001). Rather, critique is about exposing the limits of the epistemological context. In turn, recognizing the limitations of an epistemological context entails accounting for and being accountable to the relationship between the subject and objects of critique through which the context emerges. Thus critique, for Foucault (1997), is "the art of voluntary insubordination, that of reflected intractability" (p. 32) through which the critic engages in a "practice that not only suspends judgment, ... but offers a new practice of values based on that very suspension" (Butler, 2001, p. 1). This suspension is critical in both senses of the word because it asks how the very judgements we make as critical subjects are already produced and organized within the proximal and prismatic relations between our/selves as critics and the objects of critique we glorify and dismay (see also Bohm, 1996). If we engage in critique as "the movement by which the subject gives himself the right to question the truth on its effects of power and question power on its discourses of truth" (Foucault, 1997, p. 32), would we still arrive at or care to make the same judgements altogether?

Returning to the question of "what counts" as science within the context of cross-cultural science education (from Chapter 3), considering critique as prismatic opens up additional lines of questioning that can be engaged critically. While questioning might open up new lines of critical engagement, recall that questions that reveal the epistemological limitations and shape (i.e., the lines delineating the interiority and exteriority of truth) of a particular epistemic framework are for Foucault a form of critique in and of themselves. Rather than engaging in acts of judgement through which objects of critique are positioned as either *matters of fact* or *matters of fiction*, Foucault's prismatic critique invites us to consider the relationships through which these judgements are produced through optical inflection. If we revisit the earlier question regarding the matter of science education as "culturally quasi-neutral", rather than immediately framing this as good/true (i.e., a *matter of fact*) or bad/false (i.e., a *matter of fiction*), it is worth momentarily suspending judgement to ask how critique is formed with/in relation to these norms.

If we treat science as "culturally quasi-neutral" as a point of illumination, through what categories, constructs, and concepts is meaning dispersed and inflected to produce this constellation of meaning? What are the critical objects that are utilized to either uphold or question this norm? What objects are positioned as abject (e.g., how does "culture" come to be seen as problematic?) and which ones are positioned as positive levers (e.g., how does method come into stand for and as quasi-neutrality)? What optical geometries are required for these to come to be, and to be sustained as such?

Recognizing that the light/prism relationship is often interchangeable, how is it that "cultural quasi-neutrality" disperses and inflects its negative and positive objects? Treating "cultural quasi-neutrality" prismatically also invites the question of what it produces in turn when meaning is inflected through it. For example, how is it that practices deemed cultural (rather than quasi-neutral) are inflected in terms of their meaning when shone through that prism? What about sanctioned scientific methods?

What about questions of curriculum and pedagogy? When science education is filtered through the normative prism of "cultural quasi-neutrality", what kinds of pedagogies and curriculum are dispersed and inflected through the other side? Which ways of learning and ways of knowing are (re)produced? What is made possible and made impossible (e.g., What conceptions of a learner can and do emerge when cultural quasi-neutrality is part of the illumination)? What kinds of learners and learnings as well as teachers and teachings are made intelligible/unintelligible within this space?

Expanding upon the last question, how are the critics themselves, as the subjects of critique, formed with/in theses multiple relationships that often appear to be epistemologically pre-given? How are the critics inflecting this vector of light *not like that* (e.g., What facets of the optical arrangement are being deflected, inflected, inverted, and redirected; which meanings are flow through, with minor refraction at most?)? How do these inflections shape the critic whose self is in prismatic relation to them (e.g., as critics, as educators, as researchers)? When considering the ever-increasing ways in which the cross-cultural science education classroom is presented with plural(istic) ways-of-knowing-nature, processes and products that exceed the norms by which we frame science and by which we are framed as science educators, it becomes important to consider the rich luminescent web of dispersed meanings when thinking about what is (im)possible within/as science education.

BARADIAN DIFFRACTION: INCLUDING THE CRITICAL
APPARATUS IN THE PRODUCTION OF CRITIQUE

It is a well-recognized fact of physical optics that if one looks closely at an "edge," what one sees is not a sharp boundary between light and dark but rather a series of light and dark bands – that is a diffraction pattern. (Barad, 2007, p. 156)

While Barad (2007) does not explicitly make the act of critique a focus of her scholarship in *Meeting the Universe Halfway*, she does speak to critique in a more recent interview:

Critique is over-rated, over-emphasized, and over-utilized... Critique is all too often not a deconstructive practice, that is, a practice of reading for the constitutive exclusions of those ideas we can not do without, but a destructive practice meant to dismiss, to turn aside, to put someone or something down. This is a practice of negativity that I think is about subtraction, distancing and othering. (Barad, 2012a, p. 49)

Like Foucault (1997), Barad (2012a) is critical of distance and separation as the (pre-)condition under which *the* critical spirit operates. The othering of the object of critique obscures the ways in which the subject of critique is indebted to its other through its proximal and co-constitutive relationship (see also Barad, 2007, 2010); not only epistemologically, but also ethically and ontologically. For Barad (2007), such critique cannot be disassociated from "long history of using vision and optical metaphor [s] to talk and theorize about knowledge" (p. 29), particularly the "well-worn metaphor of reflection" (p. 29). For her, the commonplace understanding of reflection can be understood as such:

Mirrors reflect. To mirror something is to provide an accurate image or representation that faithfully copies that which is being mirrored. Hence mirrors are an often-used metaphor for representationalism and related questions of reflexivity. For example, a scientific realist believes that scientific knowledge accurately reflects physical reality, whereas a strong social constructivist would argue that knowledge is more accurately understood as a reflection of culture, rather than nature. (p. 86)

For Barad (2007), the act of reflection is about mirroring sameness else-where. As mentioned earlier within this chapter, this largely has to do with the ways in which the mirror metaphor produces an illusion of pre-givenness (i.e., presence) through which claims of correspondence are masked. Thus, through reflection, patterns of difference are dialectically subsumed into or sublated through sameness, making it difficult to account for and be accountable to the enactment of difference (see Chapter 2). As an alternative to reflection, Barad (2007, 2012a) proposes *diffraction* as optical metaphor that attends to relations of difference, and how they are differentially done and undone. In a nutshell, "diffraction involves reading insights through one another in a way that help illuminate differences as they emerge: how differences get made, what gets excluded, and how those exclusions matter" (Barad, 2007, p. 30).

Here, a rich possibility is offered through the metaphorical optical alternative proposed for cross-cultural and decolonizing science education. Given that cross-cultural science education endeavours to make space for ways-of-knowing-in-being that are not typically included within the curricular scope, the invitation to consider *otherness* without the necessity of bringing it into frames of *sameness* and/or being accountable to that which is exceeded by frames of sameness when applied is of importance. Too often other(ed) ways-of-knowing-in-being (e.g., Indigenous) are placed in a relationship that dialectically reduces their elements to those that they share with dominant WMS, at the expense of their uniqueness and possibilities (e.g., ethics, balance, other-than-human agency; Cajete, 1994, 2000). Furthermore, while the intent is not new, the practice of diffractive critique can bring new ways to account for and be accountable to relations of power between normative and alternative ways-of-living-with/in-nature.

However, if we continue the work of labouring the metaphor of visuality and technologies of sight with respect to critique, before applying the optical metaphor of diffraction, it is important to have an understanding of the optical referent to which it refers.[27] In short, Barad's metaphor of diffraction invites a more nuanced and complex understanding of the natural phenomena informing this referent: light. As Barad (2007) reminds us, the optical mode through which most critique operates is Euclidian (i.e., rectilinear). However, light does not always act linearly: "under certain experimental circumstances, light manifests particle-like properties [i.e., enacting Euclidian geometries], and under an experimentally incompatible set of circumstances, light manifests wave-like properties [i.e., enacting non-Euclidian geometries]" (Barad, 2000, p. 233). Accordingly, this is not an invitation to strictly consider light, and according mediated visualities as non-linear, as they are still produced as linear under particular experimental circumstances. Rather, it is an invitation to consider light and properties of mediated vision interacting as both particle and wave-like properties, as well as consider the conditions under which they become particle and wave-like. To do so, Barad (2007) introduces the physical phenomena of diffraction as useful in exploring this referent.

Before considering light as both wave and particle as a metaphorical referent for the practice of critique, it is worth quickly unpacking what it means to consider light as demonstrating wave properties before moving forward. Within classical physics, this phenomenon is called *diffraction*, and it "has to do with the way waves combine when they overlap and the apparent bending and spreading of waves that occurs when waves encounter an obstruction" (Barad, 2007, p. 74). While some physicists hold to a history in which the first phenomena (i.e., combining of waves) are referred to as *interference*, Barad (2007) reminds us that both phenomena have to do with the juxtaposition of waves. Since it is classically a property of all waves, I will give examples of both types of classical diffractions with wave phenomena that might be more familiar: sound.

To demonstrate the first definition, imagine that you are at an outdoor (soft) rock concert with two loud speakers at each end of the stage. Should you have the freedom to walk about the area, you may come to find that there are spaces where the music seems quieter, as well as spaces where the music seems louder. This has to do with the principle of wave superposition that states that when waves occupy the same position or immediate local space, their amplitudes combine to create a new wave. This new wave may be dampened through destructive interference (i.e., when the waves' amplitudes are opposite) or intensified through productive interference (i.e., when the waves' amplitudes align).

For the second instance, imagine that you are speaking into a cardboard tube. The sound that emerges from the other end does not follow the linearity of the tube, but rather spreads out. This second type for diffraction occurs when waves encounter a slit, a hole, or an obstacle whose wavelength is no greater than their own (e.g., sending light waves through the same cardboard tube would not produce any noticeable ripples). While the tube prevents the sound to exit it anywhere but the opening, each and every point along a wave can and does act as a point of origin which explains why it seems like the sound wave bends in all directions once exiting the tube.

However, recall that Barad (2000, 2007) invites us to consider light as both wave and particle. This requires us to explore diffraction as a quantum phenomenon. Here, it is productive to discuss the experiment that is emblematic of the collapse of classical Western metaphysics[28]: the two-slit experiment. In this experiment, a single particle, such as an electron is fired into the two-slit experimental apparatus that is configured to observe wave phenomena. This is significant as within classical physics, particles are largely thought to behave like other forms of matter, in mechanistic, causal, and linear manners; that unlike waves, particles as material phenomena entails spatial single occupancy (unlike the wave superposition discussed earlier). Were this simply the case, this experimental apparatus would have yielded no observation and would have been largely forgettable. However, produced through this experiment were diffraction patterns that indicate that under the right experimental conditions, particles exhibit the behaviour of waves. It is also worth noting that the corollary would also be shown to be possible as well in doing similar experiments with light waves and creating experimental conditions in which they would behave as particles. This is of deep importance as these materialities (i.e., waves and particles) exhibit and enact properties that are ontologically mutually exclusive within classical physics.

From this, a few theorists offered theories to attempt to explain this wave-particle duality. Of note is Heisenberg's uncertainty principle that posits the wave-particle duality as epistemological, as a limitation to knowability. However, Neils Bohr's theorizing, which is the commonly accepted theory, states that this duality is of ontological nature. The phenomena is not simply presenting itself again by representing its essence, but rather, the experimental conditions under which observation occur shape the properties of what the phenomena can be (see Barad, 2010, 2011).

Working with physicist Neils Bohr's journals, Barad extends his analysis by asking where the agencies of observation begin and where they end, what is included, what is excluded, what matters, and what comes to materialize. While Barad originally draws from Bohr's work to theorize materiality and materialization within the context of quantum physics, she later extends these conclusions outwards.[29] The role, the constitution, and the enactment of the apparatus is an important location where

Barad's work specifically deviates from and inflects Foucault's. However, this deviation is not one of critical negation, sublation, or subsuming (by reducing her objects of critique to *matters of fact* or *matters of critique*), but rather a diffractive reading. Recall that for Foucault (1977), apparatus such as the Panopticon are at once physical, discursive, and organizational structures, which are produced by and reproduce the capillary workings of power within society. As Barad (2007) states, "although Foucault insists that the objects (subjects) of knowledge do not pre-exist but emerge only within discursive practices, he does not explicitly analyze the insepara-bility of apparatuses and the objects (subjects)" (p. 201). In other words, while Foucault considers the ways in which apparatuses of power such as the Panopticon produces phenomena of subjectification, Barad (2007) invites a consideration of how the phenomena of subjectification comes to produce apparatuses such as Panopticon.[30]

Through a diffractive reading in which she reads Bohr's insights through Foucault's, and vice versa, Barad (2007) produces new insights. In particular, she reads Foucault's insights into societal phenomena with Bohr to postulates that the apparatus not only produces the phenomena under observation but also that the apparatus is constitutive of and consti-tuted by the phenomena as well.[31] Thus, if for Foucault subjectivity is not contained by the subject through interiority but rather a generative enactment in relation to the norms which govern the possible possibili-ties of who and what one can be (see Deleuze, 1988), Barad extends this theorizing to the apparatus as well.

Asking the question of what constitutes "an apparatus" that comes to produce and be produced by a phenomena, she reaches the conclusion that an apparatus is observed is never simply a material tool or a discursive concept through which the phenomena can be observed, but rather an entangled and enacted network of agencies at play. One example of such that Barad (2007) provides is that of the Stern-Gerlach experiment in 1922 in which Otto Stern and Walther Gerlach experimentally and empir-ically made demonstrable the theoretical concept of "space quantization". This phenomena in which electrons made quantum leaps from one discrete orbital or energy level to another within an atom was well devel-oped within theoretical atomic models, however, classical understandings of atomic configurations were reluctantly held onto until proof of some sort was given to justify the theory (or debunk the theory as a temporary stand in for another misunderstood phenomena). However, Stern and Gerlach created an instrument that, "using a particular arrangement of magnets" (p. 163), would show the ways in which a beam of silver atoms' electrons are differentially positioned, oriented, and configured within an atom through deflection: some would be deflected upwards and some would be deflected downwards. As Otto Stern recounts the experimental event,

> With Gerlach looking over my shoulder as I peered closely at the plate, we were surprised to see gradually emerge the trace of the beam... Finally we realized what [had happened]. I was the equivalent of an assistant professor. My salary was too low to afford good cigars, so I smoked bad cigars. These had a lot of sulfur in them, so my breath on the plate turned the silver into silver sulfide, which is jet black, so easily visible. It was like developing a photographic film. (Otto Stern in Barad, 2007, p. 164)

The experiment functioned. What Barad makes clear is that the very boundaries that constitute the apparatus through which phenomena stabilize and make themselves intelligible are not so easily determined, or at least enclosed within that which is usually referred to as "equipment" within a laboratory report. Here, when asking the question of what constitutes the apparatus through which the phenomena was enacted, we would necessarily have to consider not only the material agency of the cigar, but also questions of gender, class, and economics through which that particular type of cigar came to be included. As Barad (2007) cautions, this "is not to say that all relevant factors figure in the same way or with the same weight. The precise nature of this configuration (i.e., the specific practices) matters" (p. 167). Accordingly, "apparatuses are not static laboratory setups but a dynamic set of open-ended practices, iteratively refined and reconfigured". (p. 167). The apparatus is the enactment of a singular multiplicity that enfolds multiple bodies of meaning and matter that comprises each of their respective material and discursive historicities. As such, these constitutive bodies do not simply interact between one another, but rather *intra-act* within this re(con)figured body which is the experimental apparatus. Barad refers to this type of co-substantiation that occurs with/in the apparatus, as well as the phenomena under observation, as one of quantum entanglement:

> *Quantum entanglements* are generalized quantum superpositions, more than one, no more than one, impossible to count. They are far more ghostly than the colloquial sense of 'entanglement' suggests. *Quantum entanglements* are not the intertwining of two (or more)

states/entities/events, but a calling into question of the very nature of two-ness, and ultimately of one-ness as well. Duality, unity, multiplicity, being are undone. 'Between' will never be the same. One is too few, two is too many. No wonder quantum entanglements defy commonsense notions of communication 'between' entities 'separated' by arbitrarily large spaces and times. Quantum entanglements require/inspire a new sense of a-count-ability, a new arithmetic, a new calculus of response-ability. (Barad, 2010, p. 251, emphasis in original)[32]

Returning to the question of what critique is and can be, Barad's (2007) notion of diffraction invites us to consider the ways in which the very process of critique differentially produces the subject who critiques along-side the object(s) of critique (as does Foucault). Uniquely, she also invites us to consider the ways in which the norms of bodily production through which these subjects and objects come into being through the enactment of critique are not being pre-given. Just as Foucault's critique invites us to trouble the notion of an epistemological pre-givenness of the terms through which the critique operates, Barad invites us to trouble an onto-logical pre-givenness of the ways in which the usual subjects and objects of critique are segmented and separated. It is a call to consider them as superpositioned without the form of "a-count-ability" being one in which superposition entails sameness (i.e., one-ness) or radical differenti-ation (i.e., two-ness)[33]; it is a form of a-count-ability that accounts for its own ontological cuts as well as the norms of inclusion/exclusion that are shaped through this practice.

Along similar lines, Latour (2004a) states that:

The mistake we made, ... was to believe that there was no efficient way to criticize matters of fact except by moving *away* from them and directing one's attention *toward* the conditions that made them possible. But this meant accepting much too uncritically what matters of fact were. (p. 231, emphasis in original)

Latour (2004a), like Barad, reminds us here that critique need not only be about the taking apart of constructs, constraints, and consequences of particular *matters of fact* or *matters of fiction*. As these are always already the product of entanglements which are enacted, there is always the possi-bility of the very things that matter to us from being enacted in such a way that the very entanglement is re(con)figured to be a product of hegemony. If the very things we care for are constructed, it means that we are to operate with care in how we (re)enact them. Critique as diffraction, or diffractive critique, is then a process of producing, and being responsive and accountable to non-negligible patterns of difference that come to matter when two (or more) entangled material-discursive phenomena are diffracted through one another. This requires however that "we learn to tune our analytical instruments (that is our diffractive instruments) in a way that is sufficiently attentive to the details of the phenomenon we want to understand" (p. 73) and to pay attention to the fine details that would otherwise be considered negligible with/in conventional scientific and social scientific research methods.

Let us return to the question of "what counts" as science within the context of cross-cultural science education and the dominant assumption of scientific knowledge's "cultural quasi-neutrality" (from Chapter 3). Recall that like Foucault's prismatic critique, Barad's diffraction questions the a priori status of epistemology (e.g., concepts, constructs, and categories). However, the dispersal, deferral, and displacement of culture through discourse is further troubled by questioning the a priori status of ontology (e.g., space, time, matter). This entails that the ontological units onto which critical arguments are mapped are not passive (e.g., time, causality) but rather are enacted, as are the cuts by which these units come to be. Furthermore, in considering ontology as dynamic, Barad (2010) invites a reconsideration of its dualistic or dichotomized relation to epistemology; "one is too few, two is too many" (p. 251). Accordingly, culture is not only "internally" co-substantiated through superposition, but its "exteriority" (i.e., nature) is active and agentic with/in this entanglement: everything is within culture; everything is within nature (see also Latour, 1993; Kirby, 2011).

So, what does this mean for science and science education if the dominant belief that nature produces the "quasi-neutral" cultural mediations that are scientific knowledge? The lines of questioning shift from asking whether WMS *is* or *is not* a (sub-)culture and cultural production (i.e., *matters of fact* and *matters of fiction*), as well as how cultural meanings are inflected and dispersed through one another to produce a normative web. If everything is within culture and nature in their totality, and the two are co-substantiated rather than dualistic or monistic, then *everything* comes to bear in the production of Western modern scientific knowledge. To what degree do the multiple natural-cultural agents participate in the production of phenomena under observation? What ways-of-knowing, ways-of-being, or perhaps more appropriately, ways-of-knowing-in-being are enacted through such entanglements?

If we consider the ways in which the part is within the whole (see Barad, 2007; Cajete, 1994, 2000; Peat, 2002), when scientific knowledge derived from WMS is brought into the science classroom, what is produced through such diffraction? What comes to matter (and to what degree)? How do the ways-of-knowing-in-being enacted within the laboratory by scientists, technological apparatus, and by agentic matter intra-act within pedagogical entanglements with students (and other natural-cultural agents within schools) when they too are considered as a part within the whole? What occurs when other ways-of-knowing-in-being are diffracted through these normative entanglements? If entanglement does not equate equality, sameness, or uniformity, what are the patterns of difference that occur with/in? What types of negotiations, navigations, and hybrids are (im)possible?

Also, if we consider WMS to always be an enactment of knowing *with* nature rather than *about* nature (i.e., matter comes to matter within experimental conditions), what is entangled within the production that frames it to be knowing *about*? Furthermore, if knowing is always already knowing with nature if everything is always within nature, how might we engage in the multiple possible possibilities of knowing *with* nature without slipping into relativism? What are some of the systems and approaches that shape WMS and its relation with/in nature? What are other systematic and sustained engagements of learning with nature and what can be learned from/with these enactments? What can be learned from practices of a-count-ability already frame them as ways of learning *with* and *from* nature (rather than *about*; e.g., quantum physics, IWLN)? Lastly, if different ways-of-knowing-with-nature produce differently entangled possible possibilities, what might be desirable goals for science education?

CONCLUSION: RE(CON)FIGURING CRITIQUE IN SCIENCE EDUCATION

What would critique do if it could be associated with *more*, not with *less*, with *multiplication*, not *subtraction*? (Latour, 2004a, p. 237, emphasis in original)

For scholars critical engaging at the intersections of science and society (e.g., science education), critical resistance to scientific normalization through modes of exposing that "there is no such thing as natural, unmediated, unbiased access to truth, that we are always prisoners of language, that we always speak from a particular standpoint" (Latour, 2004a, p. 227) have, in the past, efficiently worked against problematic "ideological arguments posturing as matters of fact" (p. 227). However, these modes of critique have become the very tools working against critics.[34] In other words, the very tools of dismantling a normative centre have been absorbed by the centre and have been redeployed against the margins: the argument that the dominant position is but *a* situated and partial position is, through metaphoric subtraction, being applied to those who critique the norms of science. As the critical gaze is never a passive operationalization of visual metaphors (Haraway, 2001), to posit a geometric arrangement is to also put forth the very terms through which your argument can be reversed, deflected, and diverted (see also Barad, 2007).

As an emerging de/colonizing science education scholar and practitioner, my primary focus is on Indigenous science to-come: critically engaging with (re)opening the structures and strategies of science education so that Indigenous science might be other than excluded, differing, and deferred. My critical engagements with/in pedagogical practices had me slowly becoming worried about critical possibilities and the possibilities of critique (see McKinley & Aikenhead, 2005); particularly if decolonization would always be de/colonizing (i.e., always shaped in response to and slipping back into colonialism; see Carter, 2004, 2010; Higgins, 2014). Critiques through an optical geometry of *matters of fact* and *matters of fiction*, could not fully contain the ways in which one would flow into the other. When considered in tandem with an invitation to not treat one's negative objects of critique as one-dimensional and through a (never fully achievable) process of negation, I began asking if the issue at hand was not critique, but rather norms around critique which would make it appear as if there is (only) *a* way of being critical.

Summing it up, three optical technologies which metaphorically inform, shape, and (re)produce ways of being critical were explored in this chapter: namely the mirror, the prism, and the diffraction grating. While all three modes hold differential potential and promise, and the intent herein is not to prescribe one critical metaphor at the expense of another, Latour (2004a) invites us to consider the ways in which the mirror metaphor may simply have "run out of steam". While there are moments in which politically posturing as mirroring *the* truth is productive and of deep importance, it is nonetheless important to consider and confound what the metaphor makes operational.[35] By placing its objects of critique in either a fact or fiction position, the mirroring critique and critic becomes blind to the ways in which the *matters of fact* are fictional as well as how *matters of fiction* are factual. The critique itself then is not only easily taken apart by others who may not share the same point of view, but it is also always already self-rupturing through its persistent yet productive failure of containment.

The prism, informed by Foucault's theorizing of critique, is not about displacing sameness elsewhere through mirroring. Rather it is an invitation to consider the ways in which the subjects and objects of critique are dispersed through one another and, in turn, produce one another, albeit differently. Within cross-cultural science education, this was explored around questions surrounding claims of "cultural quasi-neutrality" which shape dominant approach not as something that *is* or *is not* (i.e., achieving epistemic a priori) but rather something that is (re)produced through a complex multi-linear geometry of dispersed meanings which sustain it. It becomes an invitation to think about how "cultural quasi-neutrality" is dispersed through norms which sustain it (e.g., objectivity), how these norms are in turn sustained (e.g., cultural quasi-neutrality and politics are different arenas), and how such a norm flows through prismatic spaces which might come to produce it differently (e.g., science as always already being cultural).

The diffraction grating, while sharing similarities with the prism in terms of its disruption of epistemic pre-giveness, also includes a troubling of ontological pre-giveness within the scope of what is produced by critique. Informed by Barad's quantum ontology and exploration of the undoing of classical optics, diffraction as critique is radical in its invitation to not only consider how the subjects and objects of critique are produced through their being in relation but that the very terms of a-count-ability are enacted through the critique. Neither are they *one* (i.e., monism) or *two* (i.e., dualism) prior the critique, but their entanglement of meaning and matter is qualified and enacted through the critique rather than before or after. In cross-cultural science education, this brings an important lens to consider the ways in which culture and nature, epistemology and ontology are co-substantiated without ever achieving one-ness or dualism.

As science and science education practices are always already at the interface between nature and culture, the inclusion of nature within the flux makes it of greater consequence for and to critics who would dismay cultural critiques of science and science education as not being able to account for or be accountable to nature (e.g., Matthews, 1994). Furthermore, if the ways in which we know about nature are always knowing with nature and are always ways-of-knowing-in-being: what are the entangled epistemologies and ontologies enacted through such knowing? What network of human and other-than-human agents are co-substantiated within the production of such knowledge? If WMS considers itself as "culturally quasi-neutral" and a human endeavour, what can be learned from ways-of-knowing nature that actively consider the ways which they are produced with/in culture and with/in nature?

For critique to "be associated with *more*, not with *less*, with *multiplication*, not *subtraction*" (Latour, 2004a), there is an invitation to rethink critique as the "addition" of statements of lack as a mode of engagement as nothing new is added, never augmented. Latour (2004a), by cheekily referring to this behaviour as sub-critical,[36] asks us to not take one idea and return less-than-one but rather bring it into conversation with more ideas that sustain it, and differentially shape it through and with a new network of ideas. This might allow for the multicultural science education debate to move beyond "what counts" as science (and in turn science education) towards understanding how "what counts" is produced and producible in order to (re)open the structure of science education towards Indigenous science to-come. This might include (re)considering the debate as operating through an adversary paradigm (see Chapter 3), the role of (*an*) ontology within the construction of "what counts" (see Chapters 5 and 6), or even the complex and complicated relationship between Indigenous and Western ways-of-knowing-in-being (see Chapter 7). If the goal is to augment through (re)placing our objects of critique with/in a complex and complicated web of *knowings* and *beings* rather than foreclose them as *matters of fact* or *matters of fiction*, then the use of multiple optical metaphors to achieve this purpose brings resources, not liabilities towards unsettling science education and (re)opening it towards Indigenous science to-come.

NOTES

1. An earlier and much shorter version of this chapter appears in the journal *Cultural Studies of Science Education*, volume 13, issue 1 (pp. 185–203), and is reprinted here with permission.

2. While not taken up in this chapter as such, the optical metaphors can also be thought of as analogies and articulations of the metaphysics they respectively articulate: humanism, anti-humanism, and post-humanism (see Kirby, 2011). As such, this chapter employs its own thinking as a meta-move to present entire metaphysics through their (prismatic or diffractive) articulation: the whole is in the part and the part is in the whole. Furthermore, given the adversarial nature of the multicultural science education debate (and this chapter's critique of critique as negation), presenting optical metaphors as possibilities is meant to act as an invitation rather than present entire metaphysics and traditions through lack and deficit.

3. The expression of a "thumbnail account" is a euphemism that Apffel-Marglin (2011; see also Chapter 6) often uses that is not so dissimilar from Spivak's (1976) treatment of "in a nutshell" (i.e., attempting to contain the uncontainable in the name of brevity). Here, the "thumbnail" signals that the content and issues discussed are so complex, contradictory, and convoluted that perhaps from our partial vantage points that we may never see more than a "thumbnail". Accordingly, to give anything larger than a "thumbnail" is not only impractical, but also impossible as we are always giving *an* account rather than *the* account (see also Butler, 2005). Furthermore, it is also move towards academic modesty: giving a "thumbnail account" often requires bringing together multiple in perspectives which we can never come to know fully but without which we could not piece together an account. It is a recognition of those who precede us in making an account.

4. The mantra of beginning some-where and some-time is not only a persistent reminder that we are always already within the question of Indigeneity within science education, but also an invitation to address it as such. Similarly, we must address it *some-way* and that elsewheres and elsewhens that come to bear on the very question are rife points of examination (here, in relation to criticality, we can trace back taken-for-granted understandings and enactments to 15th or 16th century Europe).

   Further, as Kuokkanen (2007) invites us to consider, in revisiting the question of "debating" the inclusion of Indigenous science from the previous chapter:

What is required is openness to and responsibility toward the "other", and this in turn requires a certain level of comprehension of indigenous epistemes. As importantly, it requires that individuals and institutions commit themselves to a critical debate about "cross-cultural" education, and that this education involves more than just integrating new material into the curriculum. The academy needs to recognize that the logic of the gift calls for changes to the ways knowledge is perceived and approached; moreover, well-intentioned individuals will need to be well equipped to deal with the complexities that emerge when different epistemes meet. (p. 108)

This is not to say that we should be debating whether or not Indigenous science is included, as is the central case of the multicultural science education debate. Rather than a call for less critical debates, it is an invitation to engage critically otherwise: moving from debates of inclusion/exclusion towards more nuanced conversations about *how* we might go about meaningfully including Indigenous science *can be* and *should be*. However, to work beyond "integrating new material into the curriculum" (p. 108) as critical engagement requires that we commit to critically examining and "debating" the very terms that constitute the contemporary conversation: the very notions of "cross-cultural education," debate, and every other relevant concept and practice that we inherit which shapes our (in)ability to engage in the question of Indigenous knowledges within science education.

5. For Foucault (1977), power is not located within subjects or objects but is rather the relation between them through which power circulates.
6. See Chapter 2 for a lengthier, generative exploration of this problematic rupture.
7. In *Practice Makes Practice*, Britzman (2003) cautions that:

> ... every curriculum, as a form of discourse, intones particular orientations, values and interests, and constructs visions of authority, power, and knowledge. The selected knowledge of any curriculum represents not only things to know, but a view of knowledge that implicitly defines the knower's capacities as it legitimates the persons who deem that knowledge important. This capacity to privilege particular accounts over others is based upon relations of power. Consequently, every curriculum authorizes relations of power... (p. 39)

To take Britzman seriously is to consider a curriculum that does not indoctrinate an impossibility as they are always within and in turn (re)produce particular relations of power. Within the context of decolonizing science education, this complex and contradictory space could be read as de/colonizing (Higgins, 2014). This, however, does not mean that every curriculum indoctrinates equally. Accordingly, like Lather (2007) and Spivak (1993/2009), I am interested in what possibilities become possible when we strive for the impossible (even when the very things we use and which use us are problematic). I recognize that this, in part, entails learning to differentially inhabit "the lines of making sense" which shape what is possible within "the arrangement of those lines" (Spivak, 1993/2009, p. 34).

8. For example, Foucault (1997) might not self-attribute to his work the prism as an optical apparatus which informs his critique. Rather, this is an insight that is offered by Deleuze (1988).

9. It is not that reflection and the mirrored apparatus it metaphorically employs is wrong in and of itself, but as Barad (2008) posits "the allure of representationalism may make it difficult to imagine alternatives" (p. 148). Mirroring has become so normalized that reflection has become taken-for-granted, sedimented into how we come to know scientific phenomena.

10. Chapter 3's dialogue in which both the cross-cultural and universalist characters were engaged in reversing the optical configuration through dialectic is exemplary of this (e.g., both work diligently to present the other's truth as a falsity).

11. As Spivak (1993/2009) suggests, the very act of using and troubling the very tools with which one labours against structures of dominance (e.g., critique) is of particular significance:

> One of Derrida's most scandalous contributions is to begin with what is very familiar in many radical positions and to take it with the utmost seriousness, with literal seriousness, so that it questions the position (de)constructively as the wholly intimate other. One is left with the useful yet semimournful position of the unavoidable usefulness of something that is dangerous. (p. 5)

As the tools with which resistance to dominance is laboured are at once inadequate yet necessary, to engage in deconstruction is to allow for the possibility of their reconstitution as something which does not (re)produce (or to a lesser extent) the very systems against which they are working. This is particularly relevant here as decolonizing tools are currently and constantly being appropriated and (re)purposed as means and ends for (neo-)colonialism (Smith, 2005).

12. Furthermore, as the signifier and signified never achieve unity (Derrida, 1976), *all* reading is *amiss* and *a miss*. I use (mis)reading here to signal the type of reading which intentionally utilizes the *play of (re)signification* as a means to leverage the space between oft-intended and more-common signified understandings and those which continue to occupy the structure of the text but hold a radical deconstructive potentiality (e.g., unintended meanings).

13. With respect to vision, Battiste, Bell, Findlay, Findlay, and Henderson (2005) state, a "Eurocentric curriculum is hidden in plain view" with/in a spectrum of educational institutions as they are often "founded on a vision and visualization of education and culture that look to Europe as the center of all knowledge and civilization" (p. 8). As vision is the primary and centered sensory medium through which not only WMS operates but also Western modern society in general (see Peat, 2002; Pink, 2006), it often smuggles in naturalized and normalized dominant theory-practices; it becomes a critical location to work within and against.

In turn, the importance of shifting the gaze from vision (i.e., sight, goals) to visualization (i.e., ways-of-seeing) cannot be understated, because of the ways in which who and what is seen, as well as how and where sight is regulated, both literally and metaphorically reinforce dominant ways-of-knowing-in-being while diminishing and denying the validity of others. For example, vision is often use as a tool of Western modernity to define itself against its otherness in an oppositional manner:

It was believed that for civilized Europeans the "higher" senses of sight and hearing were most important, in contrast associating the "lower" senses of taste, touch, and smell with animality... [as well as] "primitive" peoples [who] would show a predilection for the "lower" or "animal." (Pink, 2006, p. 5)

However, given "the plurality, hybridity, and ambiguity of visual practices," attending to vision (either literal or metaphorical) provides a significant critical and complicit location "for unpacking old and new colonialisms" (Battiste et al., 2005, p. 9).

14. Within science education (Aikenhead, 2006a; Barad, 2000; Erickson, 2000), as well as within science (Barad, 2007, 2010; Latour, 1993, 2004b), the two predominant frames through which scientific phenomena are explained, explainable, and taught are those of naturalization and socialization. In other words, science education is explained through frames in which nature and culture are the predominant and respective (but not exclusive) factors through which knowledge comes to be known (see Aikenhead, 2006a; Barad, 2000, 2007; Erickson, 2000). As

Barad (2007) explains, both naturalization and socialization are almost always premised within a nature/culture binary in which the constitutive other is treated as a passive surface upon which the dominant term is (re)presented (i.e., claims about Nature as complex are framed against a passive Culture; claims about Culture are framed against a passive Nature; see also Apffel-Marglin, 2011).

With respect to deconstruction's position (see Derrida, 1976) within the space of science, Barad (2011) mentions,

> Invoking Derrida—the 'poster boy' for social constructivism gone wild (a misguided attribution if ever there was one, but so it is), the one theorist nearly everyone but deconstructionists and poststructuralists loves to use as a foil for their own supposed reasonableness, the science warriors' darling stand-in for all that is wrong with the humanities—undercuts any pretense of a convincing straight performance. (p. 448)

In other words, because deconstruction subverts often taken-for-granted assumptions about mediated access to *an* external natural *or* cultural reality (and, more recently subverts the binary distinction between the two; see Barad, 2010, 2012b; Kirby, 2011) by subverting their stability, it is often unwelcome within science or science education (Barad, 2000, 2011).

15. The parentheses here signals that deconstruction does not deny cultural or natural reality but rather denies it stability by presenting it as vacillating between *being* and *not being* within a classical epistemology and ontology (see Barad, 2010, 2012a, 2012b, 2012c; Derrida, 1976).

16. Within Chapter 3, I address in greater length the notion that taking an oppositional stance, what Moulton (1983) refers to as the Adversary Method, is not only an ineffective mode of getting those who would disagree with you to agree, but also acts as a dialectical move that forecloses the possibility of dialogue (Bohm, 1996; Kirby, 2011), whether literal or metaphorical.

17. Latour (2004a) uses *fact* (*fait*) and *fairy* (*fée*) because of their similar etymological roots. In short, both *fait(fact)* and *fée(fairy)* share a relationship to truth. Where they differ in meaning is in the type of truth that they signal: *fact* signals a relation to observable, and verifiable truths about the natural/physical world while *fairy* signals truths that are supernatural, metaphysical, and often associated with fate. For WMS, given its complex relation to the supernatural, metaphysics, and fate (e.g., the supernatural as nature's abject other; religion, the domain of fate, as the abject other of science; see Chapter 6) as well as to truth statements that

are unverifiable through empiricism, a *fairy* might as well be a *fiction* (i.e., a non-*fact*).Furthermore, as explored in the next Chapters (5 and 6), to decry the metaphysical as *fairy* in science and science education effectively masks the ways in which both are always already metaphysical (i.e., through Cartesianism; see Barad, 2000, 2007). In turn, this obscures the workings of power that occur through the enactment of this taken-for-granted and naturalized metaphysics (see Apffel-Marglin, 2011).

18. This is a feature that is often identified within science education: positively by those who are proponents of science education "as usual" (e.g., Cobern and Loving, 2001, 2008) and often negatively by those who endeavour to open science education to cultural critiques (e.g., Aikenhead & Ogawa, 2007; Aikenhead & Michell, 2011).

19. For example, Le Grange and Aikenhead (2016) recently responded to such a claim that decolonizing scholarship operates from a "politics of resentment". Rather, they remind that they do not refuse, refute, nor resent Western knowledge traditions: "Western knowledge should become one way of knowing and not *the* way of knowing" (p. 4, emphasis in original). In turn, decolonizing is not a "politics of resentment" but rather a "pursuing of cognitive justice" (p. 6).

20. The following is a thumbnail history of the "modest witness". In Western Europe in the seventeenth century, the state required a new form of governance that was not religiously partisan as the result of many years of religious wars. Turning from the church to science to keep the peace, those working within the laboratory as third party observers – the practice of the day for experimental verification – were required to abstain from pronouncing or enacting religious affiliation when engaging in the act of observation. They were to witness the experiment "modestly". Worth considering here is that the modest witnesses were all white men of significant status, which may signal to beliefs about who was immodest "by nature" and therefore unable to participate in the cultural practice of science (see Apffel-Marglin, 2011; Haraway, 1997; Latour, 1993). For a more in-depth treatment of the "modest witness," see Chapter 6.

21. For example, in previous research projects, I have witnessed in others and in myself an inability to simply move beyond problematic and pervasive colonial norms despite knowing about them (see Higgins, 2014; Higgins & Kim, 2019; Higgins, Madden, & Korteweg, 2015). When these norms come to constitute the possible positions one can hold, they also come to relationally bear onto the ways in which they are worked against and subverted (see next section on prismatic critique). Futhermore, as Spivak (1993/2009) reminds, "merely knowing an ideology does not dissipate its effect" (p. 5).

22. As Smith (1999/2012) states, such an *enactment* of distance (via separation and seperability) cannot be torn asunder from the "specific spatial

vocabulary of colonialism which can be assembled around three concepts: (1) the line, (2) the centre, and (3) the outside" (p. 55). Not only do these three concepts that are almost always at play in (neo-)colonial logics produce notions of hierarchy through proposing a center and a margin, the drawing of the line between them signals an oppositional difference. In science education, we often see these logics atplay when the case for WMS' centrality is (increasingly implicitly) made: it is heralded as the way of knowing nature because it is framed as not being the orientations that it is defined against, such as TEK and IWLN.This is further complicated by the ways in which (neo-)colonial logics and the metaphysics of clôture simultaneously work "to make the ends coincide with the means" (Spivak, 1976, p. xx), as explored within the first chapter. Stated otherwise, the practice of distancing which precedes hierarchizing is often naturalized and normalized as being one and the same as its resulting knowledge claim of distance. Such coalesced claims inevitably suture over the relations and processes irreducibly enfolded within them. The desire to make the ends coincide with the means results in the production of cultural difference without needing to account for or be accountable to the ways in which this difference is produced (e.g., Western modern metaphysics) or what this produces in turn (e.g., Eurocentrism).

23. Reflexivity is the often-cited solution for taking into account the process through which critics set up their optical apparatus, taking into account one's own situationality (e.g., epistemology, ontology). However, it largely continues to operate through the reflective metaphor of the mirror (Barad, 2007; Haraway, 1997; Pillow, 2003). If we require an additional mirror to account for another process of reflection, then what accounts for the mirror used for accounting, another mirror? As Barad (2007) explains, "reflexivity is nothing more than iterative mimesis: even in attempts to put the investigative subject back into the picture, reflexivity does nothing more than mirror mirroring. ... Mirrors upon mirrors, reflexivity entails the same old geometrical optics of reflections" (p. 88). For an example of how this plays out in de/colonizing science and technology education, see Higgins (2014).

24. As Spivak (1976) reminds us, every translation is always already unfaithful (i.e., never achieving sameness) due to the precariousness of intertextuality.

25. While the work of Michel Foucault is relatively common within education, it is less so the case within science education (see Bazzul & Carter, 2018). As Latour (1993) reminds, post-structural approaches often deconstruct by illuminating the contingency and partiality of the very grounds upon which both *naturalizing* and *socializing* approaches to knowing are founded. Because these two approaches come to inform the

two primary research programs within science education (i.e., cognitivism and socio-constructivism; see Aikenhead, 2006b, Erickson, 2000), post-structural approaches are often un-welcome(d) (see Barad, 2000, 2011) and, accordingly, under-explored.

26. Spivak (1988) offers an important cautionary note on this subject. If for Foucault, power circulates through *all* nodes, and that accordingly, resistance to power can happen at *any* node, Spivak reminds us that power however does not circulate evenly; while resistance can happen anywhere, some locations are nonetheless more significant than others.

27. I take the time to unpack the concept of diffraction here as the impact of Karen Barad's work and interrelated web of concepts (e.g., intra-action) is only recently coming to bear on educational theory and practice (e.g., Lenz Taguchi, 2010), and even more recently within science education (e.g., Milne & Scantlebury, 2019).

28. Recall that "Derrida uses the word 'metaphysics' very simply as shorthand for any science of presence". (Spivak, 1976, p. xxi). Within the sciences, the "master-question is the same as that of all Western metaphysics: 'What is *the* being of the entity?'" (Spivak, 1976, p. xxxiii, emphasis mine). This is to say that classically within the sciences, there is, generally speaking, *a* way-of-being to the scientific phenomena under observation.However, with the two-slit experiment, *being* (and in turn, ontology) is no longer a singular affair to be observed: "so much for the solid confidence, the assured certainty, the bedrock consistency of science, at the brink of a new century... classical metaphysics has misled us" (Barad, 2010, pp. 252, 256). As a result, we are left with unsettling questions or, perhaps more productively, an invitation to consider the ways in which *nature deconstructs* extending far beyond the ways in which "nature" deconstructs (i.e., the ways in which socio-scientific cultural meanings are always-already slipping into self-transgressive moments of irruption) (see Kirby, 2011), as well as what this might mean for science education (see Wallace, Higgins, & Bazzul, 2018).

29. For Barad (2007), materialization is a complex, non-linear, and dis/continuous phenomena through which space, time, matter, and meaning are differentially enfolded. In other words, it is an ongoing process through which *everything* comes to bear and comes to be, in which the co-constituting parts do not come to act in an equal or even manner.

30. Barad (2007) invites us here to differentially consider the relation between the literal or metaphorical observer, the apparatus of observation, and the observed phenomena, not by collapsing them into one, but by considering them as co-producing and inseparable. Since apparatuses are themselves phenomena, even metaphorical apparatus that are the tools of critique

come to produce and be produced by the subject of critique (i.e., the critic) and the object of critique.

31. While beyond the scope of this chapter and book, Barad (2007) proposes that while the panopticon may be exemplary of observational technologies of the eighteenth century, ultrasound technology can be thought of as a more contemporary example of an apparatus of observation that is produced by and producing the phenomena it is meant to observe (i.e., gender*ing*).

32. While not taken up within this chapter, Barad's (2010) "new calculus of response-ability" resonates with Kuokkanen's (2007) conception of response-ability in that both invite us to respond to a world which is rendered invisible through commonplace ways-of-knowing-in-being. Reading one through the other, for diffraction patterns, invites consideration of the ways in which Western modern science's inability to respond to naturalistic phenomena that do not fit within classical metaphysics might be entangled with the inability to take seriously Indigenous ways-of-living-with-Nature above and beyond (neo-)colonial socio-cultural dynamics.

33. Similarly (but not identically), Bohm (1994) also resists the mirror metaphor of sameness and its constitutive other of pure difference by speaking to *similar differences* and *different similarities*. These similar differences and different similarities are concepts used to talk about the relations that are always already constitutive of an undivided whole(ness), as well as the impossibility of achieving the total separation required for pure sameness or difference to be achieved.

34. For me, attempts in identifying science as socially constructed through Eurocentric norms have resulted in having the same logic returned my way: through pointing out that my position too was constructed (in a society in which construction equals fabrication). In other words, opponents would receive my jabs at universalism and return them to slide debate into relativism, reversing the binary bring the possibility of critique to a standstill (see Latour, 1993, 2004a; Haraway, 2001). Chapter 3 is an example of how these logics permeate the multicultural science education debate, unproductively.

35. Spivak (1993/2009) refers to this practice as "strategic essentialism": a critical inhabitation of truth-telling and representation through essentialism that works relentlessly to undo its own essentializing. See Chapter 6 for further discussion and use thereof.

36. Thinking with Alan Turing, Latour (2004a) defines a sub-critical engagement as one in critique is done through *substraction*: "an idea presented to such a [sub-critical] mind will on average give rise to less than one idea in reply" (Turing in Latour, 2004a, p. 248). Latour (2004a), like Turing,

asks if critique can be *super*-critical, in that critique would take one idea and produces more than one rather than less than one.

References

Aikenhead, G. S. (1997). Toward a first nations cross-cultural science and technology curriculum. *Science Education, 81,* 217–238.

Aikenhead, G. S. (2001). Students' ease in crossing cultural borders into school science. *Science Education, 85,* 180–188.

Aikenhead, G. S. (2006a). *Science education for everyday life.* London, ON: Althouse Press.

Aikenhead, G. S. (2006b). Cross-cultural science teaching: Rekindling traditions for aboriginal students. In Y. Kanu (Ed.), *Curriculum as cultural practice: Postcolonial imaginations* (pp. 223–248). Toronto, ON: University of Toronto Press.

Aikenhead, G. S., & Ogawa, M. (2007). Indigenous knowledge and science revisited. *Cultural Studies of Science Education, 2*(3), 539–591.

Aikenhead, G. S., & Michell, H. (2011). *Bridging cultures: Indigenous and scientific ways of knowing nature.* Toronto, ON: Pearson Canada Inc.

Alsop, S., & Fawcett, L. (2010). After this nothing happened. *Cultural Studies of Science Education, 5*(4), 1027–1045.

Apffel-Marglin, F. (2011). *Subversive spiritualities: How rituals enact the world.* New York, NY: Oxford University Press.

Barad, K. (2000). Reconceiving scientific literacy as agential literacy. In R. Reed & S. Traweek (Eds.), *Doing science+ culture* (pp. 221–258). New York, NY: Routledge.

Barad, K. (2007). *Meeting the universe halfway: Quantum physics and the entanglement of matter and meaning.* Durham, NC: Duke University Press.

Barad, K. (2008). Posthumanist performativity: Toward an understanding of how matter comes to matter. In S. Alaimo & S. Hekman (Eds.), *Material feminisms* (pp. 120–154). Bloomington, IN: Indiana University Press.

Barad, K. (2010). Quantum entanglements and hauntological relations of inheritance: Dis/continuities, spacetime enfoldings, and justice-to-come. *Derrida Today, 3*(2), 240–268.

Barad, K. (2011). Erasers and erasures: Pinch's unfortunate 'uncertainty principle'. *Social Studies of Science.* https://doi.org/10.1177/0306312711406317.

Barad, K. (2012a). Interview with Karen Barad. In R. Dolphijn & I. van der Tuin (Eds.), *New materialism: Interviews & cartographies* (pp. 48–70). Ann Arbor, MI: Open Humanities Press.

Barad, K. (2012b). Nature's queer performativity. *Kvinder, Køn & Forskning, 1*(2), 25–53.

Barad, K. (2012c). What is the measure of nothingness? Infinity, virtuality, justice. *dOCUMENTA, 13,* 1–17.

Barnhardt, R., & Kawagley, A. (2005). Indigenous knowledge systems and Alaska native ways of knowing. *Anthropology and Education Quarterly, 36*(1), 8–23.

Barnhardt, R., & Kawagley, A. (2008). Indigenous knowledge systems and education. *Yearbook of the National Society for the Study of Education, 107*(1), 223–241.

Battiste, M., Bell, L., Findlay, I., Findlay, L., & Henderson, J. (2005). Thinking place: Animating the indigenous humanities in education. *The Australian Journal of Indigenous Education, 34,* 7–18.

Bazzul, J., & Carter, L. (2018). (Re)considering foucault for science education research: Considerations of truth, power and governance. *Cultural Studies in Science Education, 12,* 435–452.

Belczewski, A. (2009). Decolonizing science education and the science teacher: A white teacher's perspective. *Canadian Journal of Science Education, 9*(3), 191–202.

Bohm, D. (1994). *On creativity.* New York, NY: Routledge.

Bohm, D. (1996). *On dialogue.* New York, NY: Routledge.

Britzman, D. P. (2003). *Practice makes practice: A critical study of learning to teach.* Albany, NY: SUNY Press.

Butler, J. (1990). *Gender trouble.* London, UK: Routledge.

Butler, J. (2001). *What is critique? An essay on Foucault's virtue.* Retrieved online from http://eipcp.net/transversal/0806/butler/en.

Butler, J. (2005). *On giving an account of oneself.* New York, NY: Fordham University Press.

Butler, J. (2010). *Frames of war: When is life grievable?* London, UK: Verso.

Cajete, G. (1994). *Look to the mountain: An ecology of indigenous education.* Durango, CO: Kivaki Press.

Cajete, G. (2000). *Native science: Natural laws of interdependence.* Santa Fe, NM: Clear Light Books.

Carter, L. (2004). Thinking differently about cultural diversity: Using postcolonial theory to (re)read science education. *Science Education, 88*(6), 819–836.

Carter, L. (2010). The armchair at the borders: The 'messy' ideas of borders, border zones and epistemological diversity in multicultural science education. *Science Education, 94,* 1–20.

Cobern, W. W., & Loving, C. C. (2001). Defining "science" in a multicultural world: implications for science education. *Science Education, 85,* 50–67.

Cobern, W. W., & Loving, C. C. (2008). An essay for educators: Epistemological realism really is common sense. *Science & Education, 17,* 425–447.

Deleuze, G. (1988). *Foucault.* Minneapolis, MN: University of Minnesota Press.

Derrida, J. (1976). *Of grammatology.* Translated by Gayatri C. Spivak. Baltimore, MD: John Hopkins University Press.

Erickson, G. (2000). Research programmes and the student science learning literature. In R. Millar, J. Leach, & J. Osborne (Eds.), *Improving science education: The contribution of research* (pp. 271–292). Philadelphia, PA: Open University Press.

Foucault, M. (1977). *Discipline and punish: The birth of the prison*. London, UK: Allen Lane.

Foucault, M. (1997). *The politics of truth*. New York, NY: Semiotext(e).

Haraway, D. (1997). *Modest_Witness@Second_Millennium.FemaleMan_Meets_ OncoMouse*. New York, NY: Routledge.

Haraway, D. (2001). The persistence of vision. In N. Mirzoeff (Ed.), *The visual culture reader* (pp. 677–684). New York, NY: Routledge.

Harding, S. (2008). *Sciences from below: Feminisms, postcolonialities, and modernities*. Durham, NC: Duke University Press.

Higgins, M. (2011). Finding points of resonance: Nunavut students' perceptions of science. *in education, 17*(3). Retrieved from http://ineducation.ca/ article/finding-points-resonance-nunavut-students-perceptions-science.

Higgins, M. (2014). De/colonizing pedagogy and pedagogue: Science education through participatory and reflexive videography. *Canadian Journal of Science, Mathematics and Technology Education, 14*(2), 154–171.

Higgins, M. (2016). Decolonizing school science: Pedagogically enacting agential literacy and ecologies of relationships. In C. Taylor & C. Hughes (Eds.), *Posthuman research practices* (pp. 267–289). Basingstoke, UK: Palgrave Macmillan.

Higgins, M., & Kim, E. J. (2019). De/colonizing methodologies in science education: Rebraiding research theory-practice-ethics with Indigenous theories and theorists. *Cultural Studies of Science Education, 14*(1), 111–127.

Higgins, M., Madden, B., & Korteweg, L. (2015). Witnessing (the lack of) deconstruction: White teachers' "perfect stranger" position in urban Indigenous education. *Race Ethnicity and Education, 18*(2), 251–276.

Kirby, V. (2011). *Quantum anthropologies: Life at large*. Durham, NC: Duke University Press.

Kuokkanen, R. J. (2007). *Reshaping the university: Responsibility, Indigenous epistemes, and the logic of the gift*. Vancouver, BC: UBC Press.

Lather, P. (2007). *Getting lost: Feminist efforts toward a double(d) science*. New York, NY: State University of New York.

Latour, B. (1993). *We have never been modern*. Cambridge, MA: Harvard University Press.

Latour, B. (2004a). Why has critique run out of steam? From matters of fact to matters of concern. *Critical Inquiry, 30*(2), 225–248.

Latour, B. (2004b). *Politics of nature: Wow to bring the sciences into democracy*. Cambridge, MA: Harvard University Press.

Le Grange, L., & Aikenhead, G. (2016). Rethinking the 'western tradition': A response to Enslin and Horsthemke. *Educational Philosophy and Theory*, 1–7. https://doi.org/10.1080/00131857.2016.1167656.

Lenz Taguchi, H. (2010). *Going beyond the theory/practice divide in early childhood education: Introducing an intra-active pedagogy*. London, UK: Routledge.

Matthews, M. R. (1994). *Science teaching: The role of history and philosophy of science*. New York, NY: Routledge.

McKinley, E. (2001). Cultural diversity: Masking power with innocence. *Science Education, 85*(1), 74–76.

McKinley, E. (2007). Postcolonialism, indigenous students, and science education. In S. K. Abell & N. G. Lederman (Eds.), *Handbook of research on science education* (pp. 199–226). Mahwah, NJ: Lawrence Erlbaum.

McKinley, E., & Aikenhead, G. (2005). Comments on "Thinking differently about cultural diversity: Using postcolonial theory to (re)read science education". *Science Education, 89*(6), 901–906.

Milne, C., & Scantlebury, K. (Eds.). (2019). *Material practice and materiality: Too long ignored in science education*. Cham, Switzerland: Springer.

Moulton, J. (1983). A paradigm of philosophy: The adversary method. In S. Harding & M. B. Hintikka (Eds.), *Discovering reality: Feminist perspectives on epistemology, metaphysics, methodology, and philosophy of science* (pp. 149–164). Dordrecht, NL: Reidel.

Peat, D. (2002). *Blackfoot physics: A new journey into the native American universe*. Newbury Port, MA: Weiser Books.

Pillow, W. (2003). Confession, catharsis, or cure? *International Journal of Qualitative Studies in Education, 16*(2), 175–196.

Pink, S. (2006). *The future of visual anthropology: Engaging the senses*. New York, NY: Taylor & Francis.

Sammel, A. (2009). Turning the focus from 'other' to science education: Exploring the invisibility of whiteness. *Cultural Studies of Science Education, 4*, 649–656.

Siegel, H. (1997). Science education: Multicultural and universal. *Interchange, 28*, 97–108.

Siegel, H. (2001). Multiculturalism, universalism, and science education: In search of common ground. *Science Education, 86*, 803–820.

Smith, L. T. (1999/2012). *Decolonizing methodologies: Research and indigenous people*, 2nd Edition. London, UK: Zed Books.

Smith, L. T. (2005). Introduction. *International Journal of Qualitative Studies in Education, 19*(5), 549–552.

Snively, G., & Corsiglia, J. (2001). Discovering indigenous science: Implications for science education. *Science Education, 85*, 6–34.

Spivak, G. C. (1976). Translator's preface to *Of grammatology*, by Jacques Derrida. *Trans. Gayatri Chakravorty Spivak*. Baltimore, MD: Johns Hopkins University Press.

Spivak, G. (1988). Can the subaltern speak? In C. Nelson & L. Grossberg (Eds.), *Marxism and the interpretation of culture* (pp. 271–313). Urbana, IL: University of Illinois Press.

Spivak, G. C. (1993/2009). *Outside in the teaching machine*. New York, NY: Routledge.

Spivak, G. C. (1999). *A critique of postcolonial reason*. Cambridge, MA: Harvard University Press.

St. Pierre, E.A. (2011). Post qualitative research: The critique and the coming after. In N. K. Denzin & Y. S. Lincoln (Eds.), *The SAGE Handbook of Qualitative Research (4th Ed.)* (pp. 611–626). Thousand Oaks, CA: SAGE Publications.

Stanley, W. B., & Brickhouse, N. W. (1994). Multiculturalism, universalism, and science education. *Science Education, 78*, 387–398.

Stanley, W. B., & Brickhouse, N. W. (2001). Teaching sciences: The multicultural question revisited. *Science Education, 85*, 35–49.

van Eijck, M., & Roth, W. M. (2007). Keeping the local local: Recalibrating the status of science and traditional ecological knowledge (TEK) in education. *Science Education, 91*(6), 926–947.

Wallace, M. F., Higgins, M., & Bazzul, J. (2018). Thinking with nature: Following the contour of minor concepts for ethico-political response-ability in science education. *Canadian Journal of Science, Mathematics and Technology Education, 18*(3), 199–209.

Tinkering with Ontology
with/in the Multicultural Science Education Debate

Tinkering with/in the Multicultural Science Education Debate: Towards Positing *An*(Other) Ontology

The purpose of this chapter[1] is to continue working within and against the stratified and sedimented spaces of the multicultural science education debate by sustaining the deconstructive play of (re)signification of science education, labouring between what it *is*, *is not*, and *could be (come)*. This extends upon the previous chapters' work of (re)opening this debate by engaging in the play of (re)signification between the two predominant positions (i.e., cross-culturalist and universalist) as well as the modes through which the debate operates (e.g., dialectic, debate, critique as mirroring). Putting to work the alternative optical metaphors of the previous chapter (i.e., prism and diffraction grating), I consider how these common occurrences which present themselves with/in the multicultural science education debate are co-constituted by the uncommon and usually absent (see Derrida, 1976). This, in turn, allows for an exploration of the absent yet present[2] co-constitutive elements of the multicultural science education debate that produces the ways in which Indigenous science is to-come, deferring and differing its arrival.

Particularly, this chapter engages the *ontological* homework of response-ability: addressing the ways in which ontology, as absent presence, is *always already* (re)shaping science education. As signalled in Chapter 3, Cobern and Loving (2008) remind that attention to ontology is uncommon within the multicultural science education debate. Where Cobern and Loving (2008) conclude that knowing nature through WMS

© The Author(s) 2021
M. Higgins, *Unsettling Responsibility in Science Education*,
Palgrave Studies in Educational Futures,
https://doi.org/10.1007/978-3-030-61299-3_5

is universal and "common sense", I latch onto the binary co-constitution of common and uncommon, and moments in which they vacillate, as a lever to (re)open spaces of science education to other meanings (e.g., Indigenous science to-come). Drawing from an ethic of deconstructive *tinkering* (Derrida, 1976) by using concepts, categories, and constructs that are uncommon to the context of science education to explore that which is common,[3] I tinker with/in Cobern and Loving's (2008) criteria of ontological alignment to unsettle and (re)situate their claim of "common sense" towards (re)opening the logics of the multicultural science education debate.

As tinkering is central to this arc, (i.e., Chapters 5 and 6), I begin this chapter by quickly touching on what it means to engage with this process before moving to the task of tinkering with/in the multicultural science education debate.

A Preamble on Tinkering: Derrida on the Porous Dichotomy Between Bricolage and Engineering

To frame the process of tinkering within this section, I turn to Derrida's (1976) (mis)reading of Lévi-Strauss' *La pensée sauvage* in which he both separates and blurs the distinctiveness between *engineering* and *bricolage*. In short, both engineering and bricolage are processes of and for generating knowledge claims. Engineering is the movement from the ends to the means, whereby the engineer makes appropriate selections from "the discourses of formal logic, and the pure sciences" (Spivak, 1976, p. xix), picking concepts, categories, and constructs already purposed for their process. In contrast, "the bricoleur makes do with things that were meant perhaps for other ends" (Spivak, 1976, p. xix). Through tinkering, bricolage reverses the ends/means hierarchy by privileging the means over the ends or the process over the product, even if this entails the very possibility of not achieving the specified goals.[4] Significantly, as Derrida (1976) argues, the ends (i.e., knowledge, truth) and the means of knowledge production (i.e., methodology) never come to coincide. The goal of (fully) achieving knowledge is not only empirically impossible, as Lévi-Strauss posited, but also theoretically so. Thus, for Derrida (1976), "the engineer should always be a sort of bricoleur" (p. 139) while coming to recognize the very limitation of bricolage:

> The only weakness of bricolage – but, seen as a weakness is it not irre-mediable? – is a total inability to justify itself in its own discourse. The already-there-ness of instruments and of concepts cannot be undone or re-invented. (pp. 138–139)

Just as the goals of engineering never come to be, Derrida cautions against treating the bricoleur's tools as if they themselves always *were* (i.e., having reached the classical ontological status of Being). Instead, he invites consideration of the tools themselves as the productive enactments of bricolages past and to-come (see also Barad, 2010). There is always a need for "simultaneously troubling and using the concepts [and conceptual lines] we think we cannot think without..., keeping [them] as both limit and resource" (Lather, 2007, pp. 167–168). Furthermore, because engineering/bricolage is always already a porous binary, this invites a critical consideration of bricolage vis-à-vis its ends, or what it produces: "all bricolages are not equally worthwhile. Bricolage criticizes itself" (Derrida, 1976, p. 139).

HAVING AND BEING HAD BY "COMMON SENSE" DURING A SCIENCE EDUCATION PROJECT IN NUNAVUT

Because we must "begin" some-where and some-time, let's "begin" (once again)[5] in Iqualuit, Nunavut in July of 2009.[6] During the delivery of a curriculum in the Canadian arctic in which Indigenous (i.e., Inuit) and non-Indigenous youth explored differential cultural constructions of science through videography (see Higgins, 2011, 2014), I took up the call to examine and challenge the ways in which Eurocentrism "insidiously ... maintain[s] the status quo" (Sammel, 2009, p. 651; see also Belczewski, 2009; McKinley, 2001, 2007) through involving youth in revealing, (re)structuring, and (re)directing the multiple ways in which dominance is maintained. I worked towards disrupting the concepts and categories that tend to create, and are utilized to uphold, inequality within science education, as well as the systems under which these inequalities become possible. Through this project, the youth involved learned and enhanced their movie-making skills and practices in order to explore, define, and document the diverse ways-of-knowing-Nature (i.e., science) that were enacted in their community of Iqaluit, Nunavut. This participant-directed videography work took various shapes: documentary-style interviews with community members within science and technology fields, as well as their own short movies which were a form of digital story-telling. Given that such an examination must also include the many bodies that occupy and uphold these systems within science education, I decided to engage simultaneously with the oft-cited "solution" of self-reflexivity through video diaries to analyse my/self as decolonizing pedagogue. Guiding this examination were questions into the ways in which I was participating in, (re)producing, and (re)produced by the culture of power.

This was useful in creating partial and side-long glances rather than penetrating gazes, producing complex and contradictory accounts of self. I was able to reveal/learn from some of the ways in which I was produced by and producing the culture of power within science education, and inadvertently pivoting rather than "transforming" my pedagogies and self as pedagogue. Nonetheless, through revisiting this data, it became apparent that I was reifying Eurocentrism through the very process of working against it. But then, as Battiste (2005) states, we cannot treat Eurocentrism as a mere individually held prejudice that we can easily cast aside with enough knowledge as its pervasiveness renders it *common sensical*. As briefly mentioned in previous chapters, within this major research endeavour the decolonizing curriculum (e.g., border crossing) and "teacher-as-researcher" identities (e.g., culture broker) available worked both within and against a problematic centre. As such, curriculum, pedagogy, and pedagogue were exceeded in pedagogical prac-tice by the very coloniality the approach worked against, thus becoming de/colonizing (Higgins, 2014; see also Carter, 2004, 2010). One could say that the common pervaded the uncommon; I *held* but was also *being held* by "common sense".

As stated earlier, the very concepts we hold are always already exceeded by lived experience; lived experience provides deconstructive openings to think and act otherwise (Bohm, 1996; Wildcat, 2005). More than a niggling doubt about the porosity of that which earlier seemed solid, this double(d) pattern of *holding* and *being held* by "common sense" would become a location in which I would come to tinker. On *being held* by "common sense," Battiste (2005) reminds that Eurocentrism is a "consciousness in which *all* of us have been marinated" (p. 124, emphasis mine). This similarity is pronounced by difference in intensity, degree, and duration such that it comes to produce us as de/colonizing subjects differently. If how we think is part of the problem, as Lather (2007) quizzically questions us, how do we, and can we, think about how we think without using the thing with which we think? It is a project framed by at once by necessity and im/possibility: an impossibility that when creatively strived towards reconfigures what critical possibilities are possible (see Barad, 2007, 2010; Spivak, 1993/2009).

In attempting to think about how I think, one of the most significant observations that I made through this analysis was that the continued appearance of Cartesianism as a common thread. At the time, I understood Cartesianism as "both the *belief* that various meanings and materialities are discrete quantities (e.g., mind/body) as well as the process through which they are separated from that which co-constitutes them" (Higgins, 2014, p. 164, emphasis added). Like others (e.g., Pillow, 2003), I saw Cartesianism as a belief relegated to the realm of epistemology (i.e., knowledge about ontology, rather than or in addition to an enactment of ontology).

However, these qualities come to describe *ways-of-being* (i.e., ontology) instead of, or in addition to, *ways-of-knowing* (i.e., epistemology). Thus, what might it mean to take insights from the previous chapter (e.g., epistemology and ontology as co-constitutive and not pre-existing practice) and come to see Cartesianism not (only) as an epistemological facet of science education but (also) as one that is ontological? While questions of epistemology often take primacy due to uneven inter-cultural interfaces (i.e., dialectic rather than dialogue), it is often productive to tinker with/in windy, indirect, and side-long approaches when the most direct path seems over-travelled without yielding the desired outcomes. Responding to Lather's (2007) question, it might be one approach to thinking without using the very thing with which you think (when the thing with which you think is part of the problem), recognizing that such is never (fully) achieved. Thus, what might it mean to (re)consider the multicultural science education debate with ontology in mind?

TINKERING WITH/IN "COMMON SENSE," ONTOLOGY AND THE MULTICULTURAL SCIENCE EDUCATION DEBATE

Science and justice, matter and meaning are not separate elements that intersect now and again. They are inextricably fused together, and no event, no matter how energetic, can tear them asunder. (Barad, 2010, p. 242)

If we take seriously that "science and justice... are inextricably fused together" (Barad, 2010, p. 242), the ferocity with which the multicultural science education debate rattles taken-for-granted assumptions can be, to some, surprising (see van Eijck & Roth, 2007). At the epicentre of this debate are questions of "what counts" as science within school-based curriculum. Largely at stake is the inclusion or exclusion of TEK and IWLN alongside WMS, as well as the norms through which they are included, excluded, and juxtaposed. This unresolved and unresolvable debate often presents science educators with diverse and difficult queries regarding what it means to respect students and the diverse ways-of-knowing-nature that they bring with them: can science (i.e., knowing nature) and justice (i.e., respecting diverse cultural knowledges) co-exist within the science education classroom?

Between science educators who champion the inclusion of TEK and IWLN as equally valid[7] ways of knowing nature (i.e., cross-culturalists; for example, Snively & Corsiglia, 2001; Stanley & Brickhouse, 2001) and those who do not consider these ways-of-knowing-nature as *equally valid* to the "universal" standard of WMS (i.e., universalists; for example, Cobern & Loving, 2001; Siegel, 2001), this aforementioned question (and false dichotomy; see Chapters 3 and 4) continues to produce friction that is not always generative. As there continues to be an ongoing and ever-present need to respond to the conflicting and potentially incommensurable demands between epistemological validity and ethical responsiveness in science education, there have been multiple attempts to resolve the debate (see Chapter 3), by working towards producing modest intermediary positions that attempt to develop and enhance potential points of agreement between positions (Alsop & Fawcett, 2010; Cobern & Loving, 2008; van Eijck & Roth, 2007). An example of such a point of agreement is that while universalists and cross-culturalists generally do not agree whether or not science education *is* or *is not* a frequent site of scientism, indoctrination, or imposition, they both agree that it should not be. In the last few years, attempts to labour from shared assumptions towards intermediary positions have included: (a) positing an ethics of incommensurability or co-existence (e.g. El-Hani & de Ferreira Bandeira, 2008; El-Hani & Mortimer, 2007; van Eijck & Roth, 2007), (b) considering diverse and competing scientific knowledges as (re-)contextualized processes rather than inert knowledges (van Eijck & Roth, 2007), and (c) pedagogically enacting an ethics framed by the vulnerability of *not* knowing (Alsop & Fawcett, 2010).

However, in this section, I take up the possible pathway put forth by Cobern and Loving (2008). In short, Cobern and Loving (2008) posit that most proposed and partial responses to the multicultural science education debate largely centre upon questions of epistemology. Given the predominantly socio-cultural characteristics of the debate, this is not without cause. However, Cobern and Loving come at the debate sideways by tinkering with that which remains uncommon to questions of and in science education: ontology. In response to epistemic primacy, to explore how scientific knowledges might have "characteristics of verisimilitude, vis-à-vis the real world" (p. 440), Cobern and Loving suggest that the "real world" too must be seriously considered in the equation. Thus, instead, or in addition to strictly epistemological undertakings, Cobern and Loving (2008) propose that this debate be addressed through a (re)consideration of how the subject of scientific knowledge aligns with its object, or how epistemology (i.e., Culture) aligns with ontology (i.e., Nature). Through an exploration of ontological situatedness, and within the context of the multicultural science debate, the conclusion that Cobern and Loving (2008) reach is that "epistemological realism [i.e., epistemology of WMS] is literally the common ground—the common sense—we all share" (p. 443).

Drawing from insights from the previous chapters, the purpose of this chapter is not to deny the claim made by Cobern and Loving (2008) that "Epistemological Realism Really is Common Sense" (p. 425) by presenting it through a metaphorical mirror as either *fact* or *fairy* (see Latour, 2004a). Rather, it is to tinker with this uncommon approach to justify common sense; there are productive insights to be gained through using and troubling the notion that they put forth. In particular, drawing from the Chapter 4's optical metaphors, what would it mean to treat the statement as more than true or a false reflection of reality? What insights could be gained by (re)situating a "mirroring" statement (i.e., "characteristics of verisimilitude, vis-à-vis the real world" [Cobern & Loving, 2008, p. 440]) by thinking prismatically and diffractively with and about it? In particular, what would it mean to focus on how epistemological realism as "common sense" inflects and is inflected, how it is produced and what it produces? If epistemological realism is "common sense", to whom is it common (e.g., is the "we" in the "common sense – we all share" [Cobern & Loving, 2008, p. 440] in fact *all* of us or but a particular group)? For epistemic realism to be "common sense", what epistemological and ontological criteria need to be in place? How did this "common sense" come to be (made) common (i.e., through dialectic rather than dialogue)?[8] What does this "common sense" produce? Is "common sense" something we have, are had by, or a combination of both? Lastly, as "science and justice, matter and meaning are not separate elements that intersect now and again" (Barad, 2010, p. 242),[9] what does it mean to consider ethics and justice as co-constitutive elements of ontologically situated scientific epistemologies?

EPISTEMIC REALISM AND/AS "COMMON SENSE?" ONTOLOGICAL SITUATEDNESS AND/IN THE MULTICULTURAL SCIENCE EDUCATION DEBATE

Cobern and Loving (2008) state in "Epistemic Realism is Common Sense" (p. 425) that a collusion of forces have made it such that there seems to be less discussion within science education with respect to how epistemological frameworks align with the ontological reality that they attempt to represent. These factors include, but are not limited to: the move towards constructionism and socio-constructionism in science education as well as education writ large, a growing doubt towards WMS following a series of public opinion altering watershed moments (e.g., Agent Orange, napalm, Hiroshima), as well as a Kuhn's introduction of paradigms and the sudden appearance of a plenary of competing paradigms due to an increasingly multicultural social reality. The argument made is that the move to make space for what Cobern and Loving (2008) refer to "ideas that heretofore would have been called ethnoscience and folklore, pseudoscience, and even quackery" (p. 435) has meant that the focus of science education has shifted towards epistemological pluralism as a means of attempting to account for these diversely positioned ways-of-knowing-nature. While Cobern and Loving (2008) might bemoan the impact that these diverse educational and social shifts have had, and how they have reshaped science education as it is today, others have seen these as deeply productive locations and levers to pry open the spaces of science and science education (e.g., Haraway, 1997; Latour, 1993; Snively & Corsiglia, 2001). While I stand with the latter and disagree with the premise that Cobern and Loving (2008) put forth, there is nonetheless some merit in exploring their argument: not necessarily to prove it right or wrong, but rather to explore how it is produced, producible, and what it produces with/in discursive spaces of science education.

Cobern and Loving (2001, 2008) argue that questions of cross-cultural integration, interfacing, and understanding often neglect asking questions about how these meanings necessarily interface with the nature of Nature. They state, given the plurality of epistemic frameworks, that there needs to be pragmatic criteria through which systems of knowing nature (i.e., science) are deemed valid and valuable:

> Interpretations of experience are *all* one can know... One accepts the validity of interpretations in so far as they are pragmatically viable... Historically we [Western modern scientists and science educators] have believed in the ontological reality of the world and trust our epistemological efforts to describe that reality and yet always do our epistemologies fail to demonstrate that what we think we know really is what is. (Cobern & Loving, 2008, p. 433, emphasis in original)

For Cobern and Loving (2001, 2008), among many other science educators, that criterion is experience: a statement that both universalists (e.g., Matthews, 1994; Siegel, 1997, 2001) and cross-culturalists (e.g., Aikenhead & Michell, 2011; Barnhardt & Kawagley, 2005, 2008; Cajete, 1999; Snively & Corsiglia, 2001) would agree with. However, there is a significant point of deviation in how experience is both understood and enacted (e.g., experience as Truth, experience as situated and partial, having and being had by experience,[10] experience as relationality). Whether intentional or unintentional, Cobern and Loving (2008) utilize this conceptual plurality and undecidability as location to present a false dichotomy:

> It is true that we cannot know with certainty that perceptual and experiential experiences are significantly grounded in ontological reality. However, it is equally true that we cannot know for certain that perceptual and experiential experiences are not significantly grounded in ontological reality. (p. 441)

The false dichotomy presented here is either one in which epistemology is ontologically situated within "ontological reality" by utilizing an "*ontological* realism that very few educators would reject" (Cobern & Loving, 2008, p. 437, emphasis in original) or the meaning made is anti-ontological and not grounded in reality in as meaningful a way or at all. This (false) dichotomy relies upon the absent yet present a *common sense* assumption that "ontological reality" precedes meaning making, is singular, unified, and is the ontological reality defined and enacted by WMS (see Chapter 3). Thus, by (re)presenting and casting approaches that don't fall into the classical cannon of epistemic realism into (absolute) relativism, a "a pejorative as far as [they] are concerned" (p. 437), the alternative they present is but a strawman or a trick of smoke and mirrors as discussed within the previous chapter.

Further, it can be argued that the approach presented by Cobern and Loving (2008) does not bring us closer to resolving the multicultural science education debate: it simply displaces the terms of the debate elsewhere (i.e., ontology rather than epistemology). Such a displacement (re)produces a similar but different argument whose goal is to dialectically negate, subsume, or suture over the opposing view rather than create something else, something in-between through dialogue. In other words, the ontological situatedness that Cobern and Loving (2008) call for is but another means of reaffirming science, or WMS to be specific, as universal.

However, considering the notion of ontological situatedness can still bear fruit in the context of science education: what if epistemic realism best aligned with *an* ontology rather than simply "ontology"?

From Ontological Alignment to Positing an Ontology in Science Education

> How *reality* is understood matters. There are risks entailed in putting forward an ontology: making metaphysical assumptions explicit exposes the exclusions on which any given conception of reality is based. But the political potential of deconstructive analysis lies not in simply recognizing the inevitability of exclusions but in insisting on accountability for the particular exclusions that are enacted and in taking the responsibility to perpetually contest and rework the boundaries. (Barad, 2007, p. 205, emphasis in original)

Cobern and Loving (2008) highlight that "we face a metaphysical choice" (p. 441) in science education: a choice bearing significance if "how *reality* is understood matters" (Barad, 2007, p. 205, emphasis in original). This choice is situated within the realm of metaphysics as it asks us to consider the relationship between epistemology (i.e., Culture) and ontology (i.e., Nature). Within science and science education, this relationship between epistemology and ontology has been understood through competing claims of weak and strong forms of relativism and realism. In other words, the way in which this has primarily been taken up by science and science education scholars is through the critical and metaphysical questioning of the extent to which epistemology aligns with ontology.

Within the context of science, as Latour (1993) posits, most scientists reject absolute relativism (i.e., everything is cultural) as it requires the bracketing out of Nature. Similarly, scientists often also reject absolute realism (i.e., everything is natural) as it wholly brackets out Culture. Thus, as scientists generally agree that scientific knowledge is shaped by both natural and cultural factors, scientists more frequently adopt a weaker form of relativism or realism. While Nature and Culture are kept separate for both realists and relativists, there is always a relationship between Culture and Nature such that diverse cultures have different modes of accessing Nature. What is contested between the two approaches is whether knowledge about nature can be explained primarily but not exclusively through natural factors (i.e., weak realism) or through cultural factors (i.e., weak relativism). Within science education, similar discussions of realism and relativism take place (see Chapter 3) and are often included in curricula through exploration of the nature of science (NOS). In short, NOS addresses how the culture of science epistemologically understands and comes-to-know the nature of Nature, or ontology (e.g., Holbrook & Rannikmae, 2007; Plakitsi, 2010; Rudolph, 2000).

However, "as soon as Nature comes into play without being attached to a culture, a third model is always secretly used" (Latour, 1993, p. 104). Latour (1993) refers to this as "particular universalism": a framework in which Nature is stable and outside of Culture and diverse cultural positionings mediate access to knowledge about Nature. The caveat, and mean through which WMS maintains primacy, is that "one society - and it is always the Western one - defines the general framework of Nature with respect to which the others are situated" (p. 105). In other words, defining how Nature operates[11] is established as and establishes epistemic privilege for WMS. Again, as Barad (2007) states, "how *reality* is understood matters" (Barad, 2007, p. 205, emphasis in original); importantly, not only in terms of significance but also in terms of the materialization of its consequences.

Furthermore, for reasons that include but go beyond the troubling of this epistemic privilege, many critical science scholars (e.g., Barad, 2007; Kirby, 2011; Latour, 1993) have begun to examine and cast doubt upon the framework(s) through which questions of relativism and realism come to be argued. Under critical examination is the oft taken-for-granted assumption that many realist and relativist frameworks rely upon: Nature as a stable backdrop upon which Culture plays out. Critical science scholars, such as Haraway (1997) and Latour (1993), have long challenged the oft-accepted notion that "epistemology models ontology" (Polkinghorne, 1991, p. 304 in Cobern & Loving, 2008, p. 442). Returning to the metaphors explored in the previous chapter, it could be stated that the notion of epistemology *mirroring* ontology is suspended. As Barad (2007) reminds, critical science scholarship has been labouring to displace these arguments by reading them through cultural lenses to explore what it means to inflect ontology prismatically through epistemology. By showing how the culture of science has an active role in producing scientific knowledge, this opened up rich lines of questioning regarding epistemology as always situated and partial, as well as the development of an ethics that might go alongside (e.g., Harding, 1986, 1993/2004; Latour, 1993). However, as discussed in Chapter 4, what if the nature of Nature, or ontology, were not stable and passive (i.e., the condition required for both the *mirror* and the *prism*) but rather dynamic, agentic, self-differentiated, and in an ongoing state of becoming (see Barad, 2007, 2010)?[12]

Within what is being referred to as the "ontological turn," Barad (2007) draws from Neils Bohr's philosophy-physics to posit that ontology is not something that exists a priori. This is to say that scientific phenomena under observation do not pre-exist their observation, rather, they are enacted with and through observation. Ontology is not separate or separable from epistemology, but rather is always entangled, superpositioned, and diffracted. Matter comes to matter in both senses of word: it is at once important and worthy of consideration; as well as something that comes into being rather than remaining inert, static, and unagentic (see also Apffel-Marglin, 2011; Barad, 2000, 2010; Kirby, 2011).[13]

While the realm of matter and materiality (i.e., Nature) has always been the primary focus and domain of science education, a (re)consideration of how matter comes to materialize has important consequences for science education in terms of epistemology, ontology, as well as ethics (see Bang & Marin, 2015; Bazzul & Kayumova, 2016; Higgins, 2016; Kayumova, McGuire, & Cardello, 2019; Milne & Scantlebury, 2019). Of particular importance, and a focus in this chapter, is that problem and possibility that ontology is not, and has never been, a singular affair (Barad, 2007).

Rather, it is always already plural and becoming differential through the working and reworking of metaphysical cuts (e.g., the norms of bodily production—subjects and objects).

To situate science and science education ontologically requires one to posit *an* ontology, as opposed to simply situating within "ontology" (read: singular). *An* ontology is an ever-partial (i.e., having exclusions) but never relativistic accounting for an always shifting Nature. Barad (2007) reminds us that part of the positing of *an* ontology goes beyond naming *which* ontology is at work: the "accountability for the particular exclusions that are enacted" through our metaphysical choices includes "taking the responsibility to perpetually contest and rework the boundaries" (p. 205).

To posit *an* ontology thus requires, as Barad (2010) states, to consider how "science and justice" as well "matter and meaning are not separate elements that intersect now and again" (p. 242). If competing ways-of-knowing-nature are separated from their co-constitutive ontology and required to align with *an* ontology, specifically the ontology of WMS, then what does it mean to account for and being ethically accountable to this new entanglement? What is produced and what is producible?

Why Positing an Ontology Matters: Towards Accountability for How Reality Is Understood

The positing of *an* ontology and striving towards accountability for ontological enactments is of importance for scholars working within science education, as well as those which continue to inherit its legacies. To take up the call to posit of *an* ontology would necessarily require moving from questions of epistemology to questions of epistemology *and* ontology, or even onto-epistemology (i.e., the co-constitutive entanglement of knowing and being; see Barad, 2007, 2010)[14] to ask the question of how epistemology and ontology come to co-constitute one another.

Thus, returning to the "metaphysical choice" that Cobern and Loving (2008) present, one that asks which scientific epistemology best aligns and correlates with "ontology" (read: singular), it is fair to state that no choice is offered at all. They put forth, "there is simply no other rational way to account for human ability to increase instrumental epistemological power other than that knowledge has the characteristics of verisimilitude, vis-à-vis the real world" (p. 440). Here, because *an* enactment of the "real world" is already chosen by and/or for them, the "rational way" that Cobern and Loving's (2008) "common sense" metaphysical choice suggests is, again, WMS. Rather than present a modest intermediary position, I suggest that Cobern and Loving (2008) simply displace the terms of the multicultural science education debate by (re)presenting them anew, albeit elsewhere (see van Eijck & Roth, 2007).[15] However, what occurs to ways-of-knowing-nature that are not WMS when they are asked to "mirror" a Cartesian ontology?

While there have been invitations to position diverse ways-of-knowing-nature ontologically, their alignment with the ontology of WMS produces a problematic configuration. It explicitly enunciates and upholds the often-implicit message that approaches other than WMS are lesser means of knowing nature by continuing to not only centre this ontology but also failing to acknowledge that it is but one possible ontological possibility among many. Take for example Siegel's (1997) positioning of diverse ways-of-knowing-nature other than those of WMS:

> Science education must … treat members of minority, dominated cultures with respect. And it must treat the scientific ideas of these cultures with respect. But so treating these cultures and their scientific beliefs and ideas does not require those ideas be treated as correct or *as* correct as the scientific ideas of the dominant, hegemonic culture. (p. 101)

Such often unacknowledged and taken-for-granted ontological positioning and posturing continues to have adverse effects on if, and how, TEK and IWLN are included within science education (see Aikenhead & Michell, 2011; Sammel, 2009). In short, when TEK and IWLN are articulated within and/or in relation to WMS's ontology, they are not only fragmented but also potentially produced as lesser means of knowing Nature through Cartesianism.

The ontology through which WMS comes to be, Cartesianism, is the classical Western modern ontological process through which meaning and matter are individuated through separation from that which co-constitutes them (e.g., mind/body dualism). TEK and IWLN, due to their particular relational entanglements of matter and meaning, fail to (fully) fit the constructs, categories, and concepts enacted by such an ontology; making Cartesian cuts renders many aspects of TEK and IWLN absent presences (e.g., Nature as agentic; see Cajete, 2006). While many science educators have argued that TEK and IWLN stand up to the terms of WMS (e.g., validity, reliability, empirical observation, repeatability), they never stand up as well as WMS on WMS' terms (Aikenhead & Michell, 2011; Cajete, 1994, 2000). Furthermore, such a deficit-based framing (i.e., how it fails to fit WMS's epistemological and ontological enactments) obscures the importance distinctions, as well as rich contributions that TEK and IWLN have to offer from that which exceeds WMS (e.g., ethics of regeneration, spirituality; see Apffel-Marglin, 2011; Cajete, 1994, 2000).

TEK and IWLN's alignment with Cartesianism will always result in theories that are viewed as not "*as* correct as the scientific ideas of the dominant, hegemonic culture" (Siegel, 1997, p. 101, emphasis in original). They fail to cleanly fit the separate and mutually exclusive ontological and epistemological categories established by WMS because TEK and IWLN do not make such clean and clear cuts between epistemology and ontology and their constitutive domains. This is not to state that it is inherently wrong in an absolute sense to centre Cartesianism,[16] and that there are not moments in which it is an appropriate ontology to posit (e.g., when considering WMS).[17] However, to posit *an* ontology is to be held accountable to the patterns of difference, the lines of inclusion/exclusion that are produced through the "metaphysical choice[s]" that we make. Thus, to posit *an* ontology invites a differential consideration and an ongoing accounting for and ethical accountability to the ontological norms through which TEK and IWLN have been included or excluded from science education. To posit *an* ontology also invites a curricular investigation of how diverse knowledges are ontologically situated and produced, without needing to resort to relativism (see McKinley, 2007).

Positing An(Other-Than-Cartesian) Ontology: Towards Ontological Pluralism in Science Education

One does not make the subject matter relevant by starting with an unchanged traditional curriculum and coating scientific facts with "relevant examples" to make them go down easier. *In teaching for agential literacy, science is understood (not "in context") but in complex intra-action with other practices.* (Barad, 2000, p. 238, emphasis in original)

To posit *an* ontology within science education is to recognize that diverse ways-of-knowing-nature are not simply different ways of reaching the same ontological goal within the oft taken-for-granted or unacknowledged ontology of Cartesianism. Teaching a culturally pluralistic science classroom must go beyond the "candy coating" of the standard ontological account, "starting with an unchanged traditional curriculum and coating scientific facts with 'relevant examples' to make them go down easier" (Barad, 2000, p. 238). Rather, thinking with Barad (2000), if we are to teach in a way that encourages students to understand ways-of-knowing-nature as the enacted entanglement of epistemology and ontology, then we must come to understand them as complex and co-constituted practices. In part, this entails pedagogical exploration of diverse ways-of-knowing alongside the ways-of-being that co-constitute them (e.g., Barad, 2000).

To posit *an* ontology is significant within the multicultural science education debate, and beyond, because it can be said that, using the language of the ontological turn, matter has *always* mattered to Indigenous peoples in the ways in which they come to know Nature (i.e., TEK and IWLN). In other words, Indigenous peoples have never fully enacted the nature/culture binary (i.e., the mutually exclusive bracketing of nature and culture) that is commonly accepted as a defining characteristic of Western modernity, its ontology of Cartesianism, and WMS (see Bang & Marin, 2015; Cajete, 1994, 1999, 2000; Peat, 2002). For example, Cajete (2000) highlights the ways in which animal-human relationships conventionally map onto Nature/Culture breaks down within Indigenous ways-of-knowing-in-being:

> Most Native languages do not have a specific word for 'animals.' Rather, when animals are referred to they are called by their specific names. That fact that there are no specific generic words for animals underlines the extent to which animals were considered to interpenetrate with human life. (p. 152)

Beings that are often considered within the realm of Nature (e.g., animals, plants, mountains) have always been agents within the realm of Culture (see Apffel-Marglin, 2011; Bang & Marin, 2015; Barnhardt & Kawagley, 2005; Cajete, 1994, 2000). Thus, to posit *an* ontology rather than presenting ontology as singular, universal, and pre-supposed invites a differential consideration of and an ongoing accounting for, and ethical accountability to, the ontological norms through which TEK and IWLN have been included or excluded from science education.

To consider ontological plurality is not a renewed commitment to relativism. Rather, there are multiple and meaningful "points of resonance" (Peat, 2002) between WMS, TEK, and IWLN. Despite these points of resonance between ways-of-knowing-in-being, this is not to suggest that there is *an* external reality that we can differentially access through diverse cultural frames (see Latour, 1993). Again, plurality need not be though and enacted as relativism (see McKinley, 2007). Rather than relativism, to account for and to be accountable to ontological situatedness (i.e., the co-constitutive relation between *an* epistemology and *an* ontology) might be a way of enacting what Barad (2007) asks of a re(con)figured objectivity – an accounting of and for the diverse network of agents, forces, and flows which locally and globally come to produce the scientific phenomena that we seek to explore within science and science education (see also Barad, 2000).

It is nonetheless important to note that science education will be at its most fruitful when it works to engage with spaces of difference between diverse ways-of-knowing-nature to work against the ever-present risk of conflating diverse systems into sameness. This includes, but should not be limited to, the ways in which ontological sameness produces dialectic negation as discussed herein. Accounting for and being accountable to ontological difference is not only ethically significant, but epistemically generative as well: the distinctions within this plurality can help (re)shape rich and robust knowledge traditions, as well as foster the possibility of inter-cultural hybrids that bring with them the best of both worlds (e.g., Barnhardt & Kawagley, 2005).

Conclusion: Positing an Ontology as an Ethical Call

It is not possible to extricate oneself from ethical concerns and correctly discern what science tells us about the world. Realism, then, is not about representations of an independent reality but about the real consequences, interventions, creative possibilities, and responsibilities of intra-acting within and as part of the world. (Barad, 2007, p. 37)

While it is not "it is not possible to extricate oneself from ethical concerns and correctly discern what science tells us about the world" (Barad, 2007, p. 37), the relationship between science (i.e., knowing nature) and justice (i.e., respecting diverse cultural knowledges) often continues to be dichotomized within the context of science education. This make it such that the debate between the pre-dominant universalist and cross-culturalist positions rarely produces productive intermediary positions from which to fully take up both of these competing claims in the ways in which they are articulated. While there is a responsibility to be simultaneously accountable to both *how we know* and *who we teach* in science education, such should not foreclose the ability to respond. However, the entangled binary framings between Nature and Culture, epistemology and ontology, realism and ethics are notably framings: configurations open to being reconfigured.[18] As Barad (2007) invites, *realism* need not strictly be an epistemological question but can also be about the "real consequences, interventions, creative possibilities, and responsibilities" (p. 37) that come with teaching and learning within science education.

Here, part of this ontological homework of response-ability is learning to account for and being accountable to the "incarnate relation that precedes the intentionality of consciousness" (Barad, 2010, p. 265) that many science educators inherit: Cartesianism and the Nature/Culture cut that it enacts. Given that the Nature/Culture binary makes science and cultural politics incommensurable (see Latour, 2004b), response-ability requires "an iterative (re)opening up to, an enabling of responsive-ness" towards other-than-Cartesian possibilities. Thankfully, as Latour (1993) reminds, the Nature/Culture binary is never fully achieved or achievable. "We Have Never Been Modern", as the eponymous title of Latour's (1993) book proclaims, there is always already the possibility for knowing-in-being otherwise.

Towards this end within this chapter, I tinkered with Cobern and Loving's (2008) suggestion that the primary and almost exclusive focus on epistemologies within the multicultural science debate has detracted from considerations of how epistemology aligns with ontology. As is demonstrated herein, Cobern and Loving use this (re)signified natural-cultural interplay to make a stronger case for universalism (i.e., "Epistemic Realism Really is Common Sense" [p. 425]). However, in using and troubling their claim, there is the possibility for something else to emerge from this insight by differentially engaging with it, particularly if we also take

seriously the notion that ontology is not a singular affair. Such differential arrangement can support us to recognize plurality can be achieved
not by refuting Cobern and Loving's claim but rather by (re)situating it
within *a* context: *an* epistemology of epistemic realism (i.e., the epistemology of WMS) really does align best with *an* ontology of Cartesianism
(i.e., the ontology of WMS). To recognize that Cartesianism is but *an*
ontology creates space in which WMS achieves "distinction not privilege"
(Cobern & Loving, 2008, p. 444), not requiring universalists' claims of
onto-epistemic alignment to be refuted.

Notably, this potential for science without scientism requires science
educators' response-ability towards positing *an* ontology, accounting
for, as well as being accountable towards how it is produced
and what it produces. Given science education's norms and history
of inclusion/exclusion around traditional ecological knowledges and
Indigenous ways-of-living-with-nature, positing *an* ontology invites a
(re)consideration of science education's complimentary and supplemen-
tary spaces of knowing nature. Specifically, this calls for a renewed
engagement with TEK and IWLN: they have their own distinct onto-
epistemological alignments or entanglements, positing *an* ontology calls
for an ethical response-ability to account for the relational ontologies
which come to constitute them rather than requiring them to align with
Cartesianism. To engage in such ontological pluralism need not rely on
ontological or epistemological relativism as there continue to be meaningful patterns of differentiation and similarity that can be productively
engaged with (see McKinley, 2007).

I conclude this chapter with further questions about the space of
ethical response-ability offered by positing *an* ontology: if how we
think (e.g., Nature/Culture binary) is the very thing preventing forward
momentum within the multicultural science education debate, how do
we think about how we think without using the very thing with which
we think? How might science educators move towards ethical response-
ability when responsibility is not something that we simply *have* that
pre-exists our engagements but rather is also something by which *we
are had*, that is produced in its complex flow through and by us? What
further *home*work might allow us to be accountable to how we are
always already (re)produced by science education as educators? Similarly,
how can we foster response-ability to what we produce within the ever-
changing field of possible possibilities of science education (as part of
the world's ongoing becoming)? Lastly, what types of theory-practices

might facilitate the im/possible but necessary accounting for what our onto-epistemological enactments produce within science education?[19] Response-ability, as Barad (2010) reminds, is not achieved "through the realization of some existing possibility, but through the iterative reworking of im/possibility" (p. 265). While there is no singular solution to such questions, positing *an* ontology paves pathways to engage with the im/possibility of being wholly accountable and ethically responsive. To reiterate, situating science education ontologically by positing *an* ontology is not about who is right or who is wrong, nor is it about a renewed commitment to relativism. Rather, it is about coming-to-recognize a plurality of possibilities, and in turn, it means being accountable to how scientific knowledge is produced, produce-able, and what it produces in turn within and beyond the science education classroom. Because "how reality is understood matters" (Barad, 2007, p. 205), the (re)working of lines of inclusion/exclusion that such understandings and enactments produce are always already becoming something else, and this (re)working towards positing *an* ontology is an ethical call which we must all heed, albeit differently.

NOTES

1. An earlier version of this chapter appears in Milne's and Scantlebury (2019) *Material Practice and Materiality: Too Long Ignored in Science Education* (pp. 67–79), and is reprinted with permission.

2. Recall from Chapter 3 that the interplay of absence and presence signals the play of (re)signification. Absent presences are other(ed) signified meanings that might nonetheless retain the structure of the signifier that act as the constitutive otherness of that which is usually signified (see Spivak, 1993/2009). However, absent presences are not an unproblematic panacea: they are at once both unstated and un-assumed meanings (e.g., how claims of science education without scientism have not wholly done away with it) as well as those whose potentiality might allow for justice to-come (e.g., such as pedagogical plurality, allowing for Indigenous science to-come).

3. As Dr. Apffel-Marglin offers in the interview in the next chapter, it is always important to remain open to knowledge outside of one's particular disciplinary pathway. Drawing from that which is uncommon (i.e., other disciplines) can become productive tools to tinker with to (re)open the common (i.e., the occupied disciplinary space).

4. Further, if the goal is to (re)open science education, part of tinkering might entail either: (a) utilizing tools and techniques intended for science

education, but refusing to use them in intended ways; or, (b) putting to work tools and techniques not intended for use within science education (see Higgins, Wallace, & Bazzul, 2018 for exemplars of this in STEM education).

5. See Chapter 2 for a more fulsome account of this narrative vignette, albeit one differentially situated in relation to the content of that chapter (i.e., towards introducing and thinking with response-ability).

6. Once more, this mantra is to remind that with respect to the question of Indigeneity within science education, there is no "beginning": we are *always already* in relation. Accordingly, the plurality of "beginnings" from which we can account for and be accountable to are a resource rather than a liability.

   Further, the invitation within this chapter signals the ways in which metaphysics of Western modernity makes itself common, to the extent that it is often, and (too) simply, referred to as common sense. For example, the (neo-)colonial desire for and production of *origins* (see TallBear, 2013). Within the metaphysics of clôture, the process is rendered one and the same as its product (see Derrida, 1976). This is to say that origins are treated as if they were and could only singular in nature: this is their ontological status (where ontology here is to be read exclusively as the Being of beings). To say the least, this makes it difficult to account for and be accountable to the agencies that are active in the production of *an* origin. For example, TallBear's (2013) research complicates the common sensical notion that who Indigenous people *are* (Being) originates in their DNA. She does this not only by offering an Indigenous conception of Being that is (ontologically) relational, but calling into question the ways in which this "common" (read: made common through colonialism) sense shapes problematic extractivist genetic practices (i.e., taking genetic samples from Indigenous peoples to save them, rooted in an image of "vanishing Indian" that is disassociated from Western modernity's complicity in the production of this image, or the genocides which may have come to inform it).

   Which is to say, yet again, that we, who identify and are identifiable as science educators, are already within relation with Indigeneity, even if most of us are not Indigenous; that deferring and differing *an* origin is not strictly a theoretical move for theory's sake, but one that has effects that come to matter, in both senses of the word.

7. Reminder from the previous chapter that *equally valid* does not signify that TEK and IWLN achieve equivalence or sameness with WMS, and particularly not necessarily on the same terms. Rather that they offer something that is of differently significant and productive (see Aikenhead & Ogawa, 2007).

8. Latour (2004b) reminds that *common sense* and *good sense* are often opposed rather than one and the same. However, he goes on to suggest that if the "good-ness" of sense is pre-supposed, the possibility of meaning being made through *common understanding* (i.e., dialogue) rather than *made common* through dialectic is (fore)closed.

    Also, the question of how this particular "common sense" is produced and what it produces in turn is taken up in far greater detail in the latter, and more substantive, half of this section (i.e., Chapter 6).

9. This is to be differentiated from the taken-for-granted relationship that Barad (2000) alerts us to: one in which scientific literacy is perceived as always already ethical. As she states, it is often assumed that:

    > [There is an] equivalence relation between the possession of scientific knowledge and being socially responsible [that] is often implicit in discussions about scientific literacy: this is the notion that familiarity with the facts and the methods of science is all that is required for socially responsible decision-making concerning science- and technology-related issues. (Barad, 2000, p. 227)

    This ontological pre-determination of ethicality masks the ways in which the (im)possibility of ethics is never fully achieved or achievable as well as how science and science education must always remain ethically on the move and vigilant to respond to an ever-evolving set of ethical problems and possibilities (see also Roth, 2003).

10. With experience being a central concept here, as well as the larger field of education, it is important to recognize that we may not all be referring to the same understanding and enactment of the concept: experience comes to take many forms (above and beyond its doubly settled meaning).

    For example, *being had* by experience might signal at once the ways in which experience "gets the best of us," as well as the ways in which we might not "have" experiences (as possession). Rather, due to not "having" a static body of meaning or matter (i.e., the meaning and matter which makes us is never fully ours; distributed agency; see Barad, 2007; Butler, 2005; Chapter 2 on moving beyond the "I" as subject of ethics and inquiry). What remains from experiences (e.g., learnings), when we are had by them, are the marks left on our body (of meaning and matter) from the intensities and flows within the multiple assemblages we find ourselves in and the affective movement of meaning, matter, spirit with/in/around the re(con)figured body we are part of (see Apffel-Marglin, 2011).

11. Notably, particular universalism dictates how Nature operates without Nature's consent (see Barad, 2007, 2012). As Latour (1993) and Cajete (1994, 2000) state, this double(d) othering of Nature and of other-than-Western cultures are not separate nor separable enactments but rather

differential enactments of the same, intertwined through a practice of cultural "transcendence" in which WMS sees itself as neither having a culture nor being part of Nature.

12. This is of particular significance as Nature as agentic, differentiated, and fluid is important to how an Indigenous "sense of place" is understood and enacted (see Bang & Marin, 2015; Cajete, 2006).

13. Recall, from the preface, that the "ontological turn" is not wholly a *panacea* when responding to the question of Indigeneity; it is also a *poison*. As Barad (2007) invites, positing *an* ontology is also about "insisting on accountability for the particular exclusions that are enacted and in taking the responsibility to perpetually contest and rework the boundaries" (Barad, 2007, p. 205); here, the ontological turn is not excluded within the critical gaze. Specifically, this turn in educational research (and the broader humanities) often fails to acknowledge the ways in which many Indigenous peoples have been thinking about and practising with the other-than-human world since time immemorial, and also have rich conceptions of ontological fluidity and relationality. Whether intended or not, a citational politics at the ontological turn that does not acknowledge Indigeneity reproduces, albeit differently, (neo-)colonial logics by suturing over and sublating Indigenous ways-of-knowing-in-being.

At the same time, a significant part of the homework of responseability is work to within and against the language that we inherit, here it is that of science (e.g., here, drawing from critical science studies), as a means of simultaneously coming-to-know the location from which knowledge is constructed and (re)opening the ability to respond. Further, as Kuokkanen (2007) invites, including the work of (relatively) wellknown scholars into the conversation can "summon" connections and connectivities that might otherwise dismiss questions and considerations of Indigeneity. Lastly, as Spivak (1994) reminds, to denegate a particular theory or approach based on its cultural location alone can create habits in which we too simply dismiss what a particular theory might offer (particularly in terms of undoing its own situationality) while simultaneously uncritically embrace theories from the "correct" location. This has the likely unintended, but not unproblematic effect, of leaving many off-thehook if the implicit message that they are inherently incapable of escaping their own positionality: this is not insignificant considering that the field of science education is mostly based in a Western training in science and that there are many who are white (and/or are complicit in whiteness).

14. Onto-epistemology can neither be adequately referred to as both ontology *and* epistemology, nor the two as one, but rather a state of superposition and co-constitution (see Barad, 2007, 2010).

15. While this is (what I imagine to be) the intended purpose of their contribution to the science education literature, to tinker is to purpose-fully (i.e., intentionally and with purpose) (mis)use tools and concepts.

16. It is important to note however that Cartesianism and Eurocentrism co-constitute one another and that particular attention needs to be paid to how these produce science education and educator, as well what such a science curriculum and pedagogy might come to produce (see Higgins, 2014; see also Apffel-Marglin, 2011; Battiste, 2005; Cajete, 1994).

17. Part of the reason for this, if we take Barad's (2007) notion of *onto-epistemology* seriously, is that the epistemology and ontology of WMS are always already simultaneously enacted (see also Latour, 1993, 2004b). Furthermore, it has been argued that the two were also historically co-developed (see Apffel-Marglin, 2011).

18. Recall from the previous chapter that distance and separation are never neutral: they are part of the (neo-)colonial grammar and practice which comes to constitute the centre (e.g., WMS) and the margins (e.g., IWLN). In turn, these configurations are not only open to being re(con)figured, but that part of the work of hospitably receiving Indigenous ways-of-knowing-in-being in science education is to address this part of our homework: coming-to-know and address the ontological "home" which complicates and forecloses the possibility of Indigenous science to-come in science education.

19. For example, Lather (2007) suggests a double(d) reversal of the ethnographic gaze to consider not only the ways in which thought is prismatically inflected, but also the entangled apparatus through which thought it is produced and producible. Such a double(d) reversal entails both the literal reversal of studying those who do the studying (i.e., in order to reverse the direction of the ethnographic gaze), as well as the study of the way in which those who do the studying study (i.e., in order to reverse the way in which the ethnographic gaze is produced). In the second and more substantive part this section (i.e., Chapter 6), I engage with and through an interview with Dr. Apffel-Marglin who engaged in the difficult task of a double(d) reversal around how Cartesianism is produced and what it produces.

References

Aikenhead, G. S., & Michell, H. (2011). *Bridging cultures: Indigenous and scientific ways of knowing nature.* Toronto, ON: Pearson Canada Inc.

Aikenhead, G. S., & Ogawa, M. (2007). Indigenous knowledge and science revisited. *Cultural Studies of Science Education, 2*(3), 539–620.

Alsop, S., & Fawcett, L. (2010). After this nothing happened. *Cultural Studies of Science Education, 5*(4), 1027–1045.

Apffel-Marglin, F. (2011). *Subversive spiritualities: How rituals enact the world.* New York, NY: Oxford University Press.

Bang, M., & Marin, A. (2015). Nature–culture constructs in science learning: Human/non-human agency and intentionality. *Journal of Research in Science Teaching, 52*(4), 530–544.

Barad, K. (2000). Reconceiving scientific literacy as agential literacy. In R. Reed & S. Traweek (Eds.), *Doing science + culture* (pp. 221–258). New York, NY: Routledge.

Barad, K. (2007). *Meeting the universe halfway: Quantum physics and the entanglement of matter and meaning.* Durham, NC: Duke University Press.

Barad, K. (2010). Quantum entanglements and hauntological relations of inheritance: Dis/continuities, spacetime enfoldings, and justice-to-come. *Derrida Today, 3*(2), 240–268.

Barad, K. (2012). Nature's queer performativity. *Kvinder, Køn & Forskning, 1*(2), 25–53.

Barnhardt, R., & Kawagley, A. (2008). Indigenous knowledge systems and education. *Yearbook of the National Society for the Study of Education, 107*(1), 223–241.

Barnhardt, R., & Kawagley, O. (2005). Indigenous knowledge systems and Alaska Native ways of knowing. *Anthropology & Education Quarterly, 36*(1), 8–23.

Battiste, M. (2005). You can't be the global doctor if you're the colonial disease. In P. Tripp & L. J. Muzzin (Eds.), *Teaching as activism* (pp. 121–133). Montreal, QC: Queen's University Press.

Bazzul, J., & Kayumova, S. (2016). Toward a social ontology for science education: Introducing Deleuze and Guattari's assemblages. *Educational Philosophy and Theory, 48*(3), 284–299.

Belczewski, A. (2009). Decolonizing science education and the science teacher: A white teacher's perspective. *Canadian Journal of Science, Mathematics and Technology Education, 9*(3), 191–202.

Bohm, D. (1996). *On dialogue.* New York, NY: Routledge.

Butler, J. (2005). *On giving an account of oneself.* New York, NY: Fordham University Press.

Cajete, G. (1994). *Look to the mountain: An ecology of indigenous education.* Durango, CO: Kikavi Press.

Cajete, G. (1999). *Igniting the sparkle: An Indigenous science education model.* Durango, CO: Kivaki Press.

Cajete, G. (2000). *Native science: Natural laws of interdependence.* Santa Fe, NM: Clear Light Books.

Cajete, G. (2006). Western science and the loss of natural creativity. In F. Arrows (Ed.), *Unlearning the language of conquest: Scholars expose anti-Indianism in America* (pp. 247–259). Austin, TX: University of Texas Press.

Carter, L. (2004). Thinking differently about cultural diversity: Using post-colonial theory to (re)read science education. *Science Education, 88*(6), 819–836.

Carter, L. (2010). The armchair at the borders: The 'messy' ideas of borders, border zones and epistemological diversity in multicultural science education. *Science Education, 94,* 1–20.

Cobern, W. W., & Loving, C. C. (2001). Defining "science" in a multicultural world: Implications for science education. *Science Education, 85,* 50–67.

Cobern, W. W., & Loving, C. C. (2008). An essay for educators: Epistemological realism really is "common sense". *Science & Education, 17,* 425–447.

Derrida, J. (1976). *Of grammatology* (G. C. Spivak, Trans.). Baltimore, MD: John Hopkins University Press.

El-Hani, C. N., & de Ferreira Bandeira, F. P. S. (2008). Valuing Indigenous knowledge: To call it "science" will not help. *Cultural Studies of Science Education, 3*(3), 751–779.

El-Hani, C. N., & Mortimer, E. F. (2007). Multicultural education, pragmatism, and the goals of science teaching. *Cultural Studies of Science Education, 2*(3), 657–702.

Haraway, D. (1997). *Modest_Witness@Second_Millennium.FemaleMan_Meets_OncoMouse.* New York, NY: Routledge.

Harding, S. (1986). *The science question in feminism.* Ithaca, NY: Cornell University Press.

Harding, S. (1993/2004). Rethinking standpoint epistemology: What is "strong objectivity"? In S. Harding (Ed.), *The feminist standpoint theory reader: Intellectual & political controversies,* (pp. 127–140). New York, NY: Routledge.

Higgins, M. (2011). Finding points of resonance: Nunavut students' perceptions of science. *in education, 17*(3), 17–37.

Higgins, M. (2014). De/colonizing pedagogy and pedagogue: Science education through participatory and reflexive videography. *Canadian Journal of Science, Mathematics and Technology Education, 14*(2), 154–171.

Higgins, M. (2016). Decolonizing school science: Pedagogically enacting agential literacy and ecologies of relationships. In C. Taylor & C. Hughes (Eds.), *Posthuman research practices* (pp. 186–205). Basingstoke, UK: Palgrave Macmillan.

Higgins, M., Wallace, M. F., & Bazzul, J. (2018). Disrupting and displacing methodologies in STEM education: From engineering to tinkering with theory for eco-social justice. *Canadian Journal of Science, Mathematics and Technology Education, 18*(3), 187–192.

Holbrook, J., & Rannikmae (2007). The nature of science education for enhancing scientific literacy. *International Journal of Science Education, 29*(11), 1347–1362.

Kayumova, S., McGuire, C. J., & Cardello, S. (2019). From empowerment to response-ability: Rethinking socio-spatial, environmental justice, and nature-culture binaries in the context of STEM education. *Cultural Studies of Science Education, 14*(1), 205–229.

Kirby, V. (2011). *Quantum anthropologies: Life at large*. Durham, NC: Duke University Press.

Kuokkanen, R. J. (2007). *Reshaping the university: Responsibility, Indigenous epistemes, and the logic of the gift*. Vancouver, BC: UBC Press.

Lather, P. (2007). *Getting lost: Feminist efforts toward a double(d) science*. New York, NY: State University of New York.

Latour, B. (1993). *We have never been modern*. Cambridge, MA: Harvard University Press.

Latour, B. (2004a). Why has critique run out of steam? From matters of fact to matters of concern. *Critical Inquiry, 30*(2), 225–248.

Latour, B. (2004b). *Politics of nature: How to bring the sciences into democracy*. Cambridge, MA: Harvard University Press.

Matthews, M. R. (1994). *Science teaching: The role of history and philosophy of science*. New York, NY: Routledge.

McKinley, E. (2001). Cultural diversity: Masking power with innocence. *Science Education, 85*(1), 74–76.

McKinley, E. (2007). Postcolonialism, Indigenous students, and science education. In S. K. Abell & N. G. Lederman (Eds.), *Handbook of research on science education* (pp. 199–226). Mahwah, NJ: Lawrence Erlbaum.

Milne, C., & Scantlebury, K. (2019). *Material practice and materiality: Too long ignored in science education* (Vol. 18). Cham, Switzerland: Springer.

Peat, D. (2002). *Blackfoot physics: A new journey into the Native American universe*. Newbury Port, MA: Weiser Books.

Plakitsi, K. (2010). Collective curriculum design as a tool for rethinking scientific literacy. *Cultural Studies of Science Education, 5*(3), 577–590.

Pillow, W. (2003). Confession, catharsis, or cure? *International Journal of Qualitative Studies in Education, 16*(2), 175–196.

Roth, W. M. (2003). Scientific literacy as an emergent feature of collective human praxis. *Journal of Curriculum Studies, 35*(1), 9–23.

Rudolph, J. L. (2000). Reconsidering the "nature of science" as curriculum component. *Journal of Curriculum Studies, 32*(3), 403–419.

Sammel, A. (2009). Turning the focus from 'other' to science education: Exploring the invisibility of whiteness. *Cultural Studies of Science Education, 4*, 649–656.

Siegel, H. (1997). Science education: Multicultural and universal. *Interchange, 28,* 97–108.

Siegel, H. (2001). Multiculturalism, universalism, and science education: In search of common ground. *Science Education, 86,* 803–820.

Snively, G., & Corsiglia, J. (2001). Discovering Indigenous science: implications for science education. *Science Education, 85,* 6–34.

Spivak, G. C. (1993/2009). *Outside in the teaching machine.* New York, NY: Routledge.

Spivak, G. C. (1994). Responsibility. *boundary 2, 21*(3), 19–64.

Stanley, W. B., & Brickhouse, N. W. (2001). Teaching sciences: The multicultural question revisited. *Science Education, 85,* 35–49.

TallBear, K. (2013). *Native American DNA: Tribal belonging and the false promise of genetic science.* Minneapolis, MN: University of Minnesota Press.

van Eijck, M., & Roth, W. M. (2007). Keeping the local local: Recalibrating the status of science and traditional ecological knowledge (TEK) in education. *Science Education, 91*(6), 926–947. https://doi.org/10.1002/sce.20227.

Wildcat, D. R. (2005). Indigenizing the future: Why we must think spatially in the twenty-first century. *American Studies, 46,* 417–440.

Positing Cartesianism as *an* Ontology Within Science Education: Towards a More Response-Able Inheritance with Dr. Frédérique Apffel-Marglin

> Inheritance is never a *given*, it is always a task. It remains before us just unquestionably as we are heirs..., even before wanting or refusing to be, and, like all inheritors, we are in mourning... *To be*, this word... means, ...to inherit. All of the questions on the subject of being or of what is to be (or not to be) are questions of inheritance. (Derrida, 1994/2006, p. 76, emphasis in original)

Before getting on to the work of positing *an*(other) ontology that might be more commensurate with Indigenous science (to¬come) in the next chapter of this book, the purpose of this chapter is to continue the task undertaken in the previous chapter: tinkering with/in the multicultural science education debate using ontology towards a more response-able science education. As "all of the questions on the subject of being or of what is to be (or not to be) are questions of inheritance" (Derrida, 1994/2006, p. 76), the task of this chapter is to engage in the *historical* homework of response-ability by archaeologically digging into the historicity and operationalization of our ontological inheritance: Cartesianism. In order to investigate the question of *what might it mean to account for and be accountable to* an *ontology of Cartesianism?*, we must also ask productively entangled questions such as: *How, where, and when*

© The Author(s) 2021   211
M. Higgins, *Unsettling Responsibility in Science Education*,
Palgrave Studies in Educational Futures,
https://doi.org/10.1007/978-3-030-61299-3_6

did Cartesianism as "common sense" become common? How was Cartesianism produced?, How did we come to inherit Cartesianism?, and *What does it make producible?* These questions bear significance as Cartesianism is not only the ontology of WMS, but also deeply shapes the ways in which science education *attempts* to engage with the very ways-of-knowing-in-being that it continues to Other. We are its inheritors in science education, "even before wanting or refusing to be" (Derrida, 1994/2006, p. 76)—our relation to this ontology is irreducible, whether we acknowledge it or not, desire it or not. To account for and be accountable to Cartesianism is part and parcel of science education's homework of response-ability: this ontology lingers and lurks even as we attempt to move within, against, and beyond it. To leave Cartesianism unaddressed risks rendering decolonizing moves, de/colonizing.[1]

Thus, in order to continue this tinkering (i.e., using that which might be intended for other uses) with ontological absent presences, I trace our ontological *inheritance.* Specifically, how multiple *there-thens* of Cartesian "origins"[2] are haunt the *here-now* of science education: passed over pasts which constitute our inheritance. However, inheritance and responsibility are irreducibly related: "there is no inheritance without a call to responsibility" (Derrida, 1994/2006, p. 114). It is call to engage in "creative archeology, a performative reinvention, of the received wisdoms that constitute our cultural inheritance" (Kirby, 2012, p. 198) which (re)produces who science educators *are* and *can be.* It is to engage in the homework of response-ability with attentiveness to our history; particularly, the ways in which past is not yet past and should not be passed over.

Significantly, "inheritance is never *given*, it is always a task" (Derrida, 1994/2006, p. 67, emphasis in original): the task of inheritance is not one that is straightforward when our onto-epistemic inheritances present themselves as absent presences, normalized and naturalized through centuries of Eurocentric thought and colonial practice (such as is the case in science education; see Higgins & Tolbert, 2018).[3] Thus, to engage with the task of (re)opening the norms of responsiveness towards Indigenous science to-come, attention is brought to how Cartesianism plays, has played, and will play an active role in the (fore)closure of response-ability: the task of inheritance. To undertake this ontological homework of response-ability, I draw from a series of expert interviews with Dr. Frédérique Apffel-Marglin's that unpack the historical, geographical, political, economic, and religious forces of the "birth of modernity" and reveal the ways in which this "common sense" went from being uncommon to common and continues to persist. In turn, these insights are read, or more specifically diffracted, through the multicultural science education debate in order to (re)open the space for responsiveness therein. In thinking with Dr. Frédérique Apffel-Marglin, I make the argument that the *making common* of Cartesianism not only (fore)closes the possible possibilities for responding to Indigenous science to-come, but also to account for and be accountable to the ontology of WMS and what it produces.

Before this diffractive reading, I present a positional vignette to both introduce the work to come and my (co-constitutive) relation to the work herein. A methodological section on (re)thinking expert interview diffractively follows.

PATHWAYS OF CHANCE: ENCOUNTERING DR. FRÉDÉRIQUE APFFEL-MARGLIN

What if we leave room in our lives for chance and the unexpected?.... Instead of seeing' one's life planned out into the future like a vast highway, life only makes sense to us when we look back to the past and see the path we have taken with all its diversion, U-turns and side roads. There may have been no fixed plan for the future but looking back at the map of one's life journey it almost seems as if there had indeed been a goal all along, but a goal that had been concealed at every step of the way. (Peat, 2007, pp. 15–16)

Because we must "begin" some-where and some-time, let's "begin" in Lamas, Peru in June of 2013.[4] Then and there, I would have the pleasure of interviewing Dr. Frédérique Appfel-Marglin on her work around deconstructing the metaphysics of modernity as well as what can be learned through attending to ritualized action. For me, this would be what Peat (2007) refers to as a "pathway of chance".

As Peat (2007) elaborates in his "little" autobiography, our lived experiences cannot be mapped so easily: we are always walking along pathways of chance. A probabilistic approach might see us treading common paths more often than those uncommon. However, when we encounter the uncommon "do we compensate as best we can in an attempt to remain on our predetermined track?" (p. 15). Peat suggests that if "we leave room in our lives for chance and the unexpected" (p. 15), we may come to encounter new relationships as well as old relations anew. Such encounters can gently inflect our life's journey or throw into new directions altogether, into unanticipated spaces and places that we may only be able to make sense of in retrospect, if at all. Just as Peat's life was thrown into another orbit altogether through meeting his mentor-to-be David Bohm, as well as the Blackfoot peoples (and Blackfoot scholars like Leroy Little Bear), my encountering Dr. Frédérique Appfel-Marglin was a meeting of the latter pivotal type.

I first textually encountered the work of Dr. Apffel-Marglin in the fall of 2010 during an assigned reading in a graduate class on Indigenous education within her edited collection *The Spirit of Regeneration*. This opened my eyes to the incredibly useful potential of thinking about questions of (neo-)coloniality and Indigeneity anew by considering the global South and those working with the global South when thinking about and with questions of de/colonizing education. I would not meet her in person until the spring of 2012, during an invited talk at the University of British Columbia. During this talk, she was showcasing her (then recently released) book *Subversive Spiritualities: How Rituals Enact the World* that, as she states, was "the fruit of a lifetime of anthropological practice" (Apffel-Marglin, 2011, p. 7). Centring the pedagogical potentiality of ritual enactments, the purpose of the talk and of her book, as she states:

> is an effort to articulate the nature of the kinds of beings I spoke with and gave gifts to in rituals. Here I attempt to articulate for people like myself what my new manner of practicing rituals has revealed to me about our (i.e., the modernist) way of representing reality. (Apffel-Marglin, 2011, p. 8)

Her life's work has not only focused on understanding ritual enactments, but also what ritual enactments reveal about Western modern ways-of-knowing-in-being (e.g., how Western modernity produces matter and meaning and what these in turn produce). Thus, during this talk, she spoke of "twin journeys" which are highly commensurate with Battiste's (2013a, 2013b) conception of decolonizing as deconstruction and reconstruction: a double(d) reversal which she referred to as "reverse anthropology":

By "reverse anthropology," I do not mean [strictly] engaging in ethno-graphic research on some aspect of American or European culture or society. Rather, I mean an understanding of the processes that have brought about our way of seeing reality – all of it, not just a single, particular aspect. By "our way of seeing reality," I mean what we take for granted; by "we" I mean people like me, educated in the modern educa-tional system that by now is hegemonic and thus no longer tied to any specific region or ethnic group; by "reality" I mean things as fundamental as time, space, and nature as opposed to culture, the human person, and so forth. (Apffel-Marglin, 2011, p. 11)

In other words, this double(d) reversal entails both the literal reversal of studying those who do the studying (i.e., in order to reverse the direction of the ethnographic gaze), as well as the study of the way in which those who do the studying study (i.e., in order to reverse the way in which the ethnographic gaze is produced). It is a differential means of working towards thinking without the thing with which we cannot think without (Lather, 2007). Furthermore, this double(d) reversal is not without significance: to make rituals and the insights they provide "understandable and credible requires nothing less than a deconstruc-tion of the modernist onto-epistemology" (Apffel-Marglin, 2011, p. 14) which render ritual *Other*. These include addressing a constellation of interconnected concepts enacted through practice, including the separa-tion of Church and State in most Western modern contexts as well as the multiplicitous ways in which most non-Western, non-modern ritual-ized spiritual enactments are often exoticized and branded as "irrational animism" (see also Cajete, 1994, 2000; Marker, 2006). In order to make space for spirituality, she delivered a rich and robust critique of moder-nity, its teleology of progress, and its narrative of development by coming at modernity via multiple angles: social, political, religious, military, historical, and economic. From here, she laboured within, against, and beyond modernity by employing a reconstructive frame that interfaced quantum ontologies (to work within and against modernity) and Indige-nous ways-of-knowing-in-being (to work within and beyond modernity). This framework was primarily offered as a way to think and enact the world otherwise through ritualized action.

Turning towards Dr. Apffel-Marglin's work around ritualized actions that "enact the world in concert with its humans, non-humans, and other-than-humans" (Apffel-Marglin, 2011, p. 15) might seem uncommon for a science educator or within science education. Apffel-Marglin (2011), like Peat (2002, 2007), reminds that it is always important to remain open to knowledge outside of one's particular disciplinary pathway. She highlights the importance of being inter- or trans-disciplinary as "it has the distinct advantage of being less prey to the tacit knowledge hidden in entrenched fields and subfields of European studies and in particular in the boundaries creating such fields and subfields" (p. 14). However, there were far more spaces of potential connection and commonality than I had or could envision at the time of first encountering her work. For one, ritualized action has many connections to how IWLN are articulated:

> Ritual is the medium for communicating, reciprocating, creating, and working with the other-than-humans, who daily remind the humans that the world is not for humans' exclusive use. It is the other-than-humans who make clear to the humans that human desires are not the only ones. In other words, they make clear that humans are not the masters of this world, and in turn that this world is not agency-less and voiceless, a sum of natural resources to be indefinitely mined to feed the supposedly infinite desires of human beings. (Apffel-Marglin, 2011, p. 6)

These formalized processes and protocols of and for communal (inter-)being have the aim of (re)generating and the sustaining of "the iterations of the cycles of life" (p. 41), reminding that we are not passive observers outside of the world but rather active participants within. Similarly, with respect to Indigenous science, Cajete (1994) states that:

> Scientists study the tracks of subatomic particles that exist only a millionth of a second. They find the human observer influences the energy relationships and even the nature of existence of these subatomic particles. Humans do participate with everything else even at this level of natural reality. Indigenous people understood this relationship of human activity as concentric rings that extend into the spirit realm. (Cajete, 1994, p. 55)

As Cajete articulates further, "it is no accident that learning and teaching unfolded in the context of spirituality in practically every aspect of traditional American Indian [Indigenous] education" (Cajete, 1994, p. 41). If I found many productive points of resonance with Frédérique Apffel-Marglin's work, and could learn much form her, it is because the degrees of separation between the spaces we, respectively, attempt to foster were far smaller than disciplinary borders presented at first. Furthermore, and of equal importance, the locations that Dr. Apffel-Marglin laboured within and against in order to make space for Indigenous ritualized enactments were even closer: as this involved critically inhabiting and tinkering within the spaces of and related to WMS. There was much I could and can still learn from her life's labour in this arena.

Encountering Dr. Apffel-Marglin was an unexpected turn along my own personal pathways of chance; it was an encounter that subverted the very ways in which I could be in the world, I could not (not) take up what she had offered. This (re)ignited my own interest in metaphysics, the cross-cultural space between Indigenous ways-of-knowing-in-being and quantum ontologies, as well as the importance in coming-to-know the very space that I am operating within and against to allow for a possible possibility of there being a beyond. This encounter also sparked my imagination with respect to the educational possibilities laying therein, which come to inform the later chapters of the book. It so happened that Drs. Peter Cole and Pat O'Riley were hosting a University of British Columbia summer institute in Peru the summer after which Dr. Apffel-Marglin visited the University of British Columbia to give her talk. This would be located at the Sachamama Center that she ran, located in the San Martin Department of Peru and on the ancestral territories of the Kichwa-Lamista peoples. Attending this summer institute meant that there would be a possibility to engage with Dr. Apffel-Marglin in person but also a possibility of holding an expert interview.[5] Sensing an opportunity that should not be missed, I had a desire to document to allow for these rich teachings to disperse: I would interview Frédérique as part of a tinkering process in which I did and would not know the end result. In turn, I sent out an invitation to do so; the same day, she replied: "I would be delighted to participate in a dialogue or interview with you and very much look forward to it" (Dr. Frédérique Apffel-Marlin, personal communication, February 14th 2013).

Tinkering with/in Expert Interview: De/Signing Research Methodology

Wanting to recognize Dr. Frédérique Apffel-Marlin's "lifetime of anthropological practice" (Apffel-Marglin, 2011, p. 7), the methodological structure with which I sought to tinker was that of expert interview. As Bogner, Littig, and Menz (2009) state, "there is no such thing as *the* expert interview" (p. 6, emphasis in original). Rather, they are differential inflections on more general interview practices that predominantly hinge upon conceptual considerations of "what counts" as "expert", "expert knowledge", as well as what the aims and purposes of an expert interviews might be. Thus, expert interview presents itself as a methodological structure that is constituted by the ways in which the key concepts are understood and enacted. As Bogner et al. (2009) frame, a topology of expert interviews would primarily include, but would not be limited to: (a) expert interview as exploratory, to provide orientation into further inquiry; (b) expert interview as systematic retrieval, to acquire "expert knowledge" from an "expert", and; (c) expert interview as theory-generating interactive exchange.

Working towards the third type entails working within and against the second type: where notions of *fact* and *fairy* come to haunt accounts as adequately or inadequately *mirroring* reality (see Chapter 4), putting into question the status of and possibility of ever achieving "expert" and "expert knowledge".

> Regardless of what might be myth and what is reality, the anticipated promise of rapid and unproblematic access to objective data makes expert interviews an extremely appealing option for empirical social researchers. But is the expert interview method really quite so simple and uncomplicated? If so, does this then render methodological considerations superfluous? Or are expert interviews in some ways just too tempting? Do they not – in their naïve belief in the totality of expert knowledge – harbour the danger of advocating a pre-reflexive definition of what constitutes an expert? (Bogner et al., p. 2)

In turn, the practice of methodological tinkering continues to bear importance and relevance[6]: methodology is always already a space with/in which to tinker with purpose but without guarantee.

Expert interview as methodology would require tinkering with so that it was not presented as *the* truth, but rather a truth with which to think; a truth through which could be productively inflected or with which others could be diffracted (see Chapter 4). As such, I worked towards achieving this through three interconnected rounds of tinkering: before, during, and after. Before, I tinkered with the content of the expert interview. During, the expert interview tinkered with itself in its enactment. After, I tinker with the (re)presentation of the expert interview by pulling from insights generated through the theory-generating exchange.

Before: Tinkering with/in Expert Interview Content

When designing the series of expert interview questions, the original intent was to create a three-part expert interview series that would differentially represent three major themes or threads that were being pulled through Dr. Apffel-Marglin's 2011 book *Subversive Spiritualities*. Namely, these themes were: (a) her methodology of reverse anthropology; (b) working within and against Cartesianism, and lastly; (c) working beyond Cartesianism by considering the spaces of ritualized enactments. In other words, the interviews as designed were not to repeat that which was already there, but rather tinker within the already existing narratives in Frédérique's book to connect differently, more loosely, to allow for the possibility of something else to emerge.

This began with the notion that despite the cohesive and complete appearance of the narratives within her book, like any and every text, they were and continue to be open to being tinkered with/in.[7] Consisting of loosely assembled partial, situated, and relational meanings, *any* and *every* narrative is but a series of meanings that are held together by "scandalous sutures" (Derrida, 1976): what Spivak refers to as the "rage for unity" (Spivak, 1976, p. xvi). Thus, narratives are always already open to being irrupted, ruptured, and torn asunder in order to differentially sequence its frayed and fraught meanings to produce new narratives which, in turn, never achieve unity. Thus, when I was developing the questions for the interview that is featured within this chapter, I centred her work on the entangled political, economic, religious, scientific, and military practices that led to the birth and operationalization of Modernity within Western Europe. I gripped and pulled at these threads from her book *Subversive Spiritualities* to (re)constitute a narrative around this theme. However, in practice, this narrative was never fully achieved either, as the very practice of expert interview exceeds itself to produce something that is within and beyond the narrative framework available to it.

During: Expert Interview Tinkering with/in Itself

In practice, expert interviews come to be an enactment or practice that *is* and *is not* its research design: both the expected and the unexpected come into proximal relation to de/sign any possible expert interview methodology design. The arguments that scholars make in interviews are always situated within a real and/or imagined scene of address: what is said, what can be said, and how it is said is always in part produced and producible in relation to the audience to which it is intended as well as the constellation of concepts which are entangled with/in (see Chapter 4 on the prism; see also Butler, 2005; Kirby, 2011). To elaborate upon this point, Judith Butler in an interview with Vicky Kirby speaks to how possibilities of ever shifting cultural and natural norms dis/allow modes of being critical:

> I think perhaps mainly in gender Trouble I overemphasize the priority of culture over nature... At the time of Gender Trouble, ... it seemed to me that there was a cultural use of 'natural' arguments to provide legitimacy for natural genders or natural heterosexuality. But that criticism did not take into account of a nature that might be, as it were, beyond the nature/cultural divide, one that is not immediately harnessed for the aims of certain kinds of cultural legitimation practices. (Butler in Kirby, 2011, p. 93)[8]

Expert interviews are then a site for things to be thought *anew*, albeit not for the first time, within a differing web of relationships that comes to bear with differing intensities.

At the Sachamama Center in Peru, where the three interviews of roughly two hours with Frédérique occurred, each interview became an opportunity for something that *was* and *was not* her book *Subversive Spiritualities*: no two "readings" are ever the (self-)same (see Spivak, 1976). Dr. Apffel-Marglin and I, but her more so than I, narratively pulled at the anticipated threads differently in order to make new meanings: that which was already came to be in new relations by differentially centring how they were included or excluded, and to which degree. While the questions generated were derived from her book, this did not constrain the interviews, especially towards the end of the interviews. The expert interview alternated from being within the script, to being exceeded by the script, from being an expert interview to a dialogue in which I was also a participant: what Bogner and colleagues (2009) refer to as a "quasi-expert". This participation was nonetheless shaped by the relational *elsewheres* and *elsewhens* that exist within and beyond the interview (e.g., Frédérique's reading recommendations might be an inflection of a teacher–student relation held in other spaces). Furthermore, comfortably seated outside her modest abode within the Sachamama centre, Frédérique and I were also often audibly joined by other members of the ecology of relation-ships that shaped that place: the birds chirping, the dogs barking at one another through the nearby fence, the traditional Peruvian three-piece band practising down the road, the wind, among other things. These other-than-human beings that she refers to so often in her book *Subversive Spiritualities* were not simply metaphorically signified and absent, but were rather physically present and active agents in the production of the interview (Higgins & Madden, 2019; see also Barad, 2007, 2010; Cajete, 1994, 2000).

After: Tinkering with/in Expert Interview (Re)Presentation

As the interview with Dr. Apffel-Marglin presents a partial, situated, yet rich and nuanced account of how Cartesianism is produced and producible, the interview itself invites other ways of being (re)presented within, against, and beyond Cartesianism. Accordingly, I turn to Barad (2007)'s post-Cartesian extension of diffraction as metaphor for methodology to inform how I work with Frédérique's interview.

Recall that for Barad (2007), diffraction asks us to consider where the agencies of observation begin and where they end, what is included, what is excluded, what matters, and what comes to materialize. Important to note here is that, for Barad (2007), the experimental apparatus through which any phenomena manifests is never simply a material tool or a discursive concept, but also includes multiple material and discursive agencies of observation, and extends to include each of their respective material and discursive historicities. This network of bodies of meaning and matter become the experimental apparatus through a differential enactment of the norms of bodily production. As such, these bodies do not simply interact between one another, but rather *intra-act* within this re(con)figured body which is the experimental apparatus. Lastly, through intra-action, materiality and discourse, the domains of ontology and epistemology, are not mutually exclusive but rather co-constitutive. Or, as Barad (2007) states "phenomena are differential patterns of mattering ('diffraction patterns') produced through complex agential intra-actions of multiple material-discursive practices or apparatuses of bodily production" (p. 140).

Diffractive methodology is thus a practice of knowing-in-being with/in the production of patterns of difference that emerge when two or more phenomena, or entangled bodies of matter-meaning, are read through one another (Barad, 2007). This is in stark contrast to the optical metaphor of the mirror that operationalizes the production and reproduction of sameness through representationalism. Diffractive methodology is a process of producing, and being responsive and accountable to non-negligible patterns of difference that come to matter when two or more entangled material-discursive phenomena are diffracted through one another. This requires however that "we learn to tune our analytical instruments (that is our diffractive instruments) in a way that is sufficiently attentive to the details of the phenomenon we want to understand" (p. 73) and to pay attention to the fine details that would otherwise be considered negligible with/in conventional scientific and social scientific research methods. Barad (2007) poses diffractive methodology as an engaged and enacted ethico-onto-epistemological research stance that takes seriously the entanglement of ethics, materiality, and discourse as well as the patterns of difference that emerge through their ongoing diffractive intra-action.

Here, diffractive methodology comes to inform the ways in which the ways in which Dr. Apffel-Marglin's interview is read through the multicultural education debate to produce patterns of difference that might come to inform how science education might be thought otherwise. Particularly, as diffraction is about producing differences that matter, there is an emphasis on what it means to be accountable to and for ontologically positioning science with/in Cartesianism. With/in any diffractive methodology, Barad (2007) reminds that:

> We are responsible for the cuts that we help enact not because we do the choosing (neither do we escape responsibility because "we" are "chosen" by them), but because we are an agential part of the material becoming of the universe. (p. 178)

One such set of cuts enacted herein are the editorial cuts which come to produce the textual presentation of the work *as* expert interview. In producing the text as an articulation of (quasi-)expertise (e.g., expert as rational subject), many agential engagements have been excised. These include, but are not limited to, the now-absent presences of other-than-humans (e.g., dogs, wind), other humans (e.g., those labouring at the Sachamama Center), human-other-than-human hybrids (e.g., Peruvian three-piece band, blaring radio), as well as moments in which speech is interrupted by thoughtful pauses, stuttering, stammering, *ums* and *ahs*.

Both Frédérique and I have engaged in this act of narrative "cleaning" in order to produce a text that might be more cohesive, and intelligible as expert interview, to an audience to-come.[9]

Furthermore, the interview that follows is split into four parts. The cuts enacted here are emergent (i.e., a "choice" that emerges with/in relations that extend beyond me; see Barad, 2007), producing rich patterns of resonance and divergence when read through the multicultural science education debate. There is nonetheless a cautionary note that different cuts produce and make possible a different set of meaning-making phenomena, this is but one possible possibility or configuration. Each cut is divided into two smaller parts: the interview with Frédérique always preceding the later diffractive analysis.

To give a quick overview of content addressed by Dr. Apffel-Marglin, the first cut, *Nature/Culture*, addresses how she understands this dichotomy with respect to her own field of anthropology as well as how she strategically goes about addressing it. The second cut, *Descartes, Boyle, and Newton*, speaks to three important figureheads not only in the development of Cartesianism, but also the ways in which they operationalize it. The third cut, *The enclosure, the double-sided ledger, and the laboratory*, situates Cartesianism with/in the material practices of Western Modern Europe with which it could not be disentangled and which lead to its operationalization. The fourth and final cut, *The modest witness, when One Truth becomes two, and the Thirty Years' War*, positions the practices of the individual within the scientific laboratory within a larger and uneasy socio-political context.

Diffracting an Interview with Dr. Frédérique Apffel-Marglin: On the (Re)Production and Operationalization of Cartesianism and What It Produces

First Cut: Dr. Frédérique Apffel-Marglin on Nature/Culture

M. HIGGINS: Within your 2011 book *Subversive Spiritualities*, you speak to the multiple dichotomies (e.g., mind/body, male/female, colonizer/colonized) that you are working within and against in order to enact a double(d) reversal of anthropology (i.e., subverting the gaze as well as the concepts, categories, and constructs through which the anthropological gaze operates). Recognizing that to disrupt one binary is to disrupt others that are related to it (see Barad, 2010; Lather, 2007; St. Pierre, 2011a), one of the key dichotomies that you center in your work is that of Nature/Culture. Could you quickly describe how you understand Culture and Nature within this binary pairing?

F. APFFEL-MARGLIN: I have to preface what I'm going to say by saying that my understanding is totally coloured by training as an anthropologist; so what you are getting is the view that currently still, I could say, operates within anthropology. Anthropology has really made the term Culture what it is, so I should begin with that. Before anthropology defined the term Culture as we know it today during the turn from the nineteenth to twentieth century, culture was something held by a "cultured person". This usually entailed people going to the symphony, to the fine arts museum, to classical plays; that is what having culture meant. It is still used like that by people who are not reading anthropology; it is still used vernacularly like that. However, anthropology has democratized the term such that it means that everyone, anywhere in the world, has culture and that anthropologized meaning of "culture" has deeply penetrated our vernacular language. There are no people in the world that do not have culture. It is human to have culture, just as it is human to have language and to speak. Furthermore, language and culture are totally integral to one another.

Anthropology, as a discipline, takes as its object of study Culture. Studying "culture" is what anthropologists do, and the discipline has developed an understanding of Culture as an exclusive dichotomy. By exclusive dichotomy, I mean the following: that Nature is what Culture is not and Culture is what Nature is not. They determine each other negatively. It is very important to understand what that means and why it is significant. Exclusivity means that there is an absolute boundary between the two terms. While there are a plethora of dichotomies that enact a tension between two poles in which things move from one pole to the other relating to each other as a continuum, that is not the case with an exclusive dichotomy. That is a very different kettle of fish. I call the former polarities, or sometimes simply dichotomies, and the latter exclusive dichotomies. While there are exclusive dichotomies as well as plain dichotomies or polarities I reserve the term dualism to refer to exclusive dichotomies. The Nature/Culture exclusive dichotomy can be said to have dominated the field of anthropology.

M. HIGGINS: Within your book, you address the notion that the first term of any dichotomy is dominant over the latter, and that this is achieved and upheld, as you mention through a metaphysics of mutual exclusivity and separation. Usually this entails that the first term is constructed in opposition to the second, making the latter its abject other or its object of lack. What consequence does this bear upon Nature as well as those who are considered "natured" (i.e., rather than "cultured")? Furthermore, how has this been troubled?

F. *APFFEL-MARGLIN*: The troubling of the Nature/Culture dichotomy has always been important within the second wave of feminism and beyond. Feminism is an important site for this troubling because women in the Western traditions, such as Christianity and Judaism (except for the mystical traditions), have been associated with Nature (i.e., defined in opposition to Culture).

Within anthropology, this exclusive dichotomy has started to be troubled around the notion of gender, especially by a wonderful British woman anthropologist, Marylin Strathern. She does this work in her book *The Gender of the Gift* on her fieldwork within Melanesia. That was an early book that really is brilliant in troubling the relationship between gender, Culture, and Nature. Following this, more and more scholars in anthropology have troubled this exclusive dichotomy. Interestingly enough, in feminist approaches to anthropology, there is, from my point of view, a sharp divide between those who reject the troubling of that dualism and those who trouble it. The most well-known of those who reject the troubling of the Nature/Culture dichotomy is Sherry Ortner. Ortner maps the relations of male/female onto Nature/Culture. In her widely influential essay "Is Female to Male as Nature to Culture?", she simply takes it for granted that the Nature/Culture dualism is of the same kind as the female/male one; she assumes that there is such a division and that you can map gender on it.

This gender mapping has been crucial in exploring the Cartesian-Boylian-Newtonian paradigm. This is the work of Donna Haraway and, also extremely importantly, of Shapin and Schaffer on Hobbes and the air pump. So the whole Nature/Culture dualism is completely entangled with gender issues, and debates in feminism. That's why I was, from day one, deeply and personally involved.

M. *HIGGINS*: There is often a relationship between different dichotomies that allows the productive tinkering within one to have deconstructive effects within others.

F. *APFFEL-MARGLIN*: Right.

M. HIGGINS: You've begun speaking to this here with respect to the Nature/Culture binary in relation to male/female, what other dichotomies might Nature/Culture be entangled with to produce particular bodies as "natured" and what consequences does this yield?

F. APFFEL-MARGLIN: While I talked about the case of gender, the *UR* [i.e., original], the basic, fundamental trope or metaphor, even if it is not seen as that, is of Nature to be controlled by Man. I deliberately use the masculine here because men developed it. They excluded women in that newly minted epistemological paradigm known as Classical Science, an exclusion that continued until the twentieth century. It really is Man, the masculine European, Western European male. So if Nature is to be controlled by Man, this also means that all *other* humans are closer to Nature. We talked about women, but also included within what Judith Butler has called "the abject other" are those perceived as "primitive", the "savage", and lower class. Diasporic people, Indigenous people, and other colonized people are often seen as "closer to Nature" because of a perceived inability to control themselves, their biology, or their nature. These tropes are still fully operational in the world today even if not everyone subscribes to them; they unfortunately have not disappeared. In my context of working with the Kichwa-Lamista, the Indigenous peoples in what is now the Peruvian High Amazon, they are often treated as "wild people" who have no civilization or language, amongst other lacks.

M. HIGGINS: In considering exclusive dichotomies, it is always productive to look at both sides of the binary division. If those who are perceived as "being closer to Nature" are being treated in these ways, what does that mean for those who are perceived or self-perceived as "cultured?" How is the Eurocentric, patriarchal and masculine human body, or Man, and his way of being in the world shaped and affected by this exclusive dichotomy?

F. APFFEL-MARGLIN: That's a lovely question. I'm so glad that you are asking that and I would recommend a book to you. My very good friend Ashish Nandy addresses this topic in his first book called *The Intimate Enemy* published in 1983. In short, it is about colonialism but it also addresses how colonialism finally hurts the colonizer as well as the colonized. When I heard your question, I immediately thought of Ashish Nandy's work. About this, I would certainly say that this way of being is one of being *déchu* [i.e., having lost dignity]. It is an ethical fall from grace, or a form of self-perversion.

M. HIGGINS: Paulo Freire, in *Pedagogy of the Oppressed,* speaks to a similar notion: to dehumanize the Other is dehumanizing in and of (it)self through a failure to recognize the humanity of the Other.

F. APFFEL-MARGLIN: Exactly.

M. HIGGINS: This very much speaks to the notion that exclusive dichotomies are porous rather than hermetic. If we take the work of Jacques Derrida seriously, despite the prevalence and pervasiveness of this exclusive dichotomy, like other binaries, it is *déjà toujours* or always already deconstructing. In other words, it is never fully achieved as it is a myth or an imaginary. In your book *Subversive Spritualities*, you explore and leverage some examples within recent scholarship that you develop within your work that speak to the porosity of the Nature/Culture dichotomy (e.g., Cultured Nature and Natured Culture—such as cultural landscapes). Can you speak to why this is important to your work?

F. APFFEL-MARGLIN: I begin with this recent scholarship around cultural landscapes for strategic reasons. Just like the *Gedanken* experiments (i.e., thought experiments) in quantum physics have been empirically proven; the undoing of Nature/Culture has been enacted empirically and you cannot deny it. If you are using the language of the mainstream, it is harder to deny. Because scientists have already shown that the Amazon forest is anthropogenic [i.e., produced through and with human interaction], all I need to do is provide the data to make my point. I start with this because it is an easier way of entering into that topic. It is easier for people who are not inclined to hear your message [i.e., those who might disagree] because you are giving empirical evidence [i.e., and speaking on their terms]. Start where it is very hard to deny: empiricism. Because if I begin with more philosophical discussions around themes people do not want to hear, people close down before I've even begun. I have seen that happen in meetings and conferences.

M. HIGGINS: That's a very interesting strategy that you utilize throughout your book. On one hand, you're rejecting the universality and the privileging of WMS and its ongoing supersessionism (i.e., the double(d) process of WMS acting as the metre stick against which other ways-of-knowing-Nature are judged, as well the ways in which WMS supplants and displaces other ways-of-knowing-Nature; see Lewis & Aikenhead, 2001). On the other hand, you are strategically using that same structure all at the same time.

F. APFFEL-MARGLIN: Right, but only when it is to my purposes. Because of my relationships to people within the mainstream, I've learned to use these arguments. But it is important for me not to stay there, and only use it as an entry point to my argument.

First Cut—Nature/Culture and the Multicultural Science Education Debate

As Dr. Apffel-Marglin's segmented interview is diffracted with the multicultural science education debate, it comes to produce particular patterns of dynamic resonance and dissonance with one another around three key nodes: (a) Nature/Culture as mutually exclusive dichotomy; (b) problematic natural-cultural hybrids (e.g., "naturalized facts") and, (c) productively using natural-cultural hybrids.

The notion of Nature/Culture as mutually exclusive has much bearing on both the fields of anthropology and science. It is the "foundational" cut of Western modernity that puts both fields into dynamic and resonating relationality (Apffel-Marglin, 2011; Barad, 2007; Kirby, 2011; Latour, 1993, 2004). As Dr. Apffel-Marglin articulates, the mutually exclusive dichotomy between Nature and Culture can be enunciated such: "Nature is what Culture is not and Culture is what Nature is not" which is further qualified by "an absolute boundary between the two terms". As mentioned earlier within this chapter, this is deeply entangled with/in the practices of validating what one might consider valid or worthwhile (i.e., "what counts" as science) in terms of meaning-making practices with/in Nature; within this framework, all knowledge of Nature must be made from a mutually exclusively cultural position (see Barad, 2007; Latour, 1993, 2004). Through patterns of similar differences and different similarities, as Frédérique articulates, meaning-making practices within the social sciences (such as anthropology) are similarly but differently entangled with/in the Nature/Culture exclusive dichotomy. Through the capillary circulation of scientificity, meaning-making practices within the social sciences are often (re)shaped such that knowledge of Culture must be made from a mutually exclusive cultural position (see Apffel-Marglin, 2011; Lather, 2007; St. Pierre, 2011a, 2011b).

Recall that Smith (Smith 1999/2012) posits that distance is a key organizing concept for understanding how Western modernity is enacted. She expands upon this by stating that distance becomes the enactment of the "specific spatial vocabulary of colonialism which can be assembled around three concepts: (1) the line, (2) the centre, and (3) the outside" (Smith, 1999/2012, p. 55). The dividing line of mutual exclusivity is a (supposed) hermetic safeguard which masks the ways in which the outside and the inside are always already in proximal relation. Within the sciences

as well as the social sciences, one consequence of such a mutual exclusivity is the inability to account for the ways in which Nature comes
to bear upon Culture, how Culture comes to bear upon Nature, and
an extreme difficulty in occupying both spaces simultaneously (Barad,
2007, 2010; Kirby, 2011; Latour, 1993, 2004). Through scientificity, the
mutual exclusion of Nature and Culture translated into research practices
in which cultural locations of observer and observed could be conceptualized and enacted as separate and separable: discounting the ways in which
there is no outside of Culture.[10]

   With/in the multicultural science education debate, such mutual exclusivity makes difficult even the task of engaging in the question *can science
(i.e., knowing nature) and justice (i.e., respecting diverse cultural knowledges) co-exist within the science education classroom?* As Latour (2004)
reminds us, "the notions of nature and [cultural] politics had been developed over centuries in such a way as to make any juxtaposition, any
synthesis, any combination of the two terms *impossible*" (p. 3, emphasis
in original). As explored within Chapter 3, such a dichotomized relationship makes it such that one can ask ethical questions of science or scientific
questions of ethics, but never fully arriving at something that effectively
straddles and occupies both spaces simultaneously within the terms articulated by both: the production of an *ethical science* that is the product of
both yet still retains sameness. Such is, as Spivak (1976) refers, an infinite
dream of plenitude (as the concept of an ethical science would productively exceed conceptions of both ethics and science), it is an impossibility
that is nonetheless worth striving for.

   The goal of ethical justice-to-come within science education, or a way
of knowing nature that can account for or be accountable to its cultural
politics, is all the more prescient when considering what Latour (1993)
refers to as particular universalism. Recall, as Latour (1993) states, one
cultural location always comes to be deemed as *the* valid location from
which to make meaning of nature: that of Man ("the masculine European, Western European male" as Frédérique states).[11] It has been and
continues to be articulated both within the sciences and the social sciences
that the Nature/Culture binary has and upholds a Eurocentric legacy
(Apffel-Marglin, 2011; Latour, 1993). As an in-between space, the same
has been argued many times over that the multicultural science education debate is one that, at its very roots, operates within and against
the bounds of Eurocentrism (Aikenhead & Michell, 2011; Lewis &
Aikenhead, 2001; McKinley, 2000; Sammel, 2009); an entangled part

of the whole that even some scientists are beginning to comment on (e.g., Mazzocchi, 2006, 2008). Lewis and Aikenhead (2001) summarize Eurocentrism as "the idea that the people, places, and events of Western European cultures are superior and a standard against which other cultures should be judged" (p. 53). Within this debate and similar others in which epistemological claims are being made across cultural contexts, it is important to consider both the relationships, and the processes of translating, between cultural spaces. This becomes all the more pressing when the West is included within this act of translations as:

> Eurocentrism is the colonizer's model of the world in a very literal sense: it is not merely a set of beliefs, a bundle of beliefs. It has evolved, through time, into a finely sculpted model, a structured whole; in fact a single theory, a general framework for many smaller theories, historical, geographical, psychological, sociological, and philosophical. This supertheory is diffusionism. (Blaut, 1993, pp. 10–11)

As James Blaut (1993) explains, through this diffusionist[12] model of dissemination, Eurocentric modes of thought centre themselves and propagate, all the while subjecting, assimilating, and subsuming other modes of thought (see also Battiste, 2005). "Universalism", which is one of the many faces or applied strategies of Eurocentrism, is achieved through diffusionism and the ways in which Western knowledge comes to be positioned against other systems of knowledge as *the* norm. Within the multicultural science education debate this relational positioning produces "the implicit curriculum message... that the *only* science is [W]estern science" (Hodson, 1993, p. 686, emphasis in original).

This becomes all the more complex when considering that Nature/Culture as a mutually exclusive dichotomy that is but one ontological configuration among many, as explored earlier with this chapter. As Latour (1993) reminds, the Nature/Culture binary is never fully achieved or achievable. "We Have Never Been Modern" as the eponymous title of his book proclaims. There have always been natural-cultural hybrids existing in the space(s) between Nature and Culture as the result of the ways in which Nature and Culture always already flow into one another: we are always already in a space of account-ability towards and for the ways in which ontological configurations (e.g., Nature/Culture binary) coalesce with epistemology and ethics (e.g., particular universalism). Science and science educational spaces are not exempt from this occurrence. While some are problematic and some offer productive possibilities, when the world is conceived of and enacted through a mutually exclusive binary, ways-of-knowing nature cannot account for or be accountable to natural-cultural hybrids.

As Dr. Frédérique Apffel-Marglin spoke, one of the greatest consequences of the *particular universalism* (Latour, 1993) through which science endeavours to mirror Nature from its cultural location is that some bodies are proclaimed as being "closer to Nature". Such "naturalized facts" come to be double(d) diminishing: the ways of knowing articulated from these positions are denied validity (see Chapter 3), such as TEK and IWLN, as they fail to achieve the normative and "necessary" criteria of distance through which a mirroring of nature is achievable (see Chapter 4); but furthermore, it is to deny the humanity of Man's "abject other" (e.g., Indigenous peoples, women, etc.) as it relies upon a conception of Nature (i.e., ontology) as static, uniform, and unflinching such that Man's "abject other" become but uni-dimensional people (see Higgins, Wallace, & Bazzul, 2019; McKinley, 2000, 2007).

Within other fields, responses to "naturalized facts" have been to reverse the Nature/Culture binary or to abandon Nature altogether. However, such anti-ontological stances that circulated with the social sciences with a bit more ease (e.g., "too simple" readings of Butler's [1990] *Gender Trouble*)[13] would not find a welcome, frequent, or widespread home within the sciences and science education despite their overlapping spaces of meaning-making (see Barad, 2000, 2011). The reversal of the resident hierarchy between Nature and Culture would not suffice or be viable to many within science education. This can be attributed to the holding and being held by knowledge of nature as separate and separable from cultural politics, often times antagonistically; they are most commonly defined and enacted through a mutually exclusive Nature/Culture dichotomy (Latour, 2004). For many scientists and science educators, the false binary choice of renouncing science to strive for ethics produces an unintelligible scientific subject (see Chapters 3 and 5). These continue to complicate questions of accounting for and being accountable to Culture (or even culture) in the process of meaning-making with Nature.

However, as Frédérique Apffel-Marglin posits within the interview, and as explored within the Chapter 3, not all positions need to be antagonistic. There are productive intermediary positions that can operate simultaneously within and against such that it productively retains a degree of intelligibility and desirability. Thus, speaking to productive natural-cultural hybrids, recall that deconstruction is a two-part process which must necessarily begin with the reversal of the hierarchy but must then undo the hierarchy altogether by paying attention to the moments in which the hierarchy altogether vacillates between its constitutive terms. Dr. Apffel-Marglin's quick but incisive pointing towards her work, (see Apffel-Marglin, 2011) and that of others, with regards to anthropogenic (or cultural) landscapes opens up productive natural-cultural locations on terms that act as an invitation to dialogue rather than a dismissal. In other words, as anthropogenic landscapes are within the realm of scientific discovery, they cannot be so easily dismissed as being "antagonistic" cultural politics (e.g., in the case of the critical reversal of critiques of "naturalized facts"; e.g., Siegel, 2001), and furthermore act as an invitation to reconsider the very terms under which science operate within the very linguistic practices that science utilizes: it is a critical inhabitation.

Continuing this critical inhabitation, Dr. Apffel-Marglin begins to articulate in the following sections, the way in which the Nature/Culture mutually exclusive dichotomy (i.e., what is traditionally referred to simply as "ontology") is not something that strictly *is* or *is not* but rather in a state of *ongoing becoming* by highlighting processes and peoples involved in its operationalization.

Second Cut: Dr. Frédérique Apffel-Marglin on Descartes, Boyle, and Newton

M. HIGGINS: Within the Nature/Culture dichotomy, even as it always being produced anew through a continued entanglement with other binaries such as masculine/feminine, one of the common threads being pulled through is dualistic thinking. While the origin of dualistic thinking (if we can say there is one) is largely, and often, attributed to René Descartes, you extend this to also include the seventeenth century scientific revolution as well as the work of Robert Boyle and Isaac Newton. Why are these three key figures in the production of what is often referred to as a mechanical, clockwork universe, as well as dualistic thought?

F. APFFEL-MARGLIN: First, I want to preface this by saying that the new (new of course in the seventeenth century) Cartesian-Boylian-Newtonian paradigm, has of course much deeper and wider historical roots. In my book, when I discuss representationalism [i.e., the epistemological and ontological web of concepts and categories enacted and operationalized through representation], I go all the way back to Plato and the Greek alphabet. So there are deep roots that are entangled within the Cartesian-Boylian-Newtonian paradigm, even if you stick with the birth of modernity. Not as deep as Plato, but certainly a few centuries preceding the scientific revolution. Of course this is simplifying it. Nevertheless, there is something very special and key about these three. They might not have created it whole cloth [i.e., entirely responsible for its fabrication] but it would not have had the power, particularly the persuasive power, were it not for those three because at the time it was a highly debated and contested new paradigm.

Descartes was the first philosopher of materialism. He was very systematic in his approach to establishing *res cogitans* [i.e., epistemology] and, by extension through separation and mutual exclusivity, *res extensa* [i.e., ontology]. So Descartes is writing in the early 1600s, and most of his writing is around 1620–1630 and he died in 1650. Descartes was a philosopher, so he is making a thought structure; he systematized this philosophy before Boyle comes along.

Boyle is mid seventeenth century (i.e., 1650s). He is doing his experiments with the air pump but he probably only started in the 1640s; for him the key period are the 1650s. Why is Boyle important? Boyle operationalizes Descartes. How did Boyle do this? He creates the laboratory, the modest witnesses, and a literary technology, as Shapin and Schaffer have brilliantly shown. He is important because he invents these practices that constitute the scientific experimental method. How key can you be? He is operationalizing Descartes' systematization of thought by making it work in action. Boyle was an actor in the creation of *The President, Council, and Fellows of the Royal Society of London for Improving Natural Knowledge*, also known as the *Royal Society*. He did this because he had to. At the time, universities were a monopoly of the church; this new knowledge had to happen outside. At the same time as the Royal Society, there was also an Italian academy, as well as *l'Académie Française* started by Mersenne and Descartes. However, within a few decades, the Royal Society quickly became the dominant academy of science as the king of England was funding it. To open up a new academic space outside of the church was absolutely key when it came to operationalizing Cartesian materialism.

Why Newton is so important is because he brought together Descartes and Boyle by establishing the mathematical relationship between Descartes' philosophy and Boyle's practice [i.e., through what is commonly referred to as Newtonian physics today]. Canonically, the age of the scientific revolution is seen as starting with Copernicus mid-sixteenth century with his *On the Revolutions of the Heavenly Spheres* that was published posthumously in 1543. Newton, with his *Principia Mathematica* in 1687, just clinched it. He was absolutely brilliant, and also very powerful. He was lionized. He is buried where the kings are buried in England. He was the most famous man of his time. He had advised the king and he had enormous power, power that he nurtured very carefully. He knew perfectly well, that there were certain things he could not make public, such as his engagement in practices of hermeticism [i.e., non-Cartesian Western ways-of-knowing-Nature which blended scientific practices with "magical" ones such as alchemy and astrology].[14]

Second Cut: Descartes, Boyle, Newton, and the Multicultural Science Education Debate

From Dr. Apffel-Marglin's critical inhabitation of *a* historical narrative of modernity, what we begin to see emerge here is that Cartesianism is not something that *is* but is rather *becoming*. As van Eijck and Roth (2007) remind within the context of the multicultural science education debate, it is important to consider science not as knowledge but as knowledge-processes that are highly contextualized and situated. One such consideration for Cartesianism as *becoming* rather than *being* is to refuse its frequent positioning as naturalized within an ahistorical present (see Spivak, 1999). However, even if we trace its historicity, it cannot simply be referred to as knowledge originating from and obtained by René Descartes that, as *the* truth, mirrors metaphysics. Rather, it is a knowledge-process that is without *an* origin (e.g., Plato): its meaning and matter are always deferred and differing such that it forms an ongoing citational chain (see Barad, 2007, 2010; Derrida, 1976). Cartesianism, as Dr. Apffel-Marglin explains, is not created "whole cloth" by René Descartes, nor by Thomas Boyle, or Isaac Newton: Cartesianism is something that precedes, lives alongside, and outlives all three through its circulation and differential enactments across a multiplicity of spaces. Nonetheless, these three do come to be key actors in the operational-ization of Cartesianism such that it is widespread, commonplace, and "common sense"; something one has and by which one is had (see Barad, 2007; Higgins, 2014).

Diffracting Frédérique's account of Descartes, Boyle, and Newton with and through the multicultural science education debate again reinforces the notion that the Nature/Culture mutually exclusive dichotomy is but a configuration among many and that it is an onto-epistemological enactment that we are responsible for; even if, as Barad (2007) states, we choose and are chosen by such configurations (see also Butler, 2005). However, what significantly begins to resonate with this vignette is that Cartesianism, when considered through distributed agency (Barad, 2007), can be considered as a "regime of truth" (Foucault, 1977, 1979). In other words, through their respective metaphysical framework, laboratory practice, and mathematics, Descartes, Boyle, and Newton implicitly enact theory-practices that implicitly come to support one another. As Foucault (1979) explains, "'truth' is linked in a circular relation with systems of power which produce and sustain it, and to effects of power which it induces and which extend it" (p. 47). The example at the beginning of this chapter could be stated as a singular application of such a regime of truth: Cobern and Loving (2008), in their argument for (re)considering and supporting an epistemology of "epistemic realism", pivot and side-step to "ontology" (read: singular; i.e., Cartesianism) without being accountable to or for the ways in which the two share a co-constitutive onto-epistemic relationship. This is often the case when and where it comes to the defence of WMS as *the* way of knowing nature: these diverse knowledge claims which come to reinforce WMS almost always operate through the differing similarity through an implicit assumption that Cartesianism is *the* (only) ontology. It is no surprise, as Dr. Apffel-Marglin continues to expand upon this, that Cartesianism comes to be (re)presented as "common sense" (as Cobern & Loving (2008) define it), holding the ahistorical appearance of stability, neutrality, and normality.

Furthermore, as indicated here and expanded upon within the next section, knowledge and power share a co-constitutive relationship and important role in the dispersal of such knowledge-practices (Foucault, 1977, 1979). For example, the earlier footnote regarding Newton's dual positioning with respect to modern science and hermeticism reveals Newton to having, and being had, by power. He is being produced as a scientific subject that simultaneously upholds and is upheld by systems of power.[15] However, as discussed in the following section, such a dispersal of knowledge-practices comes to bear at a larger scale when considering the technologies which operationalize this way of knowing-in-being (e.g., the enclosure, the double-sided ledger, and the laboratory).

Third Cut: Dr. Frédérique Apffel-Marglin on the Enclosure, the Double-Sided Ledger, and the Laboratory

M. HIGGINS: While we are on the subject, within your book, you make clear that "the scientific revolution was not the result of decades of discoveries, but was in the making for centuries" (Apffel-Marglin, 2011, p. 33). While ideas and ways-of-thinking, like the Cartesian-Boylian-Newtonian framework, are always being (re)generated, those that "stick" (and stick around) are those that are operationalized and become embedded within the matrix of power/knowledge. It is important to speak of Descartes, Boyle, and Newton when tracing or retracing the production of the Nature/Culture exclusive dichotomy, as they were key players in making operational this dualism. However, this is not to say that the scientific revolution to which Nature/Culture is often attributed origin was not entangled with and in the social, political, religious, economic and other events, tensions, and forces that were happening at the time as well as the time which preceded it. Could you give a thumbnail account of the state of the European subject historically around and preceding the "tipping point" of modernity [i.e., the scientific revolution]?

F. APFFEL-MARGLIN: There's two ways I can answer that question. I could answer it in terms of the forces that killed hylozoism [i.e., a Western European relational way-of-knowing-in-being of the time premised around the notion that all matter is in some sense alive; shares a relation with hermeticism] and made what Descartes or Boyle called the molecular or atomic model of the world, and human-istic/materialistic dualism the dominant and winning view.

Another way to answer this question is to go historical and speak to the history of the fall of the manorial system, sometimes called the feudal system, and all the conflicts that emerge. I have to preface this by saying that I owe a serious intellectual debt to my ex-husband Stephen Marglin's work, as he's done a huge amount of historical work on the birth of capitalism. Drawing from literature on economic anthropology, historical economics, and such works, I have come to the view that the enclo-sure movement is associated with the birth of Cartesian dualism. What I argue is that it is that economic context of the emergence of mercantilism, which isn't yet capitalism, but rather proto-capitalism. It is a proto-market economy that comes to replace the manorial system that begins with the enclosure of the commons. The enclosing of the commons started in Western Europe, continued through colonialization, and now persists through globalization. It is incredibly relevant when you're dealing with Indigenous peoples, places, and protocols as this enclosure movement has continued. The enclosure of the commons is for me the key thing that gives the social, political, and religious context that gives credit and credence to Descartes', Boyle's and, Newton's intellectual work and makes it stick.

> *M. HIGGINS*: As you state in your book *Subversive Spiritualities*, understanding the enclosure movement and the encloser's advantage are important in setting the stage for and are entangled within the operationalization of the Nature/Culture dichotomy. Could you speak more to this movement and its consequences?
>
> *F. APFFEL-MARGLIN*: At the time of the enclosure movement, there was a lot of conflict. It was a time of peasant wars in France and in England. There was tremendous social and political conflict. However, the enclosure movement was simply a power play.

This plays out similarly today: governments in the global South are currently signing the Free Trade agreement with the United States in which the small print says that you have to make land available to businesses, that land being the commons for Indigenous peoples. Because Indigenous peoples don't have title to it, the governments can say "no, we're giving it to the oil companies, the mining companies, the timbering companies, the agri-chemical companies, etc.". That is currently what is going on, and it is going on everywhere in the global South.

This is what happened, first in England, France, and a few other countries in Western Europe. The rich powerful merchants simply bullied others to get their way through the local courts. Because they had money and connections, they could make it stick. They gained title to the land, and then they actually put up a fence: that's what enclosure means. So you put a fence, and say: "this is mine". Then you use the land as you see fit. At the beginning, it was because the price of wool cloth went up and the world was trading wool cloth, so they wanted to get into that trade by raising and pasturing sheep. So they simply bullied everybody with the support of the aristocracy because many merchants were aristocrats. They made it stick legally through the courts but also through direct violence.

M. HIGGINS: One of the things I appreciate in this discussion and in your book is that the enclosure is not only a figurative or epistemological form of separation but rather or also one that is literal and material. Part of this comes into play in your exploration the enclosure as an enactment through the calculation of the encloser's advantage in which you make clear that the calculation not only exerts power following its enactment, but also through its enactment; the calculation itself and that which is entangled with/in it is important.

F. APFFEL-MARGLIN: This is directly related to the beginning of banks. I've fallen by chance on this while exploring the history of mathematics. In my discussion of the whole history of mathematics within *Subversive Spritualities*, an important moment is when *one* becomes a real number. *One* is what you use for counting; when *one* becomes a number, it creates a new category, that of magnitude.

The first banks emerged in the great merchant cities of Italy. The Latin word *ratio*, while I didn't do the philology [i.e., the study of language in written historical sources located between history and linguistics] of it, I understand it to mean double-entry book-keeping from its use at the time. It also bears a relationship to rationality. Ratio becomes rationality: that's the way it is born. The double-entry ratio-based calculation [e.g., quantifiable equivalencies], that's what makes rationality possible.

To calculate one's advantage, you have to remove yourself. Remove yourself from the land and the community. The people and the land are no longer one thing, belonging to each other sharing and living together. If you go into Indigenous communities, people usually do not calculate like that, even among small farmers in India. That's where I first encountered it. In the Indigenous communities here (Peruvian High Amazon), it's glaring: no one calculates in these ways. You do things to live and to regenerate life.

It goes together with my argument about the *Burning Times* when they exterminated hylozoism, hermeticism, alchemy, and various other ways-of-knowing and -being. The peasants, the so-called witches, and the so-called magicians were very close with each other and constantly sharing knowledge through contact with each other. Therefore, calculating the encloser's advantage has political, economic, social, and spiritual consequences. They are all entangled.

The mutually exclusive Nature/Culture dichotomy has its birth in removing the people from the land, and using the land as an economic asset. Land becomes a thing, an object on or through which, depending what you do with it, one generates an economic profit based on and derived through calculations of an advantage accruing to a single (male) owner and along with it excluding the other members of the land, the non-humans, the other humans in the community and the other-than-humans, namely the earth beings. Advantage was the term that they used at the time. This is the beginning of the end.

M. *HIGGINS*: Could you speak further to the consequences of the enclo-sure movement for those who were unwillingly entangled? What did this mean for those who might not have had the privilege of being an encloser?

F. *APFFEL-MARGLIN*: The herbalists, like Indigenous peoples and their relationship to Land, depended on being able to live in the forest and knowing the forest very well, its animals and plants, in order to know how to heal people. However, once the commons were enclosed, they were no longer allowed access. They could not continue their prac-tice. The church backed the powerful by calling the commoner women witches, and by calling them heretics. At the end of the fifteenth century, in 1484, the Pope in Rome declared witches to be heretics. If they could be proven to be witches, they were to be burned. This was the Burning Times. Witches, however, had always existed, why declare them heretics at the end of the fifteenth century when they had always been around? The short answer is that they were no longer tolerated. I argue that the main reason why the inquisition set on them at that particular time in history is to be correlated with the enclosure movement.

At the time the process of enclosing the commons was going on hylozoist experimenters and Cartesian-Boylian experimenters were having debates. When discussing enclosures, you had those who were criticizing and those who were defending this practice. Critics and defenders could be from the same class; it depended on one's politics. But not unlike today, those who defended the enclosure tended to be landowners: wealthy merchants and aristocrats. While there was this tendency, you cannot say 'always' here because people did not always act according to their class.

It is not unlike what's going on today in the Global South in general. Recently, in Bagua, Peru, a similar debate was public. It was amazing to see that everyone who was criticizing the enclosure of land professed that everyone has a right to the commons. That is the meaning of a commons, belonging to everyone. The vulnerable, the old, the widows, those who are unlucky, and etc. are assured of their daily bread; they can live because they have access to the commons. The calculation of the encloser's advantage begins with the enclosure of land and thereby making land an exclusive property of the encloser.

I'm now writing for an encyclopedia of Hinduism on how food was seen in pre-colonial times in India. Everyone has a right to life by virtue of being born: that is how the polity [i.e., local government] was organized, and how people behaved. If you are alive, you deserve to eat. What emerges through the enclosure of the commons is that you eat only if you work earning wages with which to purchase food which implies the associated need of creating labour as a commodity. You have to sell your labour because it is the only way you can access food. Selling one's labour requires a previous transformation which consists in owning your individual body whose labour power is something that you can sell because you are its exclusive owner. Before the enclosure movement, and in today's Indigenous and small peasant societies, your labour is not something for which you get paid. Instead, your neighbours and your relatives get together everyday and work on your land and tomorrow you're going to work on their land—that kind of principle. You can extend it to other and bigger things, but that is how people do things; labour is not a commodity or something you sell.

> M. HIGGINS: The creation of the body as labour speaks to the operation of what Karen Barad (2007) refers to as a "metaphysics of individualism". In other words, a body, be it human, other-than-human, or more-than-human that is separate and separable. This, as you state, disrupts relationships of community between human, non-human and other-than-human. Could you elaborate more on the notion of closure and boundedness of the human body and how it ties into this separation and separability?
>
> F. APFFEL-MARGLIN: The non-porous body bounded by the skin is necessary for the emergence of labour as a commodity because you have to own your body, its power and its force, in order to sell it. The body as a biological bounded separate entity that you own so that you can sell it implies a capitalist market economy. Labour is, according to Karl Polanyi, one of the three forces of production [i.e., land, labour, and capital], without which you would not have capitalism.

If you insist on the biological boundedness of bodies and you teach that to kids, you implicitly reproduce the capitalist mode of production. This entails a devalorizing of the commons and what I call the cosmo-centric economy, the exchange of gifts within the human, non-human and other-than-human communities and everything that these represent.

Accordingly, it is a statement that Indigenous and rural peasant people are backwards and superstitious, while simultaneously opening a wide space for the entrance of a profit-motivated market economy. This closes the door to non-modern ways of life in which the commons are central. Not only commons as land but its web of interconnected commons: food as commons (i.e., one person starves, everyone starves; one person eats, everybody eats), water as commons, air as commons, knowledge as a commons.

> *M.HIGGINS*: In your book, you speak of this with respect to the spirit of the gift and communal labour. It is not the measuring or quantification of labour that is important but rather it is the act of coming together and working towards common goals that is important.
>
> *F. APFFEL-MARGLIN*: Exactly, and you can extend that. The labour is not only to cultivate the neighbour's field or your relative's field, but it is also for festivals. In medieval Europe, all the cathedrals were built that way. All the non-modern great works, like the pyramids in Egypt and in meso-America, they were built in that way. It was not paid labour because there was no labour as a commodity; you do it for the community, not only of humans but also for the non-humans and other-than-humans.

Accordingly, there are important ties between labour as commodity and land as commodity. Through the enclosure of land, land emerges as a commodity. Becoming private property, you can buy and sell land because it belongs to somebody. It is the same thing with labour as a commodity. It makes the body a form of property. You own your body; and because you own your body, you can sell your labour power on the market.

It's a profound transformation. However, it took a long time to change the habits of people, to make people develop the habits necessary to create labour as a commodity within these new norms. Nonetheless, because labour as a commodity is an abstraction, this profoundly transformed the sense of what a person is and can be.

Schooling plays an important and ongoing role with this. It was and is through schooling, especially early nineteenth century and the end of the eighteenth century during the period when schooling became obligatory, that you learn the disciplines necessary to sell your labour power. In many Indigenous communities, you typically did not or do not do this. The transitions that occurred during the modernization of Western Europe continue to be relevant today, even if these transitions happened in other parts of the world at another time.

M.HIGGINS: This is certainly a double(d) form of disciplining bodies within schools. Students are not only learning the disciplines (e.g., sciences, arts), but also learning as a form of discipline in and of itself.

F. APFFEL-MARGLIN: Of course, through many bodily cues: the bells which reinforce punctuality and a disciplining of time, bodies sitting at a desk in a row of desks for so many hours engaged in mental focus, all of these things and more.

M.HIGGINS: Foucault's *Discipline and Punish* does a terrific job of unpacking the normative organization of bodies within systems of schooling and the modernist values at play there.

Directly related to this is the question of Boyle's laboratory. While Descartes is often attributed dualistic thought (i.e., cogito ergo sum), you state that it was Boyle who operationalized it through his "technologies" of the scientific laboratory. What did these entail and how did this work towards the operationalization of the Nature/Culture binary?

F. APFFEL-MARGLIN: I have to preface this by stating that Boyle used to be an alchemist, which is a form of hylozoism. But, in the transition to the laboratory, he shifted from a hylozoist practice to a dualist one. The work of Elizabeth Potter is key here because she shows that he was a big landowner. He was politically motivated by the possibility of losing his land. That's the beauty of this kind of work. It shows that the scientific method and the experimental scientific method had everything to do with politics, economics, and religions of the time and place.

Getting back to your question, this is pure Shapin and Schaffer and the work they do in their book *Leviathan and the Air Pump*. In their work, they identify three technologies which I mentioned earlier. One was the creation of the laboratory. Prior to the scientific revolution, the hylozoist or hermeticist (known under the more general term of "occult philosopher") worked in private in a secret cabinet. There he did all sorts of alchemical experiments, as well as other kinds of things. He would use a variety of instruments to conduct experimentations. However, a crucial shift was when the lab became a public space. Why did it have to be a public space? For several reasons; one of which being that by making it a public space you are stating that you are not an occult philosopher. Making the laboratory a public space is equivalent to declaring that "I'm not an Occult Philosopher". While it was still a choice to be a hylozoist at the time since there were active debates at the time between hylozoists and mechanicists, the hylozoists were still being persecuted by the inquisitions of both Protestants and Catholics. Second, it had to be public in order to establish what Shapin and Schaffer call the social technology.

This second technology, the social technology, was utilized to establish the facts, which was called the "matter of fact" or the scientific fact, within the laboratory. The model for this system being the court,[16] and the court being a system of witnesses, facts would be established through witnessing. To do this, you had to have witnesses to the experiment and have a discussion about what was witnessed. Only when there was a consensus of what they had seen could the matter of fact be established. To achieve this what was needed was a public space with several witnesses.

Key to the act of witnessing is what Boyle called the "modest witness". The modesty, what did it entail or mean? It meant that the witnesses had to be reliable people who would not let their personal preferences, personal desires, personal biases influence what they saw, and what they would say they saw. That is what excluded women because the belief at the time was that the modesty of a woman was of the body, that she did not have modesty of the mind.

The third technology is a literary technology. It was for those who could not be there, present to witness the experiment whether they lived in some other country or simply could not come. They had to develop a technology that could describe what happened without interjecting any opinion. This literary technology came to be known as the objective style of writing. "It has happened", that type of linguistic construction. It was so that those who were not present could have what came to be called a precise and objective account of what happened. The remaining technologies had to do with the distribution of these texts. So those are the three technologies.

Third Cut: The Enclosure, the Double-Sided Ledger, the Laboratory, and the Multicultural Science Education Debate

A notion that persistently resonates as the multicultural science education debate is diffracted with this segmented cut of Dr. Apffel-Marglin's interview is that, as Aikenhead (2006) reminds us, "natural philosophy [is] the handmaiden of technology" (p. 11). Scientific technologies are never simply the applied form of science knowledge: they always come to co-constitute one another. Further, technologies are never simply neutral nor passive cultural surfaces, they are both products and producers of cultural values.

Expanding upon the lines of thinking followed within the second cut, not only is the operationalization of Cartesianism an enactment that is distributed among a series of agents, but it is also distributed among a series of increasingly "common" technologies which precede, live alongside, and succeed Boyle, Netwon, and Descartes. If we think diffractively with technologies (and not only scientific ones concerned with generating or applying knowledge *about* nature) then: Nature, metaphysics, as well as how they are understood and enacted through knowledge-practices always come to bear. Social, political, and economic technologies come to be ontological enactments; as well as epistemic ones in which social, political, and economic dimensions come to coalesce without ever fully achieving sameness (see Barad, 2007; Kirby, 2011). The technologies of the enclosure, the double-sided ledger, as well as the laboratory are all supported by particular ways-of-knowing-nature such that they are produced by and (re)produce a "metaphysics of individualism" (Barad, 2007): knowing-in-being premised upon matter and meaning being separable and individuated through mutual exclusivity. Such ontological cuts are never neutral, but are always already entangled with culture and ethics; even if the enactment is (re)producing epistemic, ontological, and ethical separation, it is nonetheless placed within *a* set of relations, even if the relationships go unaccounted for and unaccountable within said framework.

This includes, of particular significance to the multicultural science education debate, the very practices of the laboratory.[17] The practices of knowing nature within WMS are often premised upon scientists operating from a culturally unbiased, neutral, and detached position (see Aikenhead & Michell, 2011; Aikenhead & Ogawa, 2007; Siegel, 2001), and are supported by social and literary technologies which assist in the production of such (e.g., the laboratory as public space, objective writing style, and modest witness). However, as Dr. Frédérique Apffel-Marglin states, "the scientific method had everything to do with politics, economics, and religions of the time and place". As mentioned herein, Thomas Boyle, a key figure in the development of the laboratory as a technology, was also a landowner. In turn, his politics outside the laboratory came to shape his politics inside the laboratory even if the laboratory were to be promoted as an *apolitical* space; he would implicitly and explicitly be bringing in and enacting the technologies of the double-sided ledger and the enclosure within the laboratory whether they could be or were accounted for. Note that one should not confuse the practices of the laboratory with those of the enclosure or the double-sided ledger, but rather, thinking diffractively (or even with prismatic dispersal) speaks to the ways in which these practices intra-act with one another in non-negligible and dialogical manners: creating something that is not dialectically mirroring one or the other through sameness but creating a complex hybrid practice whose historical and agential constitution often goes unaccounted for.

Recall that, as mentioned earlier within the chapter, attempting to make sense of and act with TEK and/or IWLN through a Eurocentric-Cartesian framework almost always results in the perception of these knowledge-practices as lesser alternatives to WMS. Thinking with Frédérique here expands upon that notion: a Eurocentric-Cartesianism framework does not only come to devalue other ways-of-knowing-nature. When considering its operationalization as well as its entangled practices, it also comes to disrupt their (re)generation. Dr. Apffel-Marglin makes the case above that the entangled practices of the enclosure and the double-sided ledger come to produce destabilizing patterns within TEK systems within Western Europe at the turn of modernity, patterns which would and do continue through the imperial cum capitalist project of (neo-)colonialism.[18] First, through the practice of the double-sided ledger, land can be assessed as through a reductionistic logics of equivalence and sameness through quantifiability. In the short physical space between one side of the ledger and the other, the human, other-than-human, and more-than-human ecology of relationships is reduced to both a spatial area and a fiscal value which are given the status of equivalency: a veritable trick of smoke and mirrors. Secondly, by making the commons private through the enclosure, the unaccounted for relationships which shape entire lived knowledge systems are disrupted. Knowledge holders no longer have the same type of access to the landscapes with and from which they learn (see Cajete, 1994); furthermore, the relational and ethical knowledge-practices through which these ecologies are (re)generated are not sustained. Lastly, when the enclosure and the double-sided ledger are metaphorically inflected through human bodies that live(d) in and with the commons, they too become commodities. The biological boundedness of the human body and the economic logics of sameness (re)shape labour as commodity rather than something that one gives and receives unequally but rather equitably within a reciprocal gift-based economy: "You own your body; and because you own your body, you can sell your labour power on the market". As these values came to permeate the social, political, and economic facets of Western modern life, it "profoundly transformed the sense of what a person is and can be" by (re)shaping the norms of subjectification.

While these social, cultural, political, and economic values slip their way into scientific practice unnoticed and unnoticeably due to its framings as *apolitical*, the *apolitical stance* of the modest witness is also produced by the social, cultural, and political climate of the time, as explored within the next section.

Fourth Cut: Dr. Frédérique Apffel-Marglin on the Modest Witness, When One Truth Becomes Two, and the Thirty Years' War

M.HIGGINS: Regarding the modest witness and the norms that shape who can be a witness, you've mentioned that women were excluded on the basis of the gendered understandings of modesty at the time. Were there other people who were excluded and on what basis?

F. APFFEL-MARGLIN: Yes, there are absolutely others. In the laboratory's beginning, to be in the lab, you had to be independently wealthy to participate. This was for two reasons. First, you had to be wealthy because it was personally expensive: you had to have the time, as well as have education that entailed knowing mathematics and a few other things. Only the wealthy people could be educated; they were all men. As such, the "modest witness" excludes not only all women but also a lot of men such as the lower classes and the uneducated ones.

Secondly, it was systematically expensive. This was because the university was church business. One of the fundamental characteristics of the lab, as Boyle set it up, was that within that space it was absolutely forbidden to talk about religion and politics. Now why do you think that was? Why do you think he made that rule if we're talking about the seventeenth century and what was going on in the seventeenth century?

M.HIGGINS: There were still witch-hunts going on, right? There would be politico-religious consequences for those who might be hylozoists and hermeticists, which go far beyond attempting to be "modest" in one's assumptions and biases, correct?

F. APFFEL-MARGLIN: Yes, and what else?

M. *HIGGINS*: There also would have been the politics of land ownership I gather, but I'm guessing that there is something even more critical than this.

F. *APFFEL-MARGLIN*: Wars of religion. They are also key in understanding the birth and operationalization of the Boyle-Newtonian-Cartesian framework. It started with the Reformation in 1510 with Martin Luther. Protestants immediately had a following amongst the merchant and certain aristocratic classes, and even among some royal families. It divided countries between regions that were Protestant and regions that were Catholic.

The Catholic Church is a centralized church which has a pyramidal hierarchy with the Pope at the top. Because of its pyramidal and international organization, as well as its domination of education at all levels, they could decide what people should and could think and read, as well as what they should not think or read. It still exists today, the means through which a dominant curriculum was established and monopolized: what they call *Librorum Prohibitum*. It tells you what is what, what is true, what is superstitious, and what is false. They also have the power of excommunication, which was very powerful at the time. It is not like today.

The problem then and there was the following. The reformation spread very quickly. Protestants stated "we have the truth; the Catholics are mistaken; what they do is magic". In Western Europe, the *One Truth* became two. Both Catholics and Protestants each said they had the *One Truth*. It was explosive. Indeed, it created mayhem. They killed each other with gusto, brutality, and horror.

In the early seventeenth century, they assassinated Henri IV, King of France [not to be confused with Henry IV, king of England from 1399 to 1413]. He was Protestant but to be king he had to be Catholic. That is what he meant by his famous line: "la France vaut bien une messe". It was well worth going to mass in order to be the king of France. There were 8 bloody civil wars in the sixteenth century, in the 1500s, so he made a very famous law: the *Édit de Nantes* in 1598 [the Edict of Nantes]. It was a very modest proposal to live together without killing each other. In regions in which the majority was Protestant, the Catholic had the right to have their religion albeit not display it publicly. In regions in which the majority was Catholic, it was the reverse. They had the right to practice, but not to display; not rub it in the face of the other. While it was a minimalistic accommodation, it worked. However, there were 10 assassination attempts against Henry IV, and finally the last one worked: he was killed in 1610. His second successor, not the next king but the king after that, Louis XIV eventually threw all the Protestants, known as "huguenots", out of France.

This is a thumbnail outline of something that is, in my opinion, fundamental to understanding the birth of modernity in the West. This is in relation to how it related to Christianity, and the One Truth; as well as its shift to having, all of a sudden, two One Truths. But there can't be two One Truths, right?

What Boyle did with the laboratory, why it worked, and why it became the prevalent approach in that day and age by accruing the most political and economic power, was because it solved something that was unsolvable. This unsolvable issue was the One Truth becoming two truths with each religious organization saying: "we have the One Truth and you are heretics". When echoed, this led to killing each other. When Boyle created the lab and stated that you could not talk about religion within that space, he created the germ or seed of secularization. The lab, and with it all of Nature, became a neutral domain outside of religio-political domains, a new neutral domain upon which the lost certainty to which Europe was addicted could be safely reconstructed.

The lab was the first neutral, or in other words, secular place in Europe. Before that, you could not be a secular person; it did not exist in Western Europe or anywhere else. That is the birth of religion as a category. Before this, spirituality was interwoven into the fabric of life: it was one seamless way of life. That is the way it was. So this was the first time that you had a space where you could not talk about what pervades life or has pervaded life for everyone.

Speaking of religion was simply forbidden. And why was it forbidden? It was very simple: if you talked about religion, you might kill each other. The reason why politics could not be discussed within the laboratory is because politics and religion went hand in hand. They could not be separated. The Protestant had their political leaders and the Catholics had the king. Politics and religion were one and the same.

What Boyle was creating was a separation for a very pragmatic reason. He was creating a space where you could argue in a friendly, non-conflicting, and accordingly non-lethal way about other matters. What are the other matters? What you have witnessed within the laboratory. Because the model is the court, there is one more reason why he used the judiciary model of argument. Within the laboratory, everyone had a right to participate if you did not touch upon topics of politics and/or religion. Because they had a right to opine, the consensus that emerged around what happened through experimentation would be strong.

At the time, his greatest competitor, Thomas Hobbes was arguing for a new model in which people would not disagree. This was seen as dogmatic. Boyle was trying to get away from the dogmatism of the Catholic Church and the Protestants. He was attempting to get away from the idea that people have no choice because of dogmatic statements "this is how it is" whether Protestant or Catholic. He was moving away from that while strengthening truth claims through encouraging disagreement within very strict boundaries. When you allow people to disagree within bounds, the consensus is strong without being dogmatic. It became a very powerful solution that addressed and solved the issues of the times within the privileged space of the laboratory.

> M. HIGGINS: If I understand this correctly, this would also have conse-quences for what was intended by modesty at the time. As you speak to in your book *Subversive Spiritualities,* secular modesty put forward by Boyle in the laboratory is very different than the religious modesty put forth by the Christian church before it. Religious modesty entailed not being a spiritual usurper, not having too many belongings in this world, or to be a martyr of sorts. Secular modesty, in contrast, largely meant being "modest" in one's beliefs by abstaining from talking about religion or politics. However, if the secular modest person also had to be wealthy as an entry point to participation, this circumvents ethical questionings of wealth altogether.
>
> F. APFFEL-MARGLIN: You simply did not discuss those things in the lab.
>
> M. HIGGINS: The process of supersessionism, or the ways in which WMS supersedes Indigenous and other ways-of-knowing-Nature, could be said to have an almost religious quality to it. However, this is not because of ties to the church, but as you identify, this largely has to do with what you identify as an addiction to certainty and objectvity/subjectivity as one of the ways in which the Nature/Culture binary plays out. Could you elaborate upon this process?
>
> F. APFFEL-MARGLIN: As I mentioned earlier, the social-political-religious landscape of Western Europe in the seventeenth century was one of raging religious wars and the Burning Times which came to an end in the second half of the seventeenth century. As I like to say, irrev-erently, Europe was addicted to the *One Truth* because it had lived with it for 15 centuries or more with Christianity. Europe required certainty to restore law and order, as Stephen Toulmin has shown in his book *Cosmopolis.* A new neutral—and very separate—domain of inquiry,

Nature, and with it a new epistemology, were required. That new epistemology needed to have its life in a new neutral space, that of the laboratory. While there are exceptions, this is still true today. The *One Truth* is an addiction.

So in order to resolve the impossibility of two One Truths in Europe, a circumscribed space that was neutral vis-à-vis both the Protestants and the Catholics was invented. Within this space, a certainty that was not religious was created. It was desperately needed and it was a very circumscribed space. Hence, the laboratory took off.

At the time, religious wars were international. The Thirty Years' War, which lasted from 1618 to 1648, was the internationalization of what I described in France earlier. Similar things were also happening in England and Germany. The *Treaty of Westphalia* was signed to bring an end to the Thirty Years' War. This process creates, invents, and necessitates the nation state.

Immediately, the nation state needed a science to function and to govern. Here, I draw from the work of James Scott and his book *Thinking like a State* addressing why the nation state cannot function without science. I cannot simplify that story for you, but I will attempt to give a quick and partial view. In a nutshell, the state has to manage two things. First, it has to manage conflict, which was religion-based. Secondly, the state had to generate revenue. To get revenue, you have to systematize and quantify. It is like what the encloser would do when calculating his advantage but now this logic is playing out at the national level. This, of course, is a partial view.

> *M. HIGGINS*: This addiction to certainty certainly has persisted strongly within Western modern(ist) traditions—to which it has had its own internal challenges (e.g., the crisis of representation in the 1990's). However, it is not uniquely internal to Western traditions as the addiction to certainty is being exported and culturally transposed...
>
> *F. APFFEL-MARGLIN*: Through the hegemony of modern education worldwide!
>
> *M. HIGGINS*: In your book *Subversive Spiritualities*, you speak to the addiction to certainty and how WMS (i.e., agricultural models in particular) supersedes Indigenous local, longstanding, and intergenerational models due to the ways in which it is framed or enacted as holding a high degree of certainty. Could you speak to the ways in which certainty was and continues to be constructed?

F. APFFEL-MARGLIN: Here is the way I've come to see it. The founding myth of Modern science is the trial of Galileo vs. the Catholic Church. Here you have rationality and belief. Of course, this myth was carried forward into and leads to the progressive secularization of society through the laboratory later through the Nation State. This created certainty on a whole new basis: the experimental method, deduction, rationality, logical thinking, objective writing, and all; the whole package.

What I see when I take that long view is that religious certainty has been transposed into a new key, to use a musical metaphor. Scientific certainty is the same thing but, of course, the new key is crucially different. The key is crucially different but the tune is the same. It is similar in its absoluteness in the sense that its method is the only valid one, its claim to the One Truth: Science gives you the One Truth but in a new language and on a new basis. In the process it had to disenchant nature, as Weber put it, and with that move transmute all other ways of knowing-being as "pre-scientific" in need of progress. Because Western Europe was addicted to certainty and the One Truth, it had recreated it in a new secular key with Modern science.

That is why the "universal" given-ness of Nature is so pernicious. Scientific rational superiority became a deliberate part of an educational campaign to inculcate in the European youth, and to shape perceptions about those who Europe colonized. Colonialism is justified through statements such as "this is right because we are enlightening these unenlightened people; we are bringing them not just civilization, we are bringing them knowledge, True knowledge". Of course, such is predicated on seeing Nature as universal.

As Karen Barad says, you can never separate or disentangle the discursive from the material. This disentanglement, however, is key to the formulation of certainty. The issue for me is that it is very difficult to have people understand this because it goes against the grain of everything they likely have been taught.

M. HIGGINS: Dualistic thought is not something that can simply be turned on or off.

F. APFFEL-MARGLIN: Exactly. You can say it conceptually in all its implications. However, what happens in practice is otherwise and extremely difficult.

M. HIGGINS: Which dualisms are we reproducing as we try to work against others?

F. APFFEL-MARGLIN: It is always a process; an unfinished work.

Fourth Cut: The Modest Witness, When One Truth Becomes Two, the Thirty Years' War, and the Multicultural Science Education Debate

Reading the insights provided by Dr. Apffel-Marglin in this this vignette through those of the multicultural science education debate begins to reveal the ways in which the *apolitical* stance of cultural quasi-neutrality explored within chapter is a process which comes to mask the flow of common-place cultural power and politics into the laboratory, but also the ways in which this stance is produced by the social, economic, and political forces of the time. What might appear as an individualistic choice here could be (re)thought as having and being had by common sense.

This stance, as Frédérique enunciates it, is one of being a modest witness. The modest witness, as one of the multiple technologies that come to constitute laboratory practice requires that the individual whom is witnessing be educated, as well as "modest of mind". Here, we come to see the Foucaultian circularity of knowledge comes to rear its head again. First, recall that the Nature/Culture exclusive dichotomy, with/in which WMS is entangled, comes to negatively position Man's (human) Other as "closer to nature". Accordingly, women, racialized bodies, Indigenous peoples, amongst others, as being perceived as possessing "modesty of the body" rather than that of the mind,[19] were generally not included in the very processes which would implicitly (and eventually explicitly) come to position them as having lesser ability to bear witness to natural phenomena. By focusing on the Other, this has the further effect of normalizing and naturalizing the ways in which science is dominion of Man:

> In order for the modesty... to be visible, the man – the witness whose accounts mirror reality – must be invisible, that is, an inhabitant of the potent "unmarked category," which is constructed by the extraordinary conventions of self-invisibility... This self-invisibility is the specifically modern, European, masculine, scientific form of the virtue of modesty. (Haraway, 1997, p. 23)

Secondly, as Dr. Apffel-Marglin states, one must also be educated to participate. As this required access to wealth, those of lower class would also be excluded (see also Haraway, 1997; Shapin & Schaffer, 1985). While one could say that that was there-then and this is here-now, it is nonetheless important to consider how science and science education's historicity continues to shape the here-now and the there-thens to-come as we move forward. While science education is increasingly perceived as being enacted as a practice for *all* students (see Aikenhead & Elliot, 2010), it is nonetheless important to continue considering how the *all* in such statements are articulated, especially when science education scholars continue referring to WMS as *white male science* (Pomeroy, 1994; see also McKinley, 2000; Sammel, 2009).

Accordingly, while the practices of *who* gets to do science have explicitly changed, many of the implicit meanings remain when we consider the ways in which the *how* of science were and continue to be framed. As Frédérique makes explicit, the practice of being a modest witness cannot be separated from the socio-political context of the time in Western Europe: particularly from what was happening within the sphere of the Church. Following the establishment of the Protestant Church, Western Europe was thrown in turmoil: a powerful *addiction to certainty* and competing claims of holding the *One Truth* would see Catholics and Protestants kill each other "with gusto, brutality, and horror" for a thirty year period. Following the Thirty Years' War, there was need of a system of governance that would not bound to either religion: in the creation of the scientific laboratory, we would also see the birth of secularism. As Dr. Apffel-Marglin puts it, this came to shape the very practice of the laboratory in its infancy in profound ways: "it was very simple: if you talked about religion, you might kill each other". However, as mentioned earlier within this chapter, the creation of secularity as mutually exclusive from spirituality had consequences then and now: practices, like TEK and IWLN, that do not see separate spirit and matter are implicitly (and sometimes explicitly) not welcome within the formal spaces of science. This trend, while far less explicit continues to bear on educational spaces today[20]: recall from Chapter 3 that there continue to be many science education scholars, as well as science educators who frame practices other-than-WMS as less valid (e.g., Cobern & Loving, 2008; Siegel, 2001).

In these instances, it can be said that there continue to be traces of what Frédérique refers to as the *One Truth*: "religious certainty has been transposed into a new key, to use a musical metaphor".[21] The aforementioned claims of validity do not come to frame WMS as being more valid in particular moments within the multicultural science education debate, but rather as a general and a "universal" which is a priori to knowing and being. It could be stated that the transposition of power from the Church to the laboratory as instrument of the State did not address or redress Western Europe's addiction for certainty at the time, but rather resituated and differentially produced the ways in which one had access to the One Truth. It is for this reason that Spivak (1976) cautions against the (too) rapid evacuation of a concept, category, or framework: left unaddressed, its problematics often follow us in ghastly and ghostly manners to reassert themselves elsewhere albeit differently (see also Derrida, 1994/2006). The calling into question of the claim of "most valid" is not to deny it, but rather resituate it by asking the questions of when and where it is most productive; few science education scholars deny the pragmatic effectiveness of WMS in making knowledge claims about nature. However, science is indebted to and weighed down by the ways in which the One Truth continue to (re)shape it: science education continues to be a space which dialectically subsumes and sutures over TEK and IWLN as a means of resolving difference and reasserting WMS as a more nuanced version of the One Truth in an ever-diversifying world.

CONCLUSION: POSITING, ACCOUNTING FOR AND BEING ACCOUNTABLE TO AN ONTOLOGY IN SCIENCE EDUCATION

To address the past (and future), to speak with ghosts, is not to entertain or reconstruct some narrative of the way it was, but to respond, to be responsible, to take responsibility for that which we inherit (from the past and the future), for the entangled relationalities of inheritance that "we" *are*, to acknowledge and be responsive to the noncontemporaneity of the present, to put oneself at risk, to risk oneself (which is never one or self), to open oneself up to indeterminacy in moving toward what is to-come. Responsibility is by necessity an asymmetrical relation/doing, an enactment, a matter of différance, of *intra-action*, in which no one/no thing is given in advance or ever remains the same. Only in this ongoing responsibility to the entangled other, without dismissal (without 'enough already!'), is there the possibility of justice-to-come. (Barad, 2010, pp. 264–265, emphasis in original)

To posit Cartesianism as *an* ontology in science education is to inevitably speak with ghosts: that which we inherit from science's "beginnings" haunt us. Because the homework of response-ability begins from where we are, it is not only significant to engage in a creative archaeology dig into the pasts passed over that constitute us, but it is part of our responsibility as it is to "take responsibility for that which we inherit" (Barad, 2010, p. 264): what we inherit is irreducibly part of who we are and who we are becoming as science educators. Repeating a question posed earlier in this chapter, *when pasts are passed over, but still come to constitute the here-now of contemporary practice, what ghosts which continue to haunt science education are being chased away?*

As explored through tinkering with an expert interview with Dr. Apffel-Marglin, Cartesianism cannot simply be reduced to set of belief that are individually held but rather a structure: the scientific revolution operationalized and naturalized Descartes' epistemology through the technological practices of the laboratory, the modest witness, and the objective style of writing. More significantly, these practices cannot and should not be so easily disentangled from socio-material practices of Western Modern Europe which lead to and surrounded its operationalization (e.g., the enclosure movement, the Thirty Years' War). The *making common* of Cartesianism not only (fore)closes the possible possibilities for responding to Indigenous science to-come, but also the ability to account for and be accountable to the ontology of WMS and what it produces: repeating the mantra of the last chapter, "how reality is understood matters" (Barad, 2007, p. 205).

Significantly, in this contemporary moment in science education, the presented and enacted conceptions of nature-cultures makes possible and palatable the ongoing dispossession and devastation of Indigenous Land, as well as the erasure of Indigenous peoples (see Bang & Marin, 2015). Diffracting Dr. Frédérique Apffel-Marglin's account through the logics of the multicultural science education debate reveals this dispossession and devastation to have been the case from the very "beginning" of Western modernity: Indigenous erasure is not strictly a feature of contemporary science education, but rather part of the inheritance which shapes who we are and can be. The scientific revolution could not be separated from the removal of peoples from Land and (forceful) denegation of their ways-of-knowing-in-being. These logics which would later come to inform ongoing practices of (neo-)colonialism. It bears repeating, the task at hand the task at hand is not to begin a new relation but rather to engage the relation *anew*: WMS and Indigenous science are *always already* in relation.

It is for that reason that responding to Indigenous science to-come is not achieved "through the realization of some existing possibility, but through the iterative reworking of im/possibility" (p. 265): we are working against centuries of history in which Cartesianism and its consequences have been normalized and naturalized. While there is no singular solution to (re)opening the norms of responsiveness towards Indigenous science to-come, positing *an* ontology paves pathways to engage with the im/possibility of being wholly accountable to and ethically responsive by coming-to-perceive the very possibility of ontological otherness, as well as accounting for the ways-of-knowing-in-being which foreclose such possibility. To reiterate from the previous chapter, situating science education ontologically by positing *an* ontology is not about who is right or who is wrong, nor is it about a renewed commitment to relativism. Rather, it is about coming-to-recognize a plurality of possibilities (and problematics), and in turn, it means being accountable to how scientific knowledge is produced, producible, and what it produces in turn within and beyond the science education classroom. For example, thinking with Frédérique Apffel-Marglin allowed for differential considerations in the conceptual baggage that comes with the metaphysics of modernity, not to dismiss or do-away with these ways-of-knowing-in-being but rather come to work with/in the structures that (re)produce and are reproduced by science education in order to (re)open them to Indigenous science to-come. These include the ways in which the home in the homework of response-ability manifests as absent presences: "uncommon" political, religious, economic, and military practices (e.g., the enclosure) that are enfolded and co-constituting "common" scientific technologies and practices (e.g., modest witness) of a non-dissociable past and present.

As the ability to respond to Indigenous science to-come requires an "ongoing responsibility to the entangled other, without dismissal (without 'enough already!')" (Barad, 2010, pp. 264–265), the task of de/colonizing our inheritances within science education is never fully over.

Notes

1. As Spivak(1976) reminds us, the language we possess also possesses us: to (too) simply abandon a particular does not mean that it has abandoned us. If the thoughts we have and by which we are had are part of the problem, it is not so simple to think about how we think without using the thing with which we think. This is to say that science education is irreducibly related to Cartesianism and that attempts to move beyond this ontology and its consequences must simultaneously work to understand and address it, as well as its entangled productions.

2. The scare quotes here are intentional as origins are forever deferred and differing (see Derrida, 1976; Kirby, 2011). Further, as signalled in the previous chapter, the fetish for origins is irreducibly bound to (neo-)colonial ways-of-knowing-in-being (see TallBear, 2013). Dr. Apffel-Marglin makes explicitly clear that the "birth of modernity" is not *an* "origin" in the conventional sense but rather a partial and contingent origin: *a* historical moment in which modernist thinking is operationalized (see Apffel-Marglin, 2011).

3. Along similar lines, Sara Ahmed (2006) succinctly states:

   > The word inheritance includes two meanings: to receive and to possess. In a way, we convert what we receive into possessions, a conversion that often "hides" the conditions of having received, as if the possession is "already there." (p. 125)

   The task of inheritance is to also consider that which we have received which no longer appears as such, specifically that which we receive from (or, more precisely, at the expense of) the other.

4. Once again, the mantra of beginning some-where and some-time is a reminder that we are always already within the question of Indigenous ways-of-knowing-in-being within science education. In turn, there is a multiplicity of locations (both temporal and geographical) from which we can approach the question productively.

   Furthermore, this specific reminder, returning to the introduction of the very first chapter, also serves to hint at the ways in which the present is not only *here-now* but is always entangled with/in various *there-thens*: that these *elsewheres* and *elsewhens* are not strictly cognitive objects to be known (i.e., historical and geographical facts) but are ontologically co-constitutive of our contemporary present, here and now. As Karen Barad (2010) states, "to address the past (and future), to speak with ghosts, is not to entertain or reconstruct some narrative of the way it was, to respond, to be responsible, to take responsibility for that which we inherit ..." (Barad, 2010, p. 264, emphasis in original). This bears particular significance as we take up and are called to take seriously pasts already passed over which continue to haunt science education's present, as if a spectre or a ghost, reaching out to future(s)-to-come. This is to say, to take the task of attending to the absent presences within the history of science (or at least those that are usually absent within the *common sensical* narrative of science): as they also have bearing on what science education was, is, and is becoming. As the absent present relationships between science and Indigeneity is always already, the work of attending to the inheritances which haunt science education is not without significance: they are "entangled relationalities of inheritance that 'we' *are*" (Barad, 2010, p. 264, emphasis in original).

This work of inheritance as responsibility at the ontological turn points towards Derrida's (1994/2006) concept of *hauntology*, a (near-) homonym to ontology that is meant to defer and differ ontology's conventional "discourse on the Being of beings" (p. 63) to embrace that which exceeds it: the spectral. Significantly, *hauntings* are not simply concepts: Barad (2010) provides empirical evidence via quantum field theory. Hauntings are indeterminacies in the ways in which space-time-matter materializes: a particle can be in a state of superposition such that it is simultaneously here, there, now, and then (including a *there-then* in the future tense, see Barad, 2010for further). Importantly, hauntology troubles the possibility of ontology being a singular (Cartesian) affair:

> To haunt does not mean to be present, and it is necessary to introduce haunting into the very construction of a concept. Of every concept, beginning with the concept of being and time. That is what we would be calling a hauntology. Ontology opposes it only in a movement of exorcism. Ontology is a conjuration. (Derrida, 1994/2006, p. 202)

Barad further suggests that: "every concept is haunted by its mutually constituted excluded other". (Barad, 2010, p. 253). Where ontology often comes to stand in for epistemic realism, as explored within the previous chapter, it becomes useful to think with and through the notion of ontology as conjuration or exorcism. Throughout *Specters of Marx*, Derrida (1994/2006) invokes multiple meanings to the concept of conjuring. It is at once the oath solemnly sworn in secrecy to struggle against a greater power, an incantation or summoning of a spirit, as well as an exorcism. Importantly, an "effective exorcism pretends to declare the death only in order to put to death" (Derrida, 1994/2006, p. 59). When pasts are passed over, but still come to constitute the here-now of contemporary practice, it is worth asking: what ghosts might have science education been chasing away?

5. Further, it became an opportunity to experiment with more creative ways of representing, and more specifically storying, Indigenous place-based education (see Higgins & Madden, 2019).

6. As explored in Higgins, Madden, Bérard, Lenz Kothe, and Nordstrom (2017), methodological research "design" often prescriptively and prohibitively continues to act as a signifier that sutures over the signified processes of designing and doing research. "Design" often signals *a* method that is exists a priori to research, "a stand-alone, instrumental set of research practices" (St. Pierre, 2011b, p. 52). This also includes and encompasses all of its conceptual apparatus (such as, in this case, "expert" and "expert knowledge").

Rather, the practice of methodological *de/sign* differs and defers that which *design* comes to signify: design as pre-existing, design as separate or separable from other aspects of research, and design as a means to achieve and justify the ends (see also Derrida, 1976; Spivak, 1976). Because methodological fabric is also a fabrication—a performative and non-separable enactment of the interconnected space between theory, practice, and ethics—methodological design is always already open to be deconstructed and re(con)figured.This is of particular significance in spaces of de/colonizing education. Highlighting the ways in which the disciplines *discipline* what counts as knowledge and, more to the point, knowledge production processes, Smith et al. (2016) ask, "are methodologies simply new technologies of cultural assimilation?" (p. 133). To (re)open the conceptual apparatuses of methodology allows for the possibility of addressing lingering colonial referents within science education (see Higgins & Kim, 2019).

7. Such a tinkering can be understood both as tinkering *within* the narrative presented throughout her book, but also tinkering *with* it in relation to the multicultural science education debate.

8. Butler's criticism of particular cultural uses of "natural facts" still stands today as "natural facts are always informed by cultural bias" (Butler in Kirby, 2011, p. 94). As this is "one of the most important contributions that scholars such as Butler have made, ... any return to the question of Nature will need to accommodate such insights rather than put them aside" (p. 94). Nonetheless, the opening of other possibilities beyond the natural/cultural divide allowed by this minor concession sees productive uses in her later work (e.g., Butler, 2010); with an always present cautionary note against the "seductive slide that conflates representation, models, and signs that substitute for material objects, with the objects themselves" (Butler in Kirby, 2011, p. 74).

9. This can be read as a form of strategically occupying an essentializing space (see Spivak, 1993/2009). However, Spivak (1993/2009) reminds, "the strategic use of essentialism can turn into an alibi for proselytizing academic essentialisms" (p. 4) without the persistent critique of that which is being essentialized throughout, "even when it seems that to remind oneself of it is counterproductive" (p. 4). Where Frédérique's interview challenges throughout the production of the academic subject of knowledge (i.e., expert), the strategic use essentialism remains nonetheless fraught, being a *pharmakon:* at once panacea and poison (see Spivak, 1976).

10. See Chapter 7 for a lengthier discussion on the ways in which the Derridean statement that *il n'y a pas de hors-texte* is a complex one within science and science education spaces. Importantly, there not being an outside of Culture does not preclude Nature unless with subscribe to a mutual exclusivity to two (see Kirby, 2011). Rather, it invites a consideration of the ways in which *nature* is not only within the realm of Culture at the individual cognitive level and the social discursive level (via the concepts we hold that also hold us, but also that nature is rarely "pure" and beyond the effects of culture (e.g., anthropogenic landscapes) or unable to produce cultural meaning. This latter is of deep significance to (re)opening science education to Indigenous science to-come (e.g., other-than-humans as teachers).

11. It is important to distinguish between *a* Man (i.e., *a* Western European male), *the* Man, (i.e., the shadowy figure in Marxist nightmares which appears to control nearly all from behind the scenes), and *Man* as articulated here. The first two emerge from use of the representational logic of the mirror such that they are representative of *an* individual, whether real or imagined, who is a separate or separable agent (i.e., what Nietzsche refers to as the "individual of will") benefitting from the power systems articulated. The latter term speaks to the complex and capillary circulation of ways-of-knowing-in-being that uphold and operationalize systems that center *Man* and Western modernity (e.g., Eurocentrism, anthropocentrism, patriarchy).

12. Diffusion can be understood as a scientific phenomena and concept which describes the movement of a substance from an area of high concentration to an area of low concentration. Importantly, where a *diffusionist model* (e.g., Blaut, 1993) employs the concept of diffusion productively to describe the colonizing movement of Western thought outwards of Europe, it is also important to not conflate the *passive* qualities of the scientific phenomena and the ways in which Eurocentrism operate *actively*: ongoing (neo-)colonial practices are not experienced as passive "transmission" of Euro-Western values, beliefs, and traditions. As explored in the

previous chapter (i.e., the ways in which common sense is *made common*), it is important to recognize that Eurocentrism is active, forceful, and violent (both literally and otherwise); it is also agentically distributed, albeit differentially: Eurocentrism is a "consciousness in which *all* of us have been marinated" (Battiste, 2005, p. 124, emphasis mine).

13. As Colebrook (2008) suggests, Butler's (1990) *Gender Trouble* is often (too simply) read as anti-ontological as it resituates that which is problematically enacted as biological determinism (i.e., sex) into the realm of discourse (i.e., through gender performativity). Taking a cue from Foucaultian critique (see Chapter 4), it is fair to state that Butler does not wholly jettison ontology altogether, but rather ontology *like that* (i.e., a Cartesian ontology in which matter precedes meaning). In turn, Colebrook's (2008) reading of Butler suggests that materiality (i.e., ontology) emerges as co-constituted by discourse (i.e., epistemology), subverting dominant epistemological and ontological constructions and enactments rather than negating them.

14. During our interviews, Frédérique Apffel-Marglin shared the following regarding what Newton had to forego to maintain epistemic power and privilege within his geo-temporal contexts:

> It was just delightful for me to learn of that, by reading Isabelle Stengers' (2001) little play about Newton titled *La Guerre des Sciences aura-t-elle lieu?*, that one of Newton's descendants needed money in 1930 and found in Newton's attic unpublished papers, manuscripts, and other works. He auctioned them to make money. So it was then bought and studied. This work was spiritual in tone: gravity was a spiritual power. Of course, he could not go public with this. So it was not to be found out or made public until these papers were sold and people took time to study them.

Deeply intrigued, I began to read Stengers' (2001; Prirogine & Stengers, 1984) work, and deeper into the question of Newton and his complex and complicated relation to science. Elsewhere (Higgins & Tolbert, 2018), writing up nascent leads on the this, I ask:

> ... what might it mean to recast science as a practice that is not oppositionally defined against spirituality and attend to the ways in which spirituality always already been included within WMS? Prigogine and Stengers state that this is the case for one of the most emblematic men of science whose way-of-being-scientific has acted as differential mould for what it means to be a scientist: Isaac Newton. They state that many "did not know the strange

story behind the Newtonian forces! For behind Newton's cautious declaration-'I frame no hypotheses'-concerning the nature of the forces lurked the passion of an alchemist." This came to light last century when one of Newton's descendants, Gerald Wallop, the 9th Earl of Portsmouth needed funds in the midst of a costly divorce and put up Newton's unpublished papers for auction in 1936. Following their purchase and analysis, John Maynard Keynes would go on to state that Newton was "not the first of the age of reason" but "the last of the magicians." Stated otherwise, his approach to science was not devoid of spirituality, but rather the norms of becoming-scientist at the time made it such that he could not openly discuss the multilogical approach that informed and fueled his thinking (as part of the practice of modest witnessing). His unpublished work along these lines would be hidden and stored by generations of family in hopes that a time would arrive in which it might be received: a latent safeguard that he perhaps hoped would allow us to dispel the myth the Isaac Newton was Newtonian. Stengers invites us to consider if the science wars over "what counts" will have had happened if this insight and rupture would allow us to complicate science's ongoing relation to Newton by re-inserting the ways in which the man exceeded the concept. (p. 283)

Here, we have yet another rich example of how science is always already entangled with/in and prismatically producing its own norms of articulation (i.e., what can be said and done, and what cannot). Importantly, if we continue to produce scientists (as well as science educators) in the image of Newton, it bears leveraging his hope for a science-yet-to-come in which logics deemed other might come to bear on the knowledge production process. While we need unsettle our inheritances in science education, some inheritance unsettle themselves if we learn to attend to the ways in which deconstruction always already happens.

15. Furthermore, while Newton's subject position is produced by the *there-then* of the time and place, knowing Newton as scientist cannot be disentangled from the *here-now:* he continues to be, after Stengers (2001), the mould from which scientists are cast.

16. Considering science and science education as entangled with/in legal-juridical modes provides insight into the ongoing practice of dialectic rather than dialogue, as explored within Chapter 3, as scientific subjects are required to dismantle their opposition's arguments before a "court" of their peers (e.g., peer review).

17. Recall that, as explored within Chapter 4, the laboratory is always already the context in which scientific knowledge-practices are produced and enacted, whether the lab be real or imaginary. As van Eicjk and Roth

(2007) suggest, the generation of scientific knowledge requires that the context under observation be transcended and treated as if it were static, singular, and homogenous. In turn, scientific knowledge praxis requires that the context of its application also be treated as such. When considered as a practice which is also translated into science education, it is important to consider the (re)production of the double(d) subject of the laboratory: both Nature as subject of inquiry and scientist (or student-thinking-like-a-scientist) as inquiring subject.

18. Apffel Marglin (2011) states that if colonization was so effective, it was because it was perfected through its internal application within Western Europe before it was exported to Western Europe's Others. Along similar notes, during her talk titled *Four Theses on Posthuman Feminism* at UBC, Rosi Braidotti stated that the Holocaust was "colonialism coming home" (28 January 2015). The point to be made here is that (neo-)colonial systems come to affect us all, even if they do not affect us all in the same way (see Battiste, 2005, 2013b; Donald, 2012; Freire, 1973/2000): they differentially inflect our respective (re)production as de/colonizing subjects.

19. In turn, it is productive to address the ways in which the mind/body binary come to operate within science education when working within and against Cartesianism. For example, recall that in Chapter 3 you were encouraged, as a reader, to also pay attention to the affective ways (e.g., emotive, sensory) in which information registered upon their body. If science is to be framed as a human practice (see Aikenhead, 2006), it is worthwhile and important to consider, account for, and be accountable to the plural and diverse ways in which we are human.

20. While not directly discussing spaces of science education, Marker's (2006) *After the Makah Whale Hunt* is a rich exemplar of the complexities and complications which arise when bringing in Indigenous knowledges and knowledge holders into schools to discuss other-than-human members of a local ecology. As school spaces are secular (through separation of Church and State), but seen as neutral and the norm, the inclusion of IWLN and its interconnectedness of spirit and matter is brought into sharp relief and becomes a site in which Eurocentrism plays out in diverse manners (e.g., excluding knowledge and knowledge holders, only including such lesson if it omits its co-constitutive spiritual dimension).

21. Or, as Stengers (2001) puts it, while history does not repeat itself, it nonetheless repeats.

REFERENCES

Ahmed, S. (2006). *Queer phenomenology: Orientations, objects, others*. Durham, NC: Duke University Press.

Aikenhead, G. S. (2006). *Science education for everyday life*. London, ON: Althouse Press.

Aikenhead, G. S., & Elliot, D. (2010). An emerging decolonizing science education in Canada. *Canadian Journal of Science, Mathematics and Technology Education, 10*(4), 321–338.

Aikenhead, G. S., & Michell, H. (2011). *Bridging cultures: Indigenous and scientific ways of knowing nature*. Toronto, ON: Pearson Canada Inc.

Aikenhead, G. S., & Ogawa, M. (2007). Indigenous knowledge and science revisited. *Cultural Studies of Science Education, 2*(3), 539–591.

Apffel-Marglin, F. (2011). *Subversive spiritualities: How rituals enact the world*. New York, NY: Oxford University Press.

Bang, M., & Marin, A. (2015). Nature–culture constructs in science learning: Human/non-human agency and intentionality. *Journal of Research in Science Teaching, 52*(4), 530–544.

Barad, K. (2000). Reconceiving scientific literacy as agential literacy. In R. Reed & S. Traweek (Eds.), *Doing Science+Culture* (pp. 221–258). New York, NY: Routledge.

Barad, K. (2007). *Meeting the universe halfway: Quantum physics and the entanglement of matter and meaning*. Durham, NC: Duke University Press.

Barad, K. (2010). Quantum entanglements and hauntological relations of inheritance: Dis/continuities, spacetime enfoldings, and justice-to-come. *Derrida Today, 3*(2), 240–268.

Barad, K. (2011). Erasers and erasures: Pinch's unfortunate 'uncertainty principle'. *Social Studies of Science, 41*(3), 443–454.

Battiste, M. (2005). You can't be the global doctor if you're the colonial disease. In P. Tripp & L. J. Muzzin (Eds.), *Teaching as activism* (pp. 121–133). Montreal, QC: Queen's University Press.

Battiste, M. (2013a). *Deconstruction and reconstruction: Roles, responsibilities and implications of a decolonizing framework*. Retrieved from http://www.indigenouseducation.educ.ubc.ca/transformation/indigenous-perspectives/ on 11/22/2012.

Battiste, M. (2013b). *Decolonizing education: Nourishing the learning spirit*. Saskatoon, SK: Purich Publishing.

Blaut, J. (1993). *The colonizer's model of the world: Geographical diffusionism and Eurocentric history*. New York, NY: Guilford Press.

Bogner, A., Littig, B., & Menz, W. (Eds.). (2009). *Interviewing experts*. Basingstoke, UK: Palgrave Macmillan.

Butler, J. (1990). *Gender trouble*. London, UK: Routledge.

Butler, J. (2005). *On giving an account of oneself*. New York, NY: Fordham University Press.

Butler, J. (2010). *Frames of war: When is life grievable?*. London, UK: Verso.

Cajete, G. (1994). *Look to the mountain: An ecology of indigenous education*. Durango, CO: Kivaki Press.

Cajete, G. (2000). *Native science: Natural laws of interdependence*. Santa Fe, NM: Clear Light Books.

Cobern, W. W., & Loving, C. C. (2008). An essay for educators: Epistemological realism really is common sense. *Science & Education, 17*, 425–447.

Colebrook, C. (2008). On not becoming man: The materialist politics of unactualized potential. In S. Alaimo & S. Hekman (Eds.), *Material feminisms* (pp. 52–84). Bloomington, IN: Indiana University Press.

Derrida, J. (1976). *Of grammatology* (G. C. Spivak, Trans.). Baltimore, MD: John Hopkins University Press.

Derrida, J. (1994/2006). *Specters of Marx: The state of the debt, the work of mourning, & the new international* (P. Kamuf, Trans.). New York, NY: Routledge.

Donald, D. (2012). Indigenous Métissage: A decolonizing research sensibility. *International Journal of Qualitative Studies in Education, 25*(5), 533–555.

Foucault, M. (1977). *Discipline and punish: The birth of the prison*. London, UK: Allen Lane.

Foucault, M. (1979). Truth and power. In M. Morris & P. Patton (Eds.), *Power, truth, strategy* (pp. 29–48). Sydney, NSW: Feral Publications.

Freire, P. (1973/2000). *Pedagogy of the oppressed*. New York, NY: Continuum International Publishing Group.

Haraway, D. (1997). *Modest_Witness@Second_Millennium.FemaleMan_Meets_ OncoMouse*. New York, NY: Routledge.

Higgins, M. (2014). De/colonizing pedagogy and pedagogue: Science education through participatory and reflexive videography. *Canadian Journal of Science, Mathematics and Technology Education, 14*(2), 154–171.

Higgins, M., & Kim, E. J. (2019). De/colonizing methodologies in science education: Rebraiding research theory-practice-ethics with indigenous theories and theorists. *Cultural Studies of Science Education, 14*(1), 111–127.

Higgins, M., & Madden, B. (2019). Refiguring presences in Kichwa-Lamista territories: Storying with Indigenous place. In C. Taylor & A. Bayley (Eds.), *Posthumanism and higher education: reimagining pedagogy, practice and research* (pp. 293–312). Basingstoke, UK: Palgrave Macmillan.

Higgins, M., Madden, B., Bérard, M.-F., Lenz Kothe, E., & Nordstrom, S. (2017). De/signing research in education: Patchwork(ing) methodologies with theory. *Educational Studies, 43*(1), 16–39.

Higgins, M., & Tolbert, S. (2018). A syllabus for response-able inheritance in science education. *Parallax, 24*(3), 273–294.

Higgins, M., Wallace, M., & Bazzul, J. (2019). Staying with the trouble in science education. In C. Taylor & A. Bayley (Eds.), *Posthumanism and higher education: reimagining pedagogy, practice and research* (pp. 155–164). Basingstoke, UK: Palgrave Macmillan.

Hodson, D. (1993). In search of a rationale for multicultural science education. *Science Education, 77*(6), 685–711.

Kirby, V. (2011). *Quantum anthropologies: Life at large.* Durham, NC: Duke University Press.

Kirby, V. (2012). Initial conditions. *Differences, 23*(3), 197–205.

Lather, P. (2007). *Getting lost: Feminist efforts toward a double(d) science.* New York, NY: State University of New York.

Latour, B. (1993). *We have never been modern.* Cambridge, MA: Harvard University Press.

Latour, B. (2004). *Politics of nature: How to bring the sciences into democracy.* Cambridge, MA: Harvard University Press.

Lewis, B., & Aikenhead, G. (2001). Introduction: Shifting perspectives from universalism to cross-culturalism. *Science Education, 85,* 3–5.

Marker, M. (2006). After the Makah whale hunt: Indigenous knowledge and limits to multicultural discourse. *Urban Education, 41*(5), 482–505.

Mazzocchi, F. (2006). Western science and traditional knowledge. *EMBO Reports, 7*(5), 463–466.

Mazzocchi, F. (2008). Analyzing knowledge as part of a cultural framework: The case of traditional ecological knowledge. *Environments Journal, 36*(2), 39–57.

McKinley, E. (2000). Cultural diversity: Masking power with innocence. *Science Education, 85*(1), 74–76.

McKinley, E. (2007). Postcolonialism, Indigenous students, and science education. In S. K. Abell & N. G. Lederman (Eds.), *Handbook of research on science education* (pp. 199–226). Mahwah, NJ: Lawrence Erlbaum.

Peat, D. (2002). *Blackfoot physics: A new journey into the Native American universe.* Newbury Port, MA: Weiser Books.

Peat, F. D. (2007). *Pathways of chance.* Pari, IT: Pari Publishing.

Prigogine, I., & Stengers, I. (1984). *Order out of chaos.* New York, NY: Bantam Books.

Pomeroy, D. (1994). Science education and cultural diversity: Mapping the field. *Studies in Science Education, 24,* 49–73.

Sammel, A. (2009). Turning the focus from 'other' to science education: Exploring the invisibility of whiteness. *Cultural Studies of Science Education, 4,* 649–656.

Shapin, S., & Schaffer, S. (1985). *Leviathan and the air-pump.* Princeton, NJ: Princeton University Press.

Siegel, H. (2001). Multiculturalism, universalism, and science education: In search of common ground. *Science Education, 86,* 803–820.

Smith, L. T. (1999/2012). *Decolonizing methodologies: Research and Indigenous People* (2nd ed.). London, UK: Zed Books.

Smith, L. T., Maxwell, T. K., Puke, H., & Temara, P. (2016). Indigenous knowledge, methodology and mayhem: What is the role of methodology in producing Indigenous insights? A discussion from mātauranga Māori. *Knowledge Cultures, 4*(3), 131–156.

Spivak, G. C. (1976). Translator's preface. In J. Derrida, *Of grammatology* (G. C. Spivak, Trans.) (pp. ix–lxxxvii). Baltimore, MD: Johns Hopkins University Press.

Spivak, G. C. (1993/2009). *Outside in the teaching machine.* New York, NY: Routledge.

Spivak, G. C. (1999). *A critique of postcolonial reason.* Cambridge, MA: Harvard University Press.

Stengers, I. (2001) *La guerre des sciences aura-t-elle lieu?: Scientifiction.* Paris, France: Le Seuil—Les Empêcheurs de Penser en Rond.

St. Pierre, E. A. (2011a). Post qualitative research: The critique and the coming after. In N. K. Denzin & Y. S. Lincoln (Eds.), *The SAGE handbook of qualitative research* (4th ed., pp. 611–626). Thousand Oaks, CA: Sage.

St. Pierre, E. A. (2011b). Refusing human being in humanist qualitative inquiry. In N. K. Denzin & M. D. Giardina (Eds.), *Qualitative inquiry and the global crisis* (pp. 40–55). Walnut Creek, CA: Left Coast Press.

TallBear, K. (2013). *Native American DNA: Tribal belonging and the false promise of genetic science.* Minneapolis, MN: University of Minnesota Press.

van Eijck, M., & Roth, W. M. (2007). Keeping the local local: Recalibrating the status of science and traditional ecological knowledge (TEK) in education. *Science Education, 91*(6), 926–947.

Towards a Curriculum for Indigenous Science To-Come

CHAPTER 7

Response-ability Revisited: Towards Re(con)figuring Scientific Literacy

> Given the pervasiveness of assimilationism in Western science education (assimilation of all students, including the Other, into the dominant ontology, and epistemology), it is not surprising that most science education articles include the mandate of improving scientific literacy and then proceed to define it, or refer to it by way of usual contemporary science education definition. (Sammel, 2009, p. 653)

The purpose of this chapter is to revisit response-ability, with a focus on enfolding the homework of previous chapters into working towards *a* response within science education which is more hospitable to Indigenous science to-come. This response takes the form of questions of tinkering with/in curriculum and pedagogy around the singular node that is scientific literacy as "most science education articles include the mandate of improving scientific literacy and then proceed to define it, or refer to it by way of usual contemporary science education definition". (Sammel, 2009, p. 653). Recall that response-ability, in its most succinct iteration, is "an ability to respond, to respond to the world beyond oneself, as well as a willingness to recognize its existence" (Kuokkanen, 2007, p. 39). As explored within earlier chapters, there are multiple facets that shape how science education and educators are produced, producible, and thus (un)able to respond to (and enact responsibility towards) Indigenous ways-of-knowing-in-being.[1] While responsibility always precedes our coming-to-knowing-in-being, the space of response-ability from

© The Author(s) 2021
M. Higgins, *Unsettling Responsibility in Science Education*,
Palgrave Studies in Educational Futures,
https://doi.org/10.1007/978-3-030-61299-3_7

which we can account for and be accountable to these responsibilities is highly productive as it invites and requires us to consider that which shapes our very ability to respond. Yet, the space of response-ability is ever in need of an ongoing unsettling the conditions which shape our ability to respond.

However, the (re)opening of science education as a location in and from which responsibility is both perceptible and potentially enacted is not as simple as desiring it to be so. Rather, the cut between what science education *is* and *is not* must continue to be laboured in order to allow for the possibility of what science education *could become:* a "common ground and a basis for dialogue" (Cajete, 2006, p. 248). This, as Barad (2010) reminds, comes-to-be through the ongoing process of reworking the norms of im/possibility to alter or and altering the possible possibilities (see also Spivak, 1993/2009). In response to the complexities of the space between Indigenous and Western ways-of-knowing-in-being, the following questions guide ethical examination and design: *How might science education account for and be accountable to these uneven and unequal relations of power? What kind of curriculum and pedagogy might open a space of response-ability in science education towards Indigenous science to-come?* "*What kind of [curriculum and] pedagogy would help students to learn about practicing responsible science?*" (Barad, 2000, p. 239, emphasis mine). As the space of response-ability is always already at risk, I also ask the following herein: *What kind of science education might consistently rework itself to be accountable for and towards its co-constitutive exclusions?*[2]

In order to engage these questions, I braid in the work of Torres Strait Islander scholar Martin Nakata's (2007a, 2007b) theorizing of the *cultural interface*, which accounts for the ways in which hybridity between ways-of-knowing-in-being are unequal, problematic, and yet rife with possibility. Recognizing that the cultural interface is never separate from its materiality (Nakata, 2007a, 2007b), there is a continued commitment to taking seriously the role of ontology. This bears particular significance as a Cartesian ontology is rife with onto-epistemic enactments that threaten to (fore)close the ability to respond towards a space of dialogue between TEK, IWLW, and WMS. Notably, one of the consequences of Eurocentrism and Cartesianism shaping nearly *all* facets of science education is that science education is often culpable of deferring and differing its attempts to work towards inclusivity by employing its associated concepts and enactments as usually defined.

In turn, I revisit and expand upon response-ability, weaving in Karen Barad's (2010) work around this concept, as a means of further (re)opening the space of response-ability by working within, against, and beyond a primary curricular node of science education: scientific literacy. This is of particular significance as its a location in need of *unsettling*: it is at once upholding settler colonialism (e.g., Sammel, 2009) and sedimented (e.g., Bang, 2018). Hospitality requires that "we work constantly towards reconceptualizing our thinking and reconsidering our values" (Kuokkanen, 2007, p. 163), which becomes all-the-more important when concepts that are absently present (fore)close the possibility of hospitably receiving Indigenous science to-come.

Therefore, in four movements, I: (a) identify scientific literacy as a central yet uncertain concept whose critical inhabitation is ripe for other meanings and enactments; (b) explore Karen Barad's subversion of scientific literacy as *agential literacy* as a productive location to rework the connectivity towards IWLN and TEK; (c), utilize agential literacy as proximal (yet differing) relation to bring in Gregory Cajete's conception of Indigenous science as *ecologies of relationships*; and (d) explore the generative points of resonance between *agential literacy* and *ecologies of relationships*. The chapter concludes with a cautionary note on points of convergence and points of divergence, wherein I use and trouble the proximal relation between agential literacy and ecologies of relationships by suggesting that this should not be recoded as but a new location for the mirror of sameness to take hold.

Revisiting Response-Ability at the Cultural Interface

We must abandon the common (often unconscious) colonial ideas about keeping the "[I]ndigenous" (epistemes, peoples, or anything else) separate or uncontaminated to preserve its archaic nature and thereby extend its inability to intervene, dialogue, participate and disrupt. It can and it must operate within those systems, because [decolonization] *is* a theory and practice of transforming the academy at the level of its intellectual procedures and traditions. (Kuokkanen, 2007, p. 146)

Because we need to "begin" some-where and some-time, let's "begin", right here, right now, by engaging in a fulsome and differential (re)visiting of response-ability in science education.[3] In moving towards *a* response, it is important that "we abandon the common (often unconscious) colonial ideas about keeping the '[I]ndigenous' (epistemes, peoples, or anything else) separate" (Kuokkanen, 2007, p. 146) as response-ability is not simply or only the taking up of the responsibility that always already lay before us and constitute us, but also the iterative (re)opening of responsiveness towards the potentiality of perceiving and differently enacting possibilities and problematics within the distributive relations that we inherit. As a white, Euro-settler trained within the physical sciences (specifically physics) and education, I recognize the importance of not simply rejecting my tradition's epistemic, ontological, and ethical commitments and enactments even (and especially) when they become problematic as this contextual vector always comes to bear on the potential (re)opening of responsiveness: "the process of decolonization can only emerge from *within* those structures of domination, from inside the institution" (Kuokkanen, 2007, p. 146). To attempt to move beyond science education without simultaneously working within and against it runs the risk of reproducing its structures, strategies, processes, and practices elsewhere, albeit differently (see Higgins, 2014a; McKinley, 2001; Sammel, 2009; see also Higgins, Madden, & Korteweg, 2015). Response-ability is a deconstructive move, an ongoing process of accounting for and being accountable to the absent yet present knowledge-practices that continuously (re)produce educational research and science education, be they problematic or rife with possibility. Even when engaging in the work of reconstructing something that is beyond what science education *is* and *is not*, the work must nonetheless, and paradoxically, be within and against simultaneously. This is of particular significance given the deep gravitational pull of Cartesianism that makes it difficult to break from its epistemological and ontological orbits (Apffel-Marglin, 2011; Barad, 2007; Battiste, 2005; Braidotti, 2013). Further, this signal the irreducible relation between reconstruction and deconstruction: the homework of response-ability must also take seriously unsettling the very tools with which we work in designing and developing curriculum, pedagogies, and methodologies in science education (see Higgins & Kim, 2019; Higgins, Wallace, & Bazzul, 2018; Higgins et al., 2017).

As Kuokkanen (2007) identifies, one important location in engaging in the (home)work of response-ability is by disrupting and displacing the (neo-)colonial desire to keep Indigenous ways-of-knowing-in-being separate from those of educational institutions, as the consequence of this move "extend[s] its inability to intervene, dialogue, participate and disrupt" (p. 146).[4] Thankfully, as the mantra of *beginning some-where and some-time* has been ceaselessly reminding us throughout is that Indigenous and Western ways-of-knowing-in-being are always already in relation. As Blackfoot Elder and scholar Leroy Little Bear (2016) signalled in his talk, *Blackfoot Metaphysics is Waiting in the Wings*, there is no metaphysics that exists outside of its relationship to others. However, it is not enough to reverse this binary (i.e., replace absence with presence), it must be disrupted and displaced as well (see Spivak, 1976). When the relationship between Indigenous and Western ways-of-knowing-in-being is recognized, it is most often one that is troubled and troubling. Notably, this relationship is prevalently and problematically attributed to a decontextualized and ahistorical account of difference within many spaces, where dichotomous differences are conceived and Indigeneity is presented as deficient otherness (Donald, 2012; see also, Spivak, 1988a, 1999). However, neither Indigenous nor Western knowledges are "immune" to the influence of the other knowledge system (Harding, 2008; Little Bear, 2000, 2016). Even if it were organized as a dichotomy, it is porous and always already deconstructing in an ongoing cross-cultural becoming. Therefore, the task is not to place Indigenous and Western ways-of-knowing-in-being *into* relation, but rather address the (neo-)colonial structures which (re)produce this relation as one of Othering or as absent presence. Importantly, the latter is also a form of colonial containments rooted in sublating, subsuming, or suturing over Indigenous ways-of-knowing-in-being.

Further, as Tewa science educator Gregory Cajete (2006) states, "Native and Western cultures, with their seemingly irreconcilably different ways of knowing and relating to the natural world, must search for common ground and a basis for dialogue" (p. 248). There is at once a need and a possibility for dialogue across Indigenous and Western "ways of knowing and relating to the natural world" (Cajete, 2006, p. 248) despite their "very different orientations to the natural world" (Cajete 2000, p. 13). As he elaborates,

> All the basic components of scientific thought and application are metaphorically represented in most Native stories of creation and origin. Indeed, both Native science and modern science have elements of the primal human story in common. They have, however, evolved very different orientations to the natural world and very different expressions of thought regarding the role of humankind in coming to know our place

and our responsibility to the creative unfolding of the greater story of the universe. As we enter the... new millennium, Native and Western cultures and their seemingly irreconcilably different ways of knowing and relating to the natural world are finding common ground and a basis for dialogue. (Cajete, 2000, pp. 13–14)

The necessity bears repeating: the relationship between Indigenous and Western knowledge systems is often referred to as one of "Jagged World-views Colliding" (Little Bear, 2000), one that is at best as tenuous, and at worst as "seemingly irreconcilably different" (Cajete, 2000, p. 14). However, the possibility for meaningful and respectful dialogue desires further engagement, despite there already being points of resonance between IWLN, and WMS. In the next section, I think with Nakata (2007a, 2007b) in order to engage with the possibility of placing TEK, IWLN, and WMS in dialogical relation *in* and *as* science education.

Considering Methodologies and Pedagogies for/at the Cultural Interface

For spaces that are always already at the cultural interface like Indigenous knowledge systems and practices in the academy, there is perhaps a need for a "different conceptualisation of the cross-cultural space, not as a clash of opposites and differences but as a layered and very complex entanglement of concepts, theories and sets of meanings of a knowledge system" (Nakata, 2006, p. 272). While there are increasingly points of resonance within this in-between space (Peat, 2002), one should not be overly or only romantic about the possibilities (Carter, 2004, 2010). Furthermore, just as one should always be alert to the ways in which these potentially productive hybrid spaces remain contested and complicated, it is also problematic to (too easily) write them off altogether (Ahenakew, 2016; Donald, 2012; Kuokkanen, 2007; see also Spivak, 1993/2009, 1994). In other words, there continues to be a need to remain critical and complicit towards these possibilities. I agree with Nakata (2007b) who states, "not opening up theoretical positions for more complicated discussion means that the cultural interface is sutured over in favour of the Western order of things and its constitution of what an Indigenous [and ally] opposition should be" (pp. 10–11), as well as possibilities beyond opposition such as dialogue.

The cultural interface according to Nakata (2007a, 2007b) are particular discursive nodes where competing and contesting knowledge systems are positioned alongside and against each other in ways that are shaped by various discursive practices (e.g., theories, epistemic regulation, social imaginaries) that dynamically intersect with the materiality of place, space, and time. As Nakata (2007a) states, these nodes:

> inform, constrain or enable what can be seen or not seen, what can be brought to the surface or sutured over, what can be said or not said, heard or not heard, understood or misunderstood, what knowledge can be accepted, rejected, legitimized or marginalized, or what actions can be taken or not taken on both individual and collective levels. (p. 199)

Not unlike Butler's (e.g., 1993, 2005, 2010) theorization of performativity, Nakata invites us to consider that the cultural interface is not a totalized or deterministic space in which agency is foreclosed. Moreover, the interface is something you *do* (as knowledge-practice) rather than something that *is* (as knowledge) (see Chapter 3). Agency is framed by the possibilities and limitations of the cultural interface. In particular, considering agency at the cultural interface invites us to consider how the plurality of coalescing and competing forces and flows produce "the very conditions to what is possible between Indigenous and non-Indigenous positions" (Nakata, 2007b, p. 13) in their dis/continuity and uneven relationality. Negotiating these spaces is not a question of *who* can know or do, but rather *what* can be known and done through negotiating, navigating, and exploring this lived everyday tension while recognizing that:

> People's lived experience at the interface is the *point of entry for investigation, not the case under investigation*. It is to find a way to explore the actualities of the everyday and discover how to express them conceptually from within that experience, rather than depend on or deploy predetermined concepts and categories for explaining experience. (Nakata, 2007b, p. 10, emphasis mine)

Not unlike Kuokkanen's (2007) conceptualization of the homework of response-ability, the work must *begin* from who and where we are. Importantly, it must also not *end* there either. For example, drawing from personal lived experience as a point of departure, I have argued with respect to decolonizing pedagogies that every attempt to work against colonization is also within colonization and inevitably reifies (neo)colonial constructs, concepts, or structures through the process (Higgins, 2014a; see Chapter 2). In turn, the cultural interface provides a rich conceptual location to consider decolonizing pedagogies as de/colonizing to explore the (neo)colonial complexities and complications that emerge through the practice of decolonizing pedagogies (see also Carter, 2004, 2010). It has been argued that the cultural interface is an incredibly productive and apt concept for situating Indigenous learners within teaching methodologies (Nakata, 2007a, 2007b),[5] as well as non-Indigenous learners engaging with Indigeneity (McGloin, 2009).[6] Lastly, I have also argued elsewhere (Higgins, 2014b), that this co-constitutive location is also a productive site from which to consider de/colonizing research methodologies as well as pedagogies.[7]

Thus, the task of placing TEK, IWLN, and WMS in dialogical relation *in* and *as* science education requires something akin to *suspended action* to engage with the double(d) practice of deconstructing and reconstructing science education.[8] Recall from Chapter 3 that suspended action is a lived practice of dialogue and a (partial) coming-to-awareness of what we think, as well as how what we think is produced and producible (Bohm, 1996; see also Patel, 2016; Stengers, 2018). The latter entails considering how values are inflected, deferred, and deflected through our selves. This double(d) movement, which bears resemblance to Apffel-Marglin's (2011) "reverse anthropology" (see Chapter 6), might allow us to rethink and displace the self-in-relation to the norms that shape how, who, and what we can be and do with explicit attention to the ways-of-knowing-in-being which produce Indigenous science as to-come. To engage with this task of *thinking-about-how-we-think-while-we-think-it*, learnings of how WMS is co-constituted by Eurocentrism and Cartesianism through the pathways tactically wandered with/in this book are enfolded into exploring a pathway forward (e.g., the practices entangled with/in WMS such as the modest witness, the enclosure, and the double-sided ledger); we move from the homework of response-ability towards and as *a* response. Further, my partial and contingent knowledge[9] of Indigenous science also supports the (re)opening of science education for Indigenous science to-come. Such an attempt recognizes that Indigenous science and WMS are not simply different natural-cultural articulations,[10] but also within unevenly distributed relations of power.

In the next section, response-ability is revisited as a means of further labouring the interface at which pedagogical and research design occur. As a continued commitment to taking seriously the role of ontology and how it shapes the what possibilities are possible with regards to the question of Indigenous ways-of-knowing-in-being within science education, I inflect the previous exploration of response-ability (see Chapter 2) with Barad's (2010) understanding of the concept.

The Homework of Response-Ability Revisited: Towards a Reconstructive Response

To reiterate, response-ability is not responsibility in the conventional sense; it is not something that one can simply take, give, or even have. Rather, response-ability is the double(d) process of (re)opening the space of responsiveness in order to enact that responsibilities towards the co-constitutive relationships we *always already* find ourselves in. As signalled in Chapter 2, this process must entail addressing the ways in which science education is marked by epistemic ignorance. This is a relation with Indigenous knowledges that is not only marked by what science education *does not know* but also by *what it refuses to know* and *what it cannot know*. This is to say, knowledge alone is not enough when not knowing is also produced by structurally sanctioned forms of ignorance and discursively conditioned (fore)closure making coming-to-know Indigenous ways-of-knowing-in-being an impossibility (see Kuokkanen, 2007, 2010).

Entangled within questions of *epistemic ignorance* are also the questions of ontology and metaphysics. To assume that the Western modernist concepts we hold mirror the being of Nature problematically (fore)closes response-ability, not only by creating a dichotomy which perpetuates a deficit view of any other-than-Cartesian relation between epistemology and ontology, but this has negative consequences for ways-of-knowing-and-being that do not fully fit within the model. Furthermore, it (fore)closes the (re)opening of concepts we inherit that are both problems and possibilities to the task of responding to Indigenous science to-come. For example, and significantly, the "I" as the ethical subject of response and the larger metaphysics of individualism to which it adheres. Rather, response-ability invites a conception of subject that is distributed along, within, and throughout the relationships through which we are co-constituted. There is no transcendental "I" who can irrupt the space of responsibility from outside: the work of response-ability is always within, against, and beyond the co-constitutive relations of the "I". In turn, the

"I" of science education research (i.e., the researcher) cannot be thought or enacted of without the co-constitutive vectors that come to shape response-ability (i.e., the home of homework: discipline, history, culture, etc.).

In addition, as Peat (2002) reminds us, some natural-cultural locations are more in/hospitable than others when it comes to the work of response-ability (see also Kuokkanen, 2007; Spivak, 1988a). Points of resonance offer themselves, not as panaceas, but as vacillating spaces of possibility for the work of response-ability (see Ahenakew, 2016). Lastly, as response-ability is an ongoing and enacted process, it is generative to consider it with/in the space of theory-practice-ethics that is methodology. Response-ability is a methodological concept that not only allows for a productive deconstructive pause, but also offers a pathways towards *a* response: it is a rich concept with which to think in the (im)possible reconstruction of science education to allow for Indigenous science to-come.

This brings us back to and has bearing on what it means to engage in the methodological process(es) of decolonizing science education. Recall that decolonizing science education entails the double(d) process of deconstructing and reconstructing (see Battiste, 2013a, 2013b). This entails at once (re)opening (neo-)colonial structures and strategies that leverage incommensurability as a means of enacting an uneven flow of power, while simultaneously (re)constructing in a way which refuses commensurability while seeking to centre and take seriously Indigenous, diasporic, and other post-colonial ways-of-knowing-in-being in reshaping its processes and priorities. However, the very possibility of this responsibility is directly tied to the ability to respond (Kuokkanen, 2007, 2010).[11] This ability to respond is, in turn, tied to the past (and future-to-come) as inheritance: not as possession, but that by which we are possessed (Barad, 2010; Derrida, 1994/2006). It is for this reason, and worth recalling, that decolonizing science education must always be both a process of deconstructing and reconstructing (and not deconstruction and reconstruction; see Jackson & Mazzei, 2012): the very tools with which any reconstruction are engaged with must also be under erasure.

In engaging with this task of response-able reconstruction, I turn to Barad's (2010) understanding of response-ability; without too easily calling Kuokkanen's (2007, 2010) understanding of response-ability commensurate with Barad's (2010), or refusing the call to attempt to place the two in dialogical relation (as they are always already in relation). Providing rich inroads towards ethically, epistemologically, and ontologically responding (and being able to respond), Barad (2010) states that,

> Responsibility is not an obligation that the subject chooses but rather an incarnate relation that precedes the intentionality of consciousness. Responsibility is not a calculation to be performed. It is a relation always already integral to the world's ongoing intra-active becoming and not-becoming. It is an iterative (re)opening up to, an enabling of responsiveness. Not through the realization of some existing possibility, but through the iterative reworking of im/possibility, an ongoing rupturing, a cross-cutting of topological reconfiguring of the space of [response-ability]. (p. 265)

As Barad (2010) suggests, response-ability is the double(d) process of enabling responsiveness to enact the responsibilities which precedes and produces the "I" of responsibility. Response-ability, as an always iterative process without *an* origin that enfolds and unfolds the *here-now* and *there-then*, "is a relation always integral to the world's ongoing intra-active becoming and not-becoming" (Barad, 2010, p. 265). In turn, as Barad (2010) suggests, the theory-practice-ethics of response-ability never achieves the calculable prescriptivity of conventional conceptions of responsibility but rather "require[s]/inspire[s] a new sense of a-countability, a new arithmetic, a new calculus" in which "one is too few, two is too many" (p. 251).[12] She offers us methodological orientations towards doing the homework of response-ability and the means to consider it as a (deconstructive/)reconstructive methodology. Of particular significance here is the final sentence in which she states that response-ability is enacted "not through the realization of some existing possibility, but through the iterative reworking of im/possibility, an ongoing rupturing, a cross-cutting of topological reconfiguring of the space of response-ability" (p. 265). For each of these insights, I will place them in conversation with previous chapters' deconstructive approaches to differentially build upon them in relation to the ongoing project of (re)opening responsiveness towards Indigenous science to-come. Furthermore, I address how these entangled insights are taken up herein.

Response-Ability as Ongoing Rupturing. The notion of ongoing rupturing here signals, yet again, and unavoidably, deconstruction.[13]

However, in marking and making the turn towards reconstruction, it is important to note the relationship between the two while chasing the possibility that, as Barad (2010) states, *one is too few* and *two is too many*. Deconstruction "is *not* about de-construction and re-construction" (Jackson & Mazzei, 2012, p. 15, emphasis in original); it is not a process of separation in which a separate agent brings in separable constituents. Rather, it is a process of differently arranging relations of co-constitutive otherness to which there is no outside from which total separation could occur. Thus, we can engage in the "undoing yet preserving of the opposition" (Spivak, 1976, p. xix) between this deconstructive/reconstructive binary opposition as they were never separate, nor separable. Not two, yet, importantly not one: there are still differentiations that come to mark the ways in which deconstruction differentiates itself from itself (i.e., never comes to be an *is;* see Spivak, 1993/2009; St. Pierre, 2011a). As Spivak (1993/2009) suggests, such a trajectory was marked in the way in which deconstruction was always already on the move within Derrida's work: "the economy, in the early work, of protecting and preserving (*garder*) the question and, in the later, of its transformation into the call to the wholly-other (*tout-autre*)" (Spivak, 1993/2009, p. 109, emphasis in original). Following a similar yet differing trajectory within this book, response-ability as ongoing rupturing builds upon and differentially enacts the deconstructive moves of previous chapters. Notably, I differentially apply the deconstruction as critical and complicit (mis)readings put to work in the fourth chapter. Recall that critical and complicit (mis)reading is to work within and against a structure by differentially occupying it, by substituting what a concept, category, or construct *is* with what it *is (not)*.[14] By substituting what it *is (not)*, that which could be signified by a signifier but usually is not (e.g., diffraction and prismatic dispersal as metaphors that come to replace and otherwise signify the optics of critique which operate through mirroring; see Chapter 4), differential possibilities emerge which are neither same nor wholly different by retaining (a partially erased trace of) the structure (see Derrida, 1976; Spivak, 1993/2009).

Furthermore, the task of deconstructive (mis)readings requires locating self-transgressive moments (i.e., where meaning vacillates intended meanings and its excessive constitutive otherness) to pry open the structure under erasure (Spivak, 1976). As discussed in Chapter 4, such locations are ripe for (mis)readings as that which it *is (not)* already comes to bear as excess; an intentional (mis)reading becomes but a leveraging of the deconstructive possibility to (re)open the structure under erasure. As deconstructive reconstruction must inevitably be within the very structure that (re)produces me, I draw from these insights to continue to critically and complicitly occupy science education, targeting the "home" of homework that is response-ability. In particular, I look to differentially signify a key location with/in science curriculum and pedagogy that are paradoxically central yet taken-for-granted: notably scientific literacy (the most frequent curricular goal of science education). This is to be achieved by doing the labour of (mis)reading these locations by substituting that which each *is (not)*. This work begins by substituting that retains a trace of the (partially erased) structure by drawing in strongly similar yet different iterations of that which is being substituted.

However, "deconstruction is [and can be] more than working within and against a structure" (St. Pierre, 2011a, p. 613) by substituting similar yet different meanings that retain the resources of the structure; "it is also the overturning and displacement of a structure so that something(s) different can be thought/done" (St. Pierre, 2011a, p. 613). The previous deconstructive approaches are not sufficient if the reconstructive orientation is to bring in Indigenous priorities, pedagogies, and protocols so that they might come to bear; requiring of science educators that "we radically de-naturalize what we've taken for granted. Here we refuse alternatives and pursue the *supplement* [i.e., the wholly other, the to-come], what always already escapes the structure" (St. Pierre, 2011a, p. 613). Here, we can turn to Barad's second suggestion.

Response-ability as a cross-cutting of topological reconfiguring. Barad (2007) suggests that the potentiality of deconstructive work does not lay strictly in its ability to identify the constitutive otherness of concepts, categories, and constructs. Nor does its possibility wholly reside in the always already occurring rupturing and shifting of meaning. Rather, deconstruction acts as an ongoing invitation towards an engaged act of account- and response-ability towards constitutive otherness, as well as an ever-present possibility to re(con)figure the lines of inclusion/exclusion, (re)constructing with that which was excluded. As the exclusion of constitutive otherness always comes-to-be, the work of deconstructing and reconstructing continues to be a recursive, iterative, and co-constitutive process which stems from the possibilities that arise from relationality.[15]

The task of response-ability is not to place peoples, places, and processes in relation: they are always already *in* or *as relation*.[16] These relations, as Barad (2010) reminds, are shaped by a constant reworking of the norms of exclusion and inclusion that come to form a topology of cross-cuts,[17] lines that come to shape what (temporarily) *is* and *is not,* as well as *could be* and the degree to which parts of the whole come to be on another part. However, because the cuts are agential, there is always the possibility of reworking how, where, and when they are made to bring seemingly distant natural-cultural elements into a relation of proximity.

To do this (home)work of response-ability, I extend the deconstructive tinkering of Chapters 5 and 6.[18] Response-ability as reconfiguring a topology of cross-cutting encourages a (re)consideration of inside/outside: there is no "outside" from which to draw the tools with which the bricoleur (i.e., the tinkerer) labours but rather only new ways of differently connecting that which is already there. Within this chapter, I tinker *with* Indigenous-ways-of-living-with-nature not as the end result (although bricolage is always to be criticized by what is and can be engineered), but as part of the process as these ancestral and long-standing practices were never developed for the purpose of school science. Nonetheless, they offer rich tools for school science to be (re)thought and enacted otherwise.

This is of particular significance as these previous chapters also explored the technologies of power that produced metaphysical distance between that which was in a relation of natural-cultural proximity (e.g., abstracting one's own body as labour in the proto-market economy, the enclosure and the encloser's advantage; see Apffel-Marglin, 2011; Chapter 6); the (home)work of response-ability can be thought of and enacted as a practice that works to bring that which is distant into proximal relation. As Barad (2010) asks, "what if we were to recognize that differentiating is a material act that is not about radical separation [and distance], but on the contrary, about making connections and commitments?" (p. 266).

Such a deconstruction of the topological norms of inclusion/exclusion is, as explored in the previous approach for (re)opening response-ability, a way of working not only within and against a structure but also beyond (a beyond that is nonetheless with/in the whole). However, while deconstruction is always already happening (whether we witness it or not), what comes to occupy this differential opening is often but a simulacrum: a different similarity or a similar difference (see Higgins et al., 2015; Mazzei, 2007).[19] While Barad (2010) suggests that, "only in this ongoing responsibility to the entangled other, without dismissal (without 'enough already!'), is there the possibility of justice-to-come" (pp. 264–265), there is nonetheless strategic and tactical locations from which to do the (home)work of responsibility. Here, a third approach can help in reworking and leveraging this open-ended closure.

Response-Ability as the Iterative Reworking of Im/possibility. As mentioned within the very beginning of the chapter, dialogue between IWLN and WMS is in a perpetual state of im/possibility: they are not and never will be (fully) commensurate (see Cajete, 1994, 2000; Barnhardt & Kawagley, 2005, 2008; McKinley & Stewart, 2012; Peat, 2002). This, in turn, complicates the possibility of if and how IWLN and TEK are included with/in school science. Generously and generatively, Spivak (1993/2009) asks, *What becomes possible when we persistently labour the conditions of im/possibility?* In considering response-ability as the iterative reworking of im/possibility, it is important to recognize that the "field of possibilities is not static or singular but rather is a dynamic and contingent multiplicity" (Barad, 2007, p. 147). In other words, possibilities do not sit still and some possibilities are more possible than others.

Locating such possible possibilities is the work of what Cajete (1994) refers to as "creative acts of perception" (p. 19). Elsewhere, Cajete (2000) qualifies that "the idea of moving around to look from a different perspective... is contained in the creative process. Indigenous logic moves between relationships, revisiting, moving to where it is necessary to learn or to bring understandings together" (Cajete, 2000, p. 210). This idea of approaching from multiple angles provides a productive extension to the play of (re)signification that was called upon in Chapter 3. In short, the play of (re)signification signals the ways in which deconstruction is an always extended invitation to consider other possible possibilities that might take hold stemming from the inevitability of theory-practice being exceeded and self-deconstructing. Drawing from Barad, and extending my own deconstructive work that began elsewhere (see Higgins et al., 2015), we can consider deconstruction of not only binary nodes (e.g., self/other and familiar/strange), but also of binary relationships between multiple binaries (e.g., self+familiar/other+strange).

Drawing on these insights, I was on the lookout for moments in which science education's meanings vacillate in ways that might (re)open its structures through deconstruction. While on the lookout for snags within the structure of science education (eventually identified as scientific literacy and visuality), I held awareness of the possibility of a reversal of a (porous) Western/Indigenous binary. In other words, I revisited the questions, *Where might tinkering with IWLN allow for the most productive (mis)readings of science education? Where might similar yet different*

Indigenous theory-practices come to differentially occupy this (re)opening so that Indigenous science to come might take hold? I was inevitably searching for productive points of resonance that could be levered to prevent the structure from too easily suturing itself over, reverting to a state of self-sameness. However, iteratively, it is important to note the dangers of subsuming *otherness* into *sameness*:

> the wholly other, *le tout-autre,* cannot be selved or samed. It is not suscep-tible to *ipseité* or *mêmeté.* The face of the wholly-other is without a name. The "other" that we narrativize or grasp consolidates the self, through a kind of *stade du miroir* [Mirror stage[20]]. (Spivak, 1993/2009, p. 238, emphasis in original)

That which *is (not)* science, such as IWLN and TEK, simultaneously existing with/in science education and yet not, as always to-come, loses its radical potentiality when it comes to bear through dialectic rela-tions (see Chapter 3). More importantly, as Spivak (1993/2009) offers, when Otherness becomes "selved or samed," it is because it has been subsumed, sublated, or sutured over by the same (neo-)colonial systems that rendered it Other in the first place; what is known or knowable is within the (fore)closure of Western modern thinking (see also Ahenakew, 2016). Such a sameness must be used and troubled; refusing the (full) reversal of difference as (wholly) separate, separable, and outside of relationality.

SCIENCE CURRICULUM AND RESPONSE-ABILITY: RE(CON)FIGURING SCIENTIFIC LITERACY

Putting to work the above orientations, I labour the space of response-ability within, against, and beyond the primary curricular node of science education: scientific literacy. As Sammel (2009) suggests, scientific literacy presents an important location to critically inhabit as its centrality often becomes a point in which efforts to disrupt and displace science education often become re-settled in both senses of the word:

> science education may be celebrating a move towards a form of multicul-turalism where ethnic differences are maintained, supported, and welcomed within the rhetoric, but in reality the infrastructural and ideological reasons for exclusion remain unchallenged and unchanged. (Sammel, 2009, p. 253)

In order to engage in the double(d) movement of unsettling science education through disrupting and displacing scientific literacy, the aforementioned three moves are put to work. Within the first part (i.e., *response-ability as ongoing rupturing*), I identify scientific literacy as a central yet uncertain concept whose critical inhabitation is ripe for other meanings and enactments. In the second part (i.e., *response-ability as the iterative reworking of im/possibility*), I identify Karen Barad's subversion of scientific literacy as agential literacy as a productive location to rework the connectivity towards IWLN and TEK. In the third section (i.e., *response-ability as the cross-cutting of topological reconfiguring*), I utilize agential literacy as proximal (yet differing) relation to bring in Gregory Cajete's conception of Indigenous science as ecologies of relationships. Lastly (i.e., *response-ability as putting to work points of resonance*), this (re)opening of responsiveness made possible with agential literacy and ecologies of relationships is explored as a means of moving towards a more response-able science education.

Response-Ability as Ongoing Rupturing: Scientific Literacy as Central yet Uncertain

As Barad (2000) states about scientific literacy, its importance to science education is central while its very purpose is always on the move:

> There has been no shortage of rationales given on behalf of the national need for scientific literacy. Scientific literacy has been hailed as: the basis for democratic decision making about public issues; necessary for global economic competitiveness and national security; crucial for the promotion of rational thinking; a condition for cultural literacy; necessary for gainful employment in an increasingly technological world; the basis for personal decision making about health-related issues; and necessary for the maintenance of the public image of science. (Barad, 2000, p. 225)

Scientific literacy has become ubiquitous within and almost synonymous with science education (Aikenhead, 2006; Bang, 2018; Barad, 2000; Holbrook & Rannikmae, 2007; Roth, 2003). Following over four decades of use, it has become unavoidable, central, as well as the "conventional" goal of science education (Aikenhead & Elliot, 2010; Holbrook & Rannikmae, 2009). Almost anthemic, call for *scientific literacy for all* rings out across educational institutions and levels (e.g., curricular resources, policy). As Barad (2000) states above, there is no lack of reasons for which scientific literacy is leveraged and centred within science education; however, she asks: "what does it mean to be scientifically literate?" (p. 225).

As Roth (2003) states, "the concept of scientific literacy is itself not at all clear" (p. 11). This is not to say that it holds a meaning that is itself blurry, rather, one that might be described as ambivalent, pluralistic, and in ways that come to occasionally contradict themselves.[21] As Holbrook and Rannikmae (2009) state, scientific literacy is inhabited by a multiplicity of conceptual components:

> (a) Knowledge of the substantive content of science and the ability to distinguish from non-science; (b) Understanding science and its applications; (c) Knowledge of what counts as science; (d) Independence in learning science; (e) Ability to think scientifically; (f) Ability to use scientific knowledge in problem solving; (g) Knowledge needed for intelligent participation in science-based issues; (h) Understanding the nature of science, including its relationship with culture; (i) Appreciation of and comfort with science, including its wonder and curiosity; (j) Knowledge of the risks and benefits of science; and (k) Ability to think critically about science and to deal with scientific expertise. (p. 276)

The conceptual components found within this non-exhaustive list are not only differentially included/excluded but also non-uniformly enacted through following the curricular goal that is scientific literacy. It is for this reason that they state scientific literacy has been so successful in its proliferation: it "avoids the use of distracting detail and, as such, convincingly portrays a complex idea which intuitively appears to be correct" (Holbrook & Rannikmae, 2009, p. 275). Here, we find resonance with Davis (2008) who states that some of the concepts that are the most successful in achieving a degree of everywhere-ness by reaching wide academic audiences and circulation are those that are at once ambiguous and incomplete.[22]

Despite the conceptual ambiguity presented with/in scientific literacy, Barad (2000) states that, "*most commonly*, scientific literacy is thought of in terms of the successful transmission of knowledge about scientific facts and methods from knowing scientists to the ignorant masses" (p. 226, emphasis mine).[23] *Most commonly* here signals the ways in which there is *a* (i.e., singular) *common sense*: a common sense that is at once held and by which science education and educators are held (see

Chapter 5). As van Eijck and Roth (2007) remind us, drawing from Foucault (1977, 1979), the logics of science education can often be characterized as a "regime of truth". As explored in Chapters 4 and 6, regimes of truth are marked by circular relations: each "truth" is but a differential articulation of the systems of power which produces it, whose articulation in turn (re)produces the systems of power. Such a circular relation can be read in two ways: first, as signalling a hermetic circle, a (fore)closure of knowledge (see Chapters 2 and 3); and second, as the capillary circulation of power from one conceptual node to another (see Chapter 4).[24] Despite its ambivalence and the ever-present possibility of deconstructive (re)signification, it is no surprise that scientific literacy is, for the most part, (re)produced within and reproducing the norms of power of science education (see Higgins, 2014a; McKinley, 2001, 2007; Sammel, 2009; van Eijck & Roth, 2007). There has been a major paradigmatic shift in science education in the last decades which shifts the emphasis from coming to know what scientists know (i.e., cognitivism, intra-personal learning, scientific knowledge as representation of nature) to enculturation into how scientists come-to-know (i.e. socio-constructivism, inter-personal learning, scientific knowledge as representation of culture) (see Barad, 2000; Erickson, 2000; Aikenhead, 2006). As Holbrook and Rannikmae (2007) state: "no content is fundamental, but rather the content needed for enhancing scientific literacy is dependent on the culture and society in which the science education is implemented" (p. 1352). However, the "teaching practices have changed little and remain based on traditional, universalist views of science and science education" (McKinley, 2007, p. 219); *common sense* understandings of science, scientist, and science education *most commonly* come to fill these (re)opening locations (see also McKinley, 2001; Sammel, 2009).

Paradoxically, in light of the ambiguity of scientific literacy, only a small fraction of the population can be said to be scientifically literate (Barad, 2000, 2012a). Barad (2000) states:

Viewed in this way, the problem of scientific illiteracy is seen as a massive transmission failure... In light of the extraordinary monetary and intellectual resources that have been and continue to be committed to solving this problem, it is perhaps not unreasonable to ask if the metaphor itself isn't sending the wrong signal. (p. 226)

As Holbrook and Rannikmae (2009) suggest, "the core of the idea behind scientific literacy lies in its analogy with literacy" (p. 275). By invoking literacy as a metaphor, we invoke a *common sense* notion that to be scientifically literate is to be able to read, interpret, and determine the validity of a multiplicity of scientific texts: scientific documentation and instrumentation, reporting on past, current, and future socio-scientific issues, the cultural practice of science, as well as Nature itself. This, as Barad (2000) states, might be "sending the wrong signal" (p. 226): a signal which (re)produces science education as an enactment of what Barad (2007) refers to as a metaphysics of individualism (see Chapters 5 and 6). This makes scientific literacy a location that is at once inadequate yet necessary, and ripe for the ongoing rupturing of response-ability through critical inhabitation for (at least) three reasons.

First, the metaphor of literacy invites a metaphysics of individualism by calling upon reading and hermeneutics: the reader is both separate and separable from the scientific "text", both of which ontologically precede the act of reading. Furthermore, interpretive readings in/of science generally require that the reader learns through mirroring the text (when decreed as valid science) or dialectically negating the text.

Secondly, the metaphor of literacy reinforces the notion that learning to be scientifically literate is an (metaphysically) individual affair. Comparable to the ways in which we consider the act of reading as a mathematical set in which there is one reader and one text, the oft-exclaimed *scientific literacy for all* presupposes that each and every learner has an individualistic relationship with scientific literacy and that each and every text is separable and separate. As Roth (2003) states, this educational separation between both learner, text, and other texts produce the precondition for a problematic double(d) effect:

> Scientific literacy currently means to question nature in ways such that do not, reflexively, also question science and scientists. However, worse is the other part of the current rhetoric about scientific literacy—it is to be for all. All individuals (e.g. Americans), so goes the idealist rhetoric, have to learn and exhibit certain 'basic' facts and skills. Just imagine, every individual taking the same ('scientific') perspective on GMOs, genetic manipulation of the human genome, or use of drugs (such as those used to dope certain kinds of children, labelled 'ADHD' (i.e. Attention-deficit, hyperactive disorder) to make them compliant). (p. 11)

In other words, the focusing of each and every student on "one" text (i.e., the science-as-usual curriculum)[25] that is treated as separate and separable from other texts (e.g., scientific ethics) produces scientific literacy as, by design, un-self-reflexive. While the possibility of ethical self-reflexivity is not wholly foreclosed, it is evident that practices of scientific-literacy-as-usual focus primarily on the ways in which different students (through standardized testing) achieve different levels of attainment of the same curricular content. Furthermore, Roth (2003) appeals to curricular reason when he states that "we [science educators] all know that there simply exists too much specific knowledge for any individual to know the relevant facts even in more constrained contexts" (Roth, 2003, p. 19).

Third, like with reading, texts can be selected to be relevant to *each* and *every* reader in order to achieve *scientific literacy for all*. However, Barad (2000) cautions that "there's something paradoxical about the notion that something can be 'made' relevant – as if relevancy could be imposed or added onto an existing structure" (p. 221). This signals the differentiation between how relevancy often plays out between theory and practice[26]; scientific texts can become what Barad (2000) refers to as "context-coated". This, as signalled within the previous chapter, can reduce epistemic pluralism in science education as but different ways of achieving the same ontological location: knowing nature with/in Cartesianism. In turn, as Roth (2003) suggests, "there is more than one reason to rethink scientific literacy and to see it as an emergent collective praxis" (p. 21); scientific literacy need not presuppose a metaphysics of individualism from the get-go. Rather, "if we think of scientific literacy in different terms, as choreography of a particular kind in which we learn to participate by participating from the beginning, we take radically different approaches to teaching science in schools" (Roth, 2003, p. 19). Not unlike Chapter 4, in which critique is re(con)figured to address its normative enactments of distance and separation as the condition for knowledge, as well as the pre-epistemic and pre-ontological status of the object and subject of critique, I turn to Barad (2000, 2007, 2012b, 2012c) to think scientific literacy otherwise.

Response-Ability as the Iterative Reworking of Im/Possibility: Karen Barad's Shift from Scientific Literacy to Agential Literacy

Drawing from quantum physics, Barad's (2007) theory of *agential realism* provides a rich location to iteratively rework the norms of im/possibility by working within and against science and science education to produce a location that might differentially allow IWLN to take hold (i.e., in ways in which it has not already, beyond integration and tolerance). Agential realism, as Barad theorizes it, questions the humanist a priori status of Nature before Culture, as well as the anti-humanist corollary statement of Culture before Nature (Barad, 2007, 2012b, 2012c; Kirby, 2011). This work disrupts the notion that Cartesianism is *the* (only) ontology, not by negating it but rather by positioning it as one ontological configuration among many. These configurations are presented and produced as open-ended processes that are enacted rather than static. In particular, Barad's concept of intra-action enables us to gain insight into how relationality, flux, and process are conceptualized and enacted:

> The neologism 'intra-action' signifies the mutual constitution of entangled agencies. That is, in contrast to the usual 'interaction', which assumes that there are separate individual agencies that precede their interaction, the notion of intra-action recognizes that distinct agencies do not precede, but rather emerge through, their intra-action. (Barad, 2007, p. 33)

In other words, intra-action accounts for and is accountable to the various ways in which bodies of meaning (e.g., social, cultural, political, historical) and bodies of matter (e.g., biology, ecology, physics, engineering, architecture) are co-constitutive. This acts as an invitation to consider the ways in which these bodies of meaning-matter are not only produced through Cartesian norms of bodily production (i.e., subjects and objects) but also through other-than-Cartesian entanglements that would comprise and cut across multiple Cartesian subjects and objects. This is not simply a way of redrawing the lines of bodily production (e.g., researcher + instrument interaction –> researcher-instrument intra-active entanglement), it is also a (re)consideration of how they come into *being*. As Barad (2012c) states,

A quantum ontology deconstructs the classical one: there are no pre-existing individual objects with determinate boundaries and properties that precede some interaction, nor are there any concepts with determinate meanings that could be used to describe their behavior; rather determinate boundaries and properties of objects-within-phenomena, and determinate contingent meanings, are enacted through specific intra-actions, where *phenomena* are the ontological inseparability of intra-acting agencies. (pp. 6–7)

Accordingly, the production of natural-cultural bodies and their bodily norms are enacted, in flux, process-based, and performative rather than something that always already *is* (or *is not*). Because phenomena are constitutive of reality, *being* can be thought as a performative and co-constitutive *becoming*: "reality is composed not of things-in-themselves or things-behind phenomena but of things-in-phenomena" (Barad, 2007, p. 140).

The consequences of agential realism for scientific literacy are drastic. The task of epistemologically establishing a representational (i.e., humanist) relationship of equivalence with either nature (i.e., through cognitivism) or culture (i.e., through socio-constructivism) breaks down because their separation was never a priori. Rather, Barad (2000) invites us to consider how "scientific literacy becomes a matter of agential literacy – of learning how to intra-act responsibly within the world" (p. 237) around the matters of science (i.e., space, time, and matter). This is significant as agential literacy goes beyond scientific literacy's accounting for the diverse natural and cultural agents that constitute experimental phenomena studied and produced within the context of science education.[27] First, it considers the ways in which agents are always already natural-cultural. Secondly, it accounts for the ways in which these agents not only constitute but are also constituted by phenomena. Third, agential literacy ethically re(con)figures accountability as a process of not only accounting for, but also being accountable to these agents and their intra-action in the world's ongoing becoming.

Response-Ability as the Cross-Cutting of Topological Re(con)figuring: Gregory Cajete's Indigenous Ways-of-Knowing-in-Being and Science Curriculum as All my Relations

Leveraging this differential opening into WMS and, in turn, science education allows for the possibility to cross-cut the topology which presents WMS and IWLN as distant by bringing them into proximal relation. Articulating relationality, flux, and process differently and for different purposes, Tewa science educator Gregory Cajete (1994, 1999, 2000) proposes that we consider ways-of-knowing-in-being—that is, the co-substantiation of epistemology and ontology—as ecologies of relationships. These ecologies of relationships that are enacted with/in these ways-of-knowing-in-being are often referred to as both external and internal to a human(ist) subject, while noting that some of the relations external to the subject do not require a subject at all.[28] Externally, we often speak of relationships with other humans, relationships with other-than-human bodies (e.g., plants, rivers, mountains), as well as relationships with more-than-human bodies (i.e., spiritual beings) (see also Apffel-Marglin, 2011). Internally, the relationships between heart, mind, body, and spirit are often called upon.

Furthermore, the boundary between exteriority and interiority is one that is porous, and it is this porosity that allows us to be with/in relation. This ontological porosity extends to space and time to make being in the world a question of process, flux and holistically being *of* the world rather than *in*. As Cajete (1994) states, "a constant building upon earlier realities is a basic characteristic of Indigenous processes... [in which] we engineer the new reality built upon earlier ones, while simultaneously addressing the needs, and acting in the sun, of our times" (p. 27). The intentionality here signals that Cajete's ecology of relationships (sometimes referred to as "sense of place") is not simply a way-of-knowing-in-being in which the world is enacted through the flux of relationships, but that there is also an ongoing accounting for and accountability to the ecology of relationships such that it is (re)generated and sustained.[29] It is for this reason that Cajete (2000) reminds us that within many Indigenous languages there is an expression akin to "all my relations" (e.g., *Mitakuye Oyasin* in Lakota). "All my relations" is an epistemological, ontological, and ethical accounting for and being accountable to the ecologies of relationships we find ourselves in and constituted by which extends beyond the immediate present to include generations past and those still yet-to-come. It is a metaphysical principle through and by which Indigenous "people understood [and understand] that all entities of nature – plants, animals, stones, trees, mountains, rivers, lakes and a host of other living entities – are embodied [and co-constitutive] relationships that must be honoured" (Cajete, 2000, p. 178).

An Indigenous science education curriculum of "all my relations" has been in place since time immemorial in the form of land- or place-based education (Cajete, 1994, 1999, 2000; see also Aikenhead & Michell, 2011; Kawagley, 2006). Despite disruption by ongoing (neo-)colonial practices, Cajete (2000) reminds us that Indigenous knowledge holders continue to engage these traditional yet ever-evolving contemporary ways-of-knowing-in-being by "seeking, making, sharing, and celebrating" (p. 178) the ecological relationships they find themselves with/in.[30] Accordingly, with/in Cajete's (1994, 1999, 2000) conception of Indigenous science education, scientific literacy would not simply be a task of knowing *about* nature but rather knowing-in-being *with* nature as an inseparable and co-constitutive part of the ecologies of relationships in order to learn "the subtle, but all important, language of relationship" (Cajete, 2000, p. 178).

This teaching of knowing-in-being *with* is woven into and enacted through traditional Indigenous approaches to teaching and learning, such as Indigenous storywork (e.g., Archibald, 2008). As Barnhardt and Kawagley (2008) remind us, while Indigenous stories hold rich representations of nature (i.e., knowledge *about* nature when read with/in Cartesian representationalism), their potential lies in honouring a knowing-in-being *with* the plants, the animals, and a wide range of other-than-human bodies that are teachers with/in the ecologies of relationships particular to a place. As a pedagogy through which Indigenous peoples "came [and come] to perceive themselves as living in a sea of relationships" (Cajete, 2000, p. 178), it is a way to witness already existing relations and foster the possibility of new ones.

This is significant within science education wherein, as explored within the previous curricular section, to be scientifically literate largely becomes a function of being able to "read" nature (and others' accounts of nature). However, to know *with* nature rather than *about* as a pedagogical framing (re)opens the space of response-ability for both educators and students, albeit a potentiality that is not always actualized.

Response-Ability as Putting to Work Points of Resonance: (Re)Thinking Scientific Literacy at the Cultural Interface

Through the potentiality of this proximal relation between Barad's agential realism and Cajete's ecologies of relationships, it becomes productive to explore and leverage both their *similar differences* and *differing similarities* (see Bohm, 1994) in working towards *a* response, a more response-able science education.[31] Notably, thinking with both, and the points of resonance between, encourages considering the ways in which bodies that are typically considered natural, rather than cultural, to have agency. To begin to consider other-than-human beings as agentic is a deeply productive step in the direction of taking seriously Indigenous science to-come, particularly the Indigenous notion that the plurality of other-than-human bodies such as animals, plants, rocks, rivers, constitute a sentient landscape which is always already teaching us, should we choose to and/or be able to listen *with* (Cajete, 1994, 1999, 2000; Marker, 2015). As Leroy Little Bear puts it "trees talk to you, but you don't expect them to speak in English or Blackfoot" (in Peat, 2002, p. 288). Articulated otherwise, place can be thought of as anthropogenic (i.e., cultural landscapes that are not "pure" of human interaction; e.g., spectres of humanity across the Amazon rainforest; see Apffel-Marglin, 2011); but it is important to simultaneously recognize that its agency in producing meanings and matterings does not solely reside in its hybrid human-other-than-human relations. This is significant to taking seriously as a "sense of place" is central to Indigenous science to-come: landscapes remember as they bear the markings of their own entangled becomings (see Apffel-Marglin, 2011; Barad, 2007; Cajete, 2000). "Hearing" the stories that place might tell requires attunement to the unique relationships that shape the ecologies of relationships we find ourselves in, an attunement Indigenous ways-of-knowing-in-being that have practiced in places that have co-constituted these practices since time immemorial (Cajete, 2000, 2006).

Further, thinking with ecologies of relationships and agential literacy disrupts and displaces the metaphysics of individualism. Again, it is through the metaphysics of individualism that the humanist subject maintains and (re)produces power through distance and separation, as well as how humanism separates, distances, and organizes space, time, and matter (i.e., Nature). Significantly, it obscures and obfuscates ongoing accounting for and accountability to the ways in which we are always already iteratively, epistemologically, and ontologically co-constituted (Barad, 2007, 2010, 2012b). Thinking with both Cajete and Barad invites us to think about the ways in which agency is distributed and enacted intra-actively across *all my relations*. Barad (2007) refers to the "doing" and "undoing" of intra-action as *posthumanist performativity*.[32] In turn, this grants materiality a similar flux and undecidability, and in the process extends the range as to which bodies can and do engage in performativity, as well as the norms by which bodies come into being.[33] As Barad (2007) reminds us, the ways in which we enact our intra-actions matter because "each one reconfigures the world in its becoming – and yet they never leave us; they are sedimented into our becoming, they become us" (p. 394). In other words, it speaks to *all my relations*: the ways in which we enact our ecologies of relationships leave their marks upon the bodies connected with/in the entanglement. Importantly, disrupting and displacing the metaphysics of individualism is refusing and resisting its (fore)closure of ongoing and ever-needed possibilities for ethics (see also Apffel-Marglin, 2011; Cajete, 1994, 1999, 2000; Peat, 2002).

Further, thinking with posthumanist performativity helps us think about non-linear temporality and its subversive potentiality through ontological indeterminacy. Rather than using the language of predictability and certainty implied through a linear and causal relationship between past natural and cultural events towards an ever-certain present, considering time as an ontological indeterminacy invites thinking with a natural-cultural future as a possible possibility that shapes the present with/in a non-linear causal relationship: *Does the past produce the present? Does the present shape the past? What about the relationships with the future?* On time, Barad (2007) states:

> Time is not a succession of evenly spaced individual moments. It is not simply there as substance of measure, a background uniformly available to all beings as a reference or an ontological primitive against which change and stasis can be measured. (p. 180)

For Barad, time is performative and comes into dis/continuous being through its enactment. This dis/continuous being, or to vacillate between *being* and *not being*, is, in short, what it means to be ontologically indeterminate. If even the past is open to being re(con)figured (e.g., quantum tunneling; see Barad, 2007) in the present, then what happens to the temporal linear causality that WMS relies upon to make knowledge claims? What if time were always already an entangled variable to account for and be accountable to rather than a control (or controllable substance)?

This resonates with Indigenous ways-of-knowing-in-being that recognize that the world itself is in flux and in process such that it might be more appropriate to state that it is ontologically *becoming* rather than *being*. Such ontological indeterminacy has significant consequences for pedagogy. For Cajete (1994), "learning involves a transformation that unfolds through time and space" (p. 54) and that enfolds space and time (see also Peat, 2002). This is significant as it makes space for a plurality of ways-of-knowing-in-being to include other ways of enacting temporality such as Indigenous forms and flows of time such as non-Euclidian circularity (Cajete, 1994, 2000; Peat, 2002). Also, considerations of time as enfolded and time as always already more than an inert, immutable, and linear backdrop upon which nature and culture play-out invites an ongoing consideration of the ways in which time makes itself intelligible through its entangled performativity with other agencies. This becomes all-the-more meaningful when considering the ways in which a future temporalities are often overcoded by a singularizing (neo-)colonial settler futurity.

The ways in which multiple space-time-matterings make their presence known in singular instances in bi-directional causal ways invites us to not only consider how the past shapes the present and the futures-to-come, but also how the plurality of undeterminable futures shape the present, as well as the past. This non-linear causality invites us not only to consider how we are shaped by potential futures-to-come, but more importantly, how we are always already ethically bound to these potentialities that we can never fully come to know. As Cajete (1994) states, "everything leaves a track, and in the track is the story: the state of being of each thing in its interaction with everything else" (pp. 55–56). Potential futurities are always already with/in us. In turn, there is always an ethical hope in the subversive potentiality of the future as it is always at once yet-to-come and not-yet-to-come.

Conclusion: Response-Able Design as a Hospitable Move Towards Indigenous Science to-Come

> The question of hospitality will never come to a close, nor should it – the moment we consider the problem solved, we arrive at a totalizing closure.... Hospitality is a productive crisis in which we work constantly towards reconceptualizing our thinking and reconsidering our values – in other words, we move beyond the disruptive, hegemonic, and exploitative exchange paradigm and its priorities towards a new relationship in which the academy is compelled to recognize and accept its responsibility toward the "other." (Kuokkanen, 2007, p. 163)

The potentiality of deconstructive work lays neither strictly in its ability to identify the constitutive otherness of concepts, categories, and constructs, nor in meaning that inevitably ruptures and shifts what science *is* and *is (not)*. Rather, the key ongoing possibility towards an engaged act of account- and response-ability towards constitutive otherness is an ever-present invitation to reconstruct with that was excluded. In other words, after Kuokkanen (2007), it is an ongoing call to hospitality, one which "will never come to a close" (p. 163). As we continue moving towards curricular and pedagogical practices of science education that productively and respectfully inhabits the cultural interface between WMS and IWLN, it is important to remember that pluralism in science education does not simply imply there are different epistemological means of approaching the same ontological reality to generate similar results. To this, Latour (1993) reminds that some epistemological means are nonetheless positioned as superior to others through particular universalism (some usually being Western modern epistemologies). As St. Pierre (2011a) suggests:

> We move away from Plato's gift of ontological determination, a logic of identity [i.e., self-sameness] and prediction – *science is this; science is not that* – toward a logic of the "and" – *This and this and this and this...* (p. 613, emphasis in original)

We can move away from ontology (as the Being of beings) not through negation, but as a coming to recognize it as but one ontological possibility among many. It is for this reason that Barad (2000) claims adaptation of science education is never simply one of "candy coating" the content with relevant material; rather, "questions of relevancy are intertwined with questions of subjectivity and epistemic responsibility" (p. 222).

However, responsibility and responsiveness are not enough for hospitality to take hold. Rather, responsibility and responsiveness also require the ability to respond. This insight became the methodological drive herein via the concepts of the *cultural interface* (Nakata, 2007a, 2007b) and, again, *response-ability* (see Barad, 2010; Kuokkanen, 2007, 2010; Spivak, 1994). Braiding in Barad's (2010) conception of response-ability, as well as the previous deconstructive approaches in previous chapters, (re)opening the space of response-ability was explored through three co-constitutive strategies: (a) response-ability as ongoing rupturing; (b) response-ability as the cross-cutting of topological re(con)figuring; and (c) response-ability as the iterative reworking of im/possibility. These three approaches were then leveraged to re(con)figure curriculum and pedagogy around the central yet conceptually ambiguous node of *scientific literacy*; a generative snag in the fabric of science education.

Identifying scientific literacy's necessary yet inadequate, prevalent yet ambiguous, status within science education made it a rich curricular location from which to engage in the work of response-ability. As "learning and teaching are occurring at all times, at all levels, and in a variety of situations" (Cajete, 1994, p. 40), science curriculum could be then re(con)figured with/in the relational ways-of-knowing-in-being that co-substantiate it. Agential realism and Indigenous science here provide important insights into the entanglement of knowing and being, as well as how we might imagine them otherwise in the pursuit of knowing nature (i.e., scientific literacy).

There are rich points of convergence between Barad's *agential realism* and Cajete's *ecologies of relationships* which offer themselves as locations to labour in the name of hospitality towards Indigenous science to-come. Namely, both are ways-of-knowing-in-being that are shaped by ethics, accountability, and responsibility to the (re)generation of that which we, as humans, co-constitute and are co-constituted by. However, to only focus on commensurability is to employ the metaphor of the mirror; commensurability then becomes an act of mirroring sameness elsewhere, dialectically subsuming into or sublating through sameness patterns of difference, as well as making it difficult to account for and be accountable to the enactment of difference. Further, while there are deep and productive points of resonance between quantum and Indigenous ontologies, there are still patterns of difference that matter.

Significantly, where quantum physics is a recent phenomenon in which WMS is irrevocably facing its ontological limits and limitations (see Barad, 2007, 2010; Peat, 2002), Indigenous ways-of-knowing-in-being have been developed, practiced, and honoured since time immemorial with a built-in ethic of (re)generation and sustainability (see Cajete, 1994; Kawagley, 2006). Constitutive of these ways-of-knowing, and also another difference that matters, is the relationship to spirituality and more-than-human beings. Placing the two in relation without conflating them has great consequences for what scientific literacy *is*, *is not*, and, perhaps most importantly, *could be*: a site which recognizes that "the question of hospitality will never come to a close" (Kuokkanen, 2007, p. 163).

NOTES

1. Notably explored herein: (a) enactments of adversarial dialectics that place science and ethics as binary opposites (see Chapter 3); (b) the mirror metaphor that comes to mask epistemic and ontological becomings, while protecting (from) critical engagement (see Chapter 4); as well as (c) the ways in which "common sense" create a circular logic that comes to simultaneously uphold and conceal *an* ontology of Cartesianism and its inseparable ethical and epistemological assumptions and enactments (Chapters 5 and 6).

2. Significantly, asking questions of ethics before-the-fact (i.e., what *would* be or *might* be responsible) rather than after-the-fact (i.e., what *is* or *has been* responsible) is in alignment with being accountable to the agential cuts that we enact (see Chapter 6) and attempts to heed Barad's (2012a) cautionary note that the "notion of consequences is [often] based on the wrong temporality: asking after potential consequences is too little, too late" (p. 53). To uphold these commitments to ethical questions focus on pedagogical design, rather than simply or only those of delivery.

   Nonetheless, delivery is a component of any curriculum or pedagogy: the lived enactment of either always exceeds the ways in which they are planned for (see Aoki, 1991/2005). Part of the challenge is designing pedagogies that work towards (re)opening spaces that are (fore)closed so that there may be the possibility of radical hospitality in the lived encounter.

   Furthermore, as explored in Chapter 6, "design" (i.e., research design) is a notion that is often overcoded by a theory/practice binary (which is co-constituted with/by a Nature/Culture binary) in which design often pre-exists practice, rather than always already being entangled within its production. In turn, this chapter is not about translating "theory into practice" (which presupposes separation and separability) but rather translating the theory-practice of earlier chapters into a differentially contextualized theory-practice (see Higgins, Madden, Berard, Lenz Kothe, & Nordstrom, 2017).

3. Whereas the refrain of beginning some-where and some-time has been utilized throughout the book to generally signal the ways in which we are all always already within the question of Indigeneity and that there are a multiplicity of spatial and temporal locations to critically inhabit, there are moments in which right-here, right-now is also a generative point of engagement that should not be dismissed, particularly if it accounts for and is accountable to a futurity which cannot yet be anticipated, that which is *to-come*.

With respect to higher education generally, Kuokkanen (2007) states, "if the academy only welcomes what it is ready to welcome – that is, what it recognizes and what it considers it must welcome – it is not hospitality" (p. 131). Rather, what is needed is an *unconditional* welcome. Such an unconditional welcome is not only the possibility of (re)opening the norms of responsiveness towards a (co-)constitutive yet othered body of knowledge (here, Indigenous ways-of-living-with Nature), but also the very possibility of a continued existence: If disciplinary spaces such as science education profess an "unlimited commitment to the truth", (Kuokkanen, 2007, p. 139) it cannot do so without explicitly recognizing the plurality of ways-of-knowing-in-being and taking seriously what they have to offer in terms of pursuing truth-finding. Here, the temporality of a future *to-come* marks not only an arrival that cannot be predicted (which will require radical hospitality), but also the temporality of the continued existence of disciplinary spaces:

> The ethics and the future of the academy require hospitality. Without openness to the "other," responsibility toward the "other," there can be no future of and in the academy. The future of the university will be found in its openness to the "other." This openness will have to involve more than merely opening doors to [I]ndigenous peoples while dismissing or failing to recognize their epistemes. (Kuokkanen, 2007, pp. 139–140)

Again, for this work of hospitality to manifest, there is homework to be done to (re)open spaces foreclosed by epistemic ignorance. The work of *a* response in a future-to-come cannot be separated from the homework of response-ability; in a future-to-come in which homework has been engaged with, there will be no possibility of disavowal. That part of this (home)work is coming to know the limits to what is and can be known from within our spaces: the threshold of knowledge. As Kuokkanen (2007) states, "in order to have a future, the academy will have to acknowledge the threshold; only in this way will it accept its responsibility and be able to respond" (p. 141). Similarly, within science education, our ability to respond to Indigenous ways-of-knowing-in-being shapes our very ability to have a future: the irreducible link to Indigeneity through its othering also binds the disciplinary space to what it can be and become. The homework of response-ability is not only a radical reversal of the gaze, but also a commitment to the sustaining of *a* home (and also one that might be hospitable towards Indigenous science to-come).

4. Recall from Chapter 3, and its cultural homework of response-ability, that one of the ways in which scientific knowledge could be (re)considered as *a* knowledge-practice (among many) was to extend science's internal dynamism and heterogeneity outwards by considering the complex and conflicting ways in which WMS, IWLN, and TEK interact and interface. As a knowledge-practice, this interface is not a location that simply *is* but rather is a *doing*. It is an enactment that is often marked by the assumption that cultural traits are separate and separable. This renders invisible: (a) the many elements of WMS that stem from non-Western cultural sciences and vice versa, as well as their respective historicities; (b) the complex co-constitutive and cross-cultural scientific knowledge production processes; and, (c) the complexities that occur at the currently lived and differently situated cultural interfaces that both they and their students occupy (see Belczewski, 2009; Snively & Corsiglia, 2001; van Eijck & Roth, 2009; see also Kuokkanen, 2007).

5. For Nakata (2007a), this is the work of engaging with *Indigenous standpoint theory*,

> a method of inquiry, a process for making more intelligible 'the corpus of objectified knowledge about us [Indigenous peoples]' as it emerges and organizes understanding of our lived realities. [It is] theorizing knowledge from a particular and interested position — not to produce the 'truth' of the Indigenous position but to better reveal the workings of knowledge and how understanding of Indigenous people is caught up and implicated in its work. (p. 215)

6. McGloin (2009) adds the rejoinder that the labouring of one's position vis-à-vis the cultural interface is a task that befits not only Indigenous peoples as the tensions at the interface affect each and everyone one of us (as we are always already in relation), albeit differentially. Particularly, she states that for Western academics and educators, forging a standpoint by criticaclly engaging with the ways in which the self at the cultural interface

is produced and producible allows for the possibility of differently inhabiting this Indigenous-Western interface that is often characterized by a unidirectional dialectic negation.

7. Specifically, I argue that the cultural interface is a useful concept for considering similarly situated research methodologies because the suturing over at the cultural interface does not only occur on any particular (human) body but also occurs upon many bodies of knowledge. As these include bodies of methodological knowledge, the complex and complicating ways in which this over-writing occurs need to be worked within and against (see also Higgins & Kim, 2019).

8. While not dismissing the importance of finding ways of relating Indigenous and Western ways-of-knowing-in-being through complementarity rather than contradiction (i.e., dialogue), and highlighting the need for suspended action, Ahenakew (2016) cautions that this "becomes a problem when we cannot recognize what is lost in translation" (Ahenakew, p. 333). To animate translation as a significant, yet complex and complicated node of the cultural interface, Ahenakew (2016) uses the metaphor of grafting:

> Grafting is used in biology as the process of transplanting something from one organism into another (e.g., hybrid plants or cell/skin implants). Grafting, in itself, is neither good nor bad. Indeed, hybridity can be a generative process. However, in the context of grafting Indigenous knowledges into non-Indigenous ways of knowing, we are operating with severely uneven environments shaped by historical circumstances where the grafting/hybridizing does not happen as a mutual exercise, but as assimilation. Grafting, in this sense, can further contribute to the elimination of Indigenous peoples as distinct Indigenous peoples both in their relationship with the state, in their relation to the land, and in terms of the perceived worth of their knowledge. (Ahenakew 2016, p. 325)

For example, the presupposition that ontology precedes epistemology (so that all knowledge claims map onto the same ontological reality) would act as a form of knowledge-as-ignorance. This would result in an enactment of "a kind of grafting that seems to maintain primary loyalty to accepted notions of time, progress, reality, and being" (Ahenakew, 2016, p. 333). In turn, Ahenakew's (2016) concept of grafting invites an ongoing rethink(ing) of the very possibility of braiding together these paradigms with decolonizing goals in mind, without abandoning the project.

Rather, there is a need to *(at)tend* to the graft. Here, the word play is productive, because not only does Ahenakew invite us to *attend to* the ways in which grafting becomes a site in which (neo-)coloniality plays out, but also signals the ways in which there is simultaneously a *tendency towards* these dynamics as well. Importantly, this impossibility does not let us off the hook (as what becomes possible when we strive towards the impossible?). Rather, it is a call to be cognizant of "the utilitarian risk to all-too-quickly instrumentalize and embrace Indigenous research methodologies as quick-fix solutions to or escapes from deep-rooted and ongoing (neo)colonial thinking" (Ahenakew, 2016, p. 323).

What would it mean to engage in a less innocent cross-cultural methodological design that accounted for and was accountable to what is lost through grafting in science education? Ahenakew (2016) suggests transforming this impossibility into a generative indeterminacy:

> For those of us writing within academia, the first small step we need to take is to make grafting visible. Making grafting visible means writing in a way that makes what is invisible noticeably absent so that it can be remembered and missed. (Ahenakew, p. 333)

It is a call to not make visible that which is absent (and made absent through colonial logics) but rather make visible its absence so that it can be missed and remembered, as the attempt to narrativize the un-narrativizable (within the grammar of Western modernity) only traps us in positions where we think we have represented but rather we have lost things in translation. (At)tending to the graft is a form of response-ability that accounts for and is accountable to the very limits of the systems that we inherit and inhabit.

9. Partial knowledge does not reduce or dismiss responsibility: all knowings are partial and contigent. Furthermore, partial knowings are nonetheless knowings (see Butler, 2005).

10. Further, it bears repeating that they are also not differing cultural articulations of Nature, particularly as this framework almost always becomes what Latour (1993) refers to as *particular universalism*: a relativist approach in which one position (that of Western modernity) comes to be the (most) correct one from which to observe the phenomena. Further, it dictates Nature's operations without Nature's consent. This matters greatly in learning (from) Indigenous science: coming-to-know partially entails learning from what other-than-humans have to teach (e.g., Cajete, 1994, 2000).

11. Recall from the first chapter that fully embracing incommensurability brought with it its own set of challenges: the possibility of response requires a certain commitment to a certain degree of commensurability (e.g., dialoguing across difference). As Kuokkanen (2007) puts it: "even if one is sometimes tempted to embrace more pessimistic view that modern and Indigenous epistemes are incommensurable, the academy will move forward only by committing itself to responsibility and thus, responsiveness" (p. 101).

12. Here, the sense of one being too few and two too many is a criticism of the ways in which Western modern thinking, and even some of its responses, has a difficult time conceptualizing a response beyond wholistic one-ness and dichotomous two-ness such that more subtle and nuanced forms of difference might exist such that multiple partial connections exist and proliferate.

13. *Ongoing rupturing,* here, also signals the ways in which deconstruction *happens*: the structure under erasure is always already in a state of ongoing rupturing (Jackson & Mazzei, 2012). As such, the structure of science education cannot and should not be recognized as a closed form whose knowability precedes rupturing (despite claims otherwise). However, as with other deconstructive moves which stem from witnessing deconstruction, it is to pay attention to the deconstructive potentiality of irruptions as not all irruptive possibilities are desirable and not all are critically generative. This is why Derrida (1976) reminds us that deconstruction critiques itself.

14. Recall that *is (not)* signifies that which is otherwise unintended but still retains, albeit differently, the structure being critically inhabited (see Chapter 4).

15. On the subject of relationality, Kirby (2011) reminds us that:

> Relationality is not an "in-between" the *de-tailing* of entities. If the Earth's grammar is necessarily internal, a shifting algorithm, than any "part" of the Earth would be a virtual geometry with hologrammatic resonance rather than a separated entity, broken off from its larger and now absent, or perhaps still attached totality. (p. 39, emphasis in original)

In *Quantum Anthropologies,* Vicky Kirby (2011) re-reads Derrida's (1976) iconic statement that "there is nothing outside the text" (Derrida, 1976, p. 163) by juxtaposing it to the earlier one that there is "[no] outside of metaphysics" (Derrida, 1976, p. 19) to consider the ways in which "there is no outside of Nature" (Kirby, 2011, p. 38). This is of particular significance within science education where post-structuralism's primary focus on Culture and cultural indetermination have made it such that a *too simple* reading of these approaches can be read as a jettisoning of the very concept of science: knowing nature, as well as knowing Nature (i.e., the metaphysical relation between space, time, and matter; see Barad, 2011). This is "not to suggest that we need to 'get real' and add Nature's authorship to this strange text as if Culture's inadequacies might be healed by a natural supplement" (p. 13). Rather as in the above block quotation by Kirby, it is to explore a relational conception of the world in which Derridean relationality does not preclude Nature. By refusing to revert Nature to a pre-critical status and to consider Nature *and* Culture as separate and separable, Kirby (2011) suggests Nature-Culture as a constitutive wholeness that never achieves *one*-ness. From this, relationality emerges not as the by-product of entities (i.e., de-tailing) but is the always already active constitution of a co-constitutive part/whole (see also Barad, 2007, 2010; Cajete, 1994; Peat, 2002).

16. As Donald (2012) remarks, because there already exists an Indigenous-Western relationality which is often but not always marked by coloniality, there is always the possibility of enacting it anew, differently, with a decolonizing ethic. Furthermore, as explored within the previous chapter, even the "One Truth" of science is inevitably related to the very things that it oppositionally defines itself against (e.g., Indigenous science, metaphysics).

17. This is not unlike the current geopolitical lines that are entangled with/in nationhood. These, as Marker (2015) reminds by drawing from Coast Salish peoples' navigation of the US–Canada border, are cultural enactments that do not transcend culture but are rather entangled with/in culture (i.e., Coast Salish people enact a differential spatial enactment of who they are which exceeds and is exceeded by current geopolitical conditions).

18. To quickly recap (from Chapter 5), deconstructive tinkering is first and foremost a process of reversing and (re)opening the engineering/bricolage binary through: (a) the use of tools otherwise unintended for the task at hand; (b) using intended tools in ways they were not intended; and privileging the process over the product as the "product" of knowledge creation never (fully) comes to be (see Derrida, 1976).

19. While one of the ways in which the deconstruction that is always already happening can be read is in the inevitable rupturing of any and every structure, it does not always come to present the norms of power through which the structure comes to (re)produce itself as a simulacrum of its former self. Elsewhere (Higgins et al., 2015), I noticed the ways in which educators who began the inclusion of Indigenous materials in their teaching practice differently articulated coloniality (e.g., presenting an image of Indigenous peoples as pan-Indigenous and of a past already past). Such rearticulates the need to continuously focus on a process as

opposed to a product (e.g., a "decolonized" curriculum which may come to mask the ways in which it continues to uphold colonialism; see also Carter, 2004, 2010).

20. The mirror stage, drawing from Lacanian psycho-analytics, is when the Imaginary we hold dialectically (in)takes the Real, producing cuts which prevent the wholly other from being anything more than what can already be known. This is of particular relevance considering the ways in which IWLN and TEK are often only considered science when they fit the criteria of "valid" science (which often happens to be that of WMS).

21. So much so that Holbrook and Rannikmae (2009) go on to state that despite or because of the prevalence of the term (i.e., scientific literacy), there are science educators who would prefer the adoption of more conceptually precise terms in its stead.

22. Davis (2008) makes the case for the popularity and pervasiveness of the concept of *intersectionality* in the social sciences.

23. Not only this, but Roth (2003) reminds us that these facts are often but "a faint and distorted image of scientists' science". (p. 10; see also Aikenhead & Michell, 2011; Aikenhead & Ogawa, 2007; McComas, 1998). Furthermore, scientific literacy is complicated by the notion that "it is assumed that it is the scientists' place to define what should be known in the field. Why should this be when at other times and places scientists claim a lack of interest in how science is used and taught?" (Roth, 2003, p. 12; see also Barad, 2000). The stance of Modest Witness (see Chapter 6) that is often held by the subject of scientific inquiry is diametrically at odds with the subject of education to whom they are accountable to(wards).

24. It is not uncommon to see statements like the following: "A common rationale given for studying science subjects in school is the achievement of scientific literacy" (Holbrook & Rannikmae, 2007, p. 1347). However, when the most common understanding of scientific literacy is "studying science", such statements achieve a certain circularity (in both senses) by distributing its collective meaning across its similar yet different articulations of *science-as-usual* (Roth, 2003; van Eijck & Roth, 2007).

   With respect to circularity, Bang (2018) suggests that scientific literacy has become so dogmatic that it is not only circular, but also ouroborossified (in the image of a snake eating its own tail), only accepting to intake itself in a monstrous cycle of associated with "greed, appetite, self-destruction, and endlessness" (p. 809) as well as an ahistorical self-genesis (p. 809). Further, Bang (2018) makes the case that scientific literacy is an example what Deleuze (1994) refers to as a "dogmatic image of thought". In other words, "what has developed sociohistorically to represent thought and stop people from thinking beyond a sedimented, stagnant, and stratified notion of the 'actual'" (Bazzul, Wallace, & Higgins, 2018, p. 824). Resonant with the post-colonial notion of foreclosure (e.g., Spivak, 1999), the result of such an image of thought is that "these unchallenged, basic definitions [of scientific literacy] exist even in (the most) well-intended science curriculum" (Bazzul et al., 2018, p. 826).

25. However, even a science-as-usual curriculum fails to achieve *one*-ness as it is always already co-constituted by a variety of texts: both intentionally (e.g., sources) and unintentionally (e.g., the relational act of reading is inter-textual and brings other texts to bear on the meaning made).

26. Or more appropriately how, as Lenz Taguchi (2010) suggests, theory is always already entangled within practice (see also Carter, 2010; Spivak, 1988a).

27. It is worth noting that Barad (2007) encourages anthropomorphism (i.e., attributing cultural values to otherwise deemed acultural bodies) if it can be put to the service of working against anthropocentrism (i.e., the centring of humans).

28. As Cajete (2000) states, Indigenous ways-of-knowing-in-being entail a "reciprocal compact of care and responsibility" (p. 183) that is an ongoing enactment (rather than an ontologically pre-existing quality) that is deeply creative, co-constitutive, and relational. However, this enactment is not one that is *only* human, or necessarily co-constituted *with* humans:

> Creative use of the environment guaranteed its continuity, and Indigenous peoples understood the importance of allowing their land its rich life because they believed their land understood the value of using humans. If humans could use the land, the land would also use them to enrich it and keep it alive. They and the place they lived were equal partners in life. (Cajete, 2000, p. 204)

It is one that is also co-constitutively enacted by the various beings, both other-than-human and more-than-human, which come to inhabit an ecology of relationships (see also Apffel-Marglin, 2011; Cajete, 1994).

29. As Cajete (2000) suggests, such accounting for and accountability to the ecology of relationships has much to do with the notion that these relations are the precondition for Indigenous ways-of-knowing-in-being rather than simply a choice that is chosen or taken up by separate and separable entities:

> The land nurtures humans and humans nurture the land, the foundation of a reciprocal compact of care and responsibility, which is continually reaffirmed through the various expressions of Native technology. Given this special relationship, the separation of culture and nature would be considered unnatural. Likewise, the separation

of humankind from nature and the creation of discrete categories for viewing nature inherent in [most of] the disciplines of Western science would be viewed as equally unnatural and arbitrary (p. 183).

Extending this, the Nature/Culture binary is not only a (neo-)colonial imposition, but furthermore, as mentioned throughout, reading the world through such a binary fails to account for the ways in which knowing (i.e., epistemology) and being (i.e., ontology) are entangled within Indigenous ways-of-knowing-in-being (Cajete, 2000; see also Apffel-Marglin, 2011; Bang & Marin, 2015).

30. Castellano (2000) outlines three foundations of Indigenous knowledge: *traditional* knowledge (intergenerational), *empirical* knowledge (gained through careful observation), and *revealed* knowledge (acquired through dreams, visions, and intuitions). To these three Cajete (2009) added a fourth: *modern* knowledge. This last foundation involves the participation in "modern" practices so that it may complement the other foundations. It is important to recall that modernity does not always entail Western modernity. As scientific and technological products and processes are taken up, the often-held assumption is that they remain unaffected. Instead, what traditions these scientific technologies suture over and, more important, how these technologies are shaped by these traditions should be considered (see also Harding, 2008; Nakata, 2007a).

31. Importantly, Kuokkanen (2007) suggests that the homework of response-ability is marked by "more than new pedagogy, although pedagogical changes would follow as a result" (Kuokkanen, 2007, p. 153). For example, these insights were woven into a pedagogical response in the form of an interdisciplinary curricular inquiry that is beyond the scope of this book. Named *Visually Storying Relationships with Nature*, this was a pedagogical project involved two middle-school classes in an urban school in Metro Vancouver for a one-month period. It encouraged participants not to "read" nature through scientific literacy but rather to narrate *with* nature as a form of agential literacy that fostered an ecology of relationships: producing photography-assisted comic books which told such relational stories *with* Nature (see Higgins, 2016a; see also Higgins, 2014b, 2016b).

32. This extends the Butlerian notion that epistemology is always already performative. Performativity, a persistent theme through Butler's work, is the anti-ontological doing and undoing of epistemological categories, concepts, and conditions such as identity (e.g., Butler, 1990), ethics (e.g., Butler, 2005), framings (e.g., Butler, 2010). For Butler, there is no doer behind the deed or foundational essence behind epistemology, but rather, knowing and ways-of-knowing are always enactments within a citational

chain. Barad (2007) extends Butler's notion of performativity by including materiality as performative and co-constitutive of discourse.

33. Although, as Barad (2007) reminds us, while everything comes to matter, not everything matters and comes to matter equally.

REFERENCES

Ahenakew, C. (2016). Grafting Indigenous ways of knowing onto non-Indigenous ways of being. *International Review of Qualitative Research, 9*(3), 323–340.

Aikenhead, G. S. (2006). *Science education for everyday life*. London, ON: Althouse Press.

Aikenhead, G. S., & Elliot, D. (2010). An emerging decolonizing science education in Canada. *Canadian Journal of Science, Mathematics and Technology Education, 10*(4), 321–338.

Aikenhead, G. S., & Michell, H. (2011). *Bridging cultures: Indigenous and scientific ways of knowing nature*. Toronto, ON: Pearson Canada Inc.

Aikenhead, G. S., & Ogawa, M. (2007). Indigenous knowledge and science revisited. *Cultural Studies of Science Education, 2*(3), 539–591.

Apffel-Marglin, F. (2011). *Subversive spiritualities: How rituals enact the world*. New York, NY: Oxford University Press.

Archibald, J. (2008). *Indigenous storywork: Educating the heart, mind, body, and spirit*. Vancouver, BC: UBC Press.

Aoki, T. T. (1991/2005). Teaching as indwelling between two curriculum worlds. In W. F. Pinar & R. L. Irwin (Eds.), *Curriculum in a new key: The collected works of Ted T. Aoki* (pp. 413– 423). Mahwah, NJ: Erlbaum.

Bang, L. (2018). In the maw of the Ouroboros: an analysis of scientific literacy and democracy. *Cultural Studies of Science Education, 13*(3), 807–822.

Bang, M., & Marin, A. (2015). Nature–culture constructs in science learning: Human/non- human agency and intentionality. *Journal of Research in Science Teaching, 52*(4), 530–544.

Barad, K. (2000). Reconceiving scientific literacy as agential literacy. In R. Reed & S. Traweek (Eds.), *Doing Science+Culture* (pp. 221–258). New York, NY: Routledge.

Barad, K. (2007). *Meeting the universe halfway: Quantum physics and the entanglement of matter and meaning*. Durham, NC: Duke University Press.

Barad, K. (2010). Quantum entanglements and hauntological relations of inheritance: Dis/continuities, spacetime enfoldings, and justice-to-come. *Derrida Today, 3*(2), 240–268.

Barad, K. (2011). Erasers and erasures: Pinch's unfortunate 'uncertainty principle'. *Social Studies of Science, 41*(3), 443–454.

Barad, K. (2012a). Interview with Karen Barad. In R. Dolphijn & I. van der Tuin (Eds.), *New materialism: Interviews & cartographies* (pp. 48–70). Ann Arbor, MI: Open Humanities Press.

Barad, K. (2012b). Nature's queer performativity. *Kvinder, Køn & Forskning, 1*(2), 25–53.

Barad, K. (2012c). What is the measure of nothingness? Infinity, virtuality, justice. *dOCUMENTA, 13*, 1–17.

Barnhardt, R., & Kawagley, A. (2005). Indigenous knowledge systems and Alaska Native ways of knowing. *Anthropology and Education Quarterly, 36*(1), 8–23.

Barnhardt, R., & Kawagley, A. (2008). Indigenous knowledge systems and education. *Yearbook of the National Society for the Study of Education, 107*(1), 223–241.

Battiste, M. (2005). You can't be the global doctor if you're the colonial disease. In P. Tripp & L. J. Muzzin (Eds.), *Teaching as activism* (pp. 121–133). Montreal, QC: Queen's University Press.

Battiste, M. (2013a). *Deconstruction and reconstruction: Roles, responsibilities and implications of a decolonizing framework.* Retrieved from http://www.indigenouseducation.educ.ubc.ca/transformation/indigenous-perspectives/ on 11/22/2012.

Battiste, M. (2013b). *Decolonizing education: Nourishing the learning spirit.* Saskatoon, SK: Purich Publishing.

Bazzul, J., Wallace, M. F., & Higgins, M. (2018). Dreaming and immanence: Rejecting the dogmatic image of thought in science education. *Cultural Studies of Science Education, 13*(3), 823–835.

Belczewski, A. (2009). Decolonizing science education and the science teacher: A white teacher's perspective. *Canadian Journal of Science Education, 9*(3), 191–202.

Bohm, D. (1994). *On creativity.* New York, NY: Routledge.

Bohm, D. (1996). *On dialogue.* New York, NY: Routledge.

Braidotti, R. (2013). *The posthuman.* Cambridge, UK: Polity.

Butler, J. (1990). *Gender Trouble.* London, UK: Routledge.

Butler, J. (1993). *Bodies that matter: On the discursive limits of "sex".* New York, NY: Routledge.

Butler, J. (2005). *On giving an account of oneself.* New York, NY: Fordham University Press.

Butler, J. (2010). *Frames of war: When is life grievable?.* London, UK: Verso.

Cajete, G. (1994). *Look to the mountain: An ecology of indigenous education.* Durango, CO: Kivaki Press.

Cajete, G. A. (1999). *Igniting the sparkle: An Indigenous science education model.* Durango, CO: Kivaki Press.

Cajete, G. (2000). *Native science: Natural laws of interdependence.* Santa Fe, NM: Clear Light Books.

Cajete, G. (2006). Western science and the loss of natural creativity. In F. Arrows (Ed.), *Unlearning the language of conquest: Scholars expose anti-Indianism in America* (pp. 247–259). Austin, TX: University of Texas Press.

Cajete, G. (2009, May). *Re-building sustainable Indigenous communities.* Keynote address at Dream Catching conference, Winnipeg, MB, Canada.

Carter, L. (2004). Thinking differently about cultural diversity: Using post-colonial theory to (re)read science education. *Science Education, 88*(6), 819–836.

Carter, L. (2010). The armchair at the borders: The 'messy' ideas of borders, border zones and epistemological diversity in multicultural science education. *Science Education, 94*, 1–20.

Castellano, M. (2000). Updating Aboriginal traditions of knowledge. In G. J. Dei, B. L. Hall, & D. Goldin Rosenberg (Eds.), *Indigenous knowledges in global contexts: Multiple readings of our world* (pp. 21–36). Toronto, ON: University of Toronto Press.

Davis, K. (2008). Intersectionality as buzzword: A sociology of science perspective on what makes a feminist theory successful. *Feminist Theory, 9*(1), 67–85.

Deleuze, G. (1994). *Difference and repetition.* New York, NY: Columbia University Press.

Derrida, J. (1976). *Of grammatology* (G. C. Spivak, Trans.). Baltimore, MD: John Hopkins University Press.

Derrida, J. (1994/2006), *Specters of Marx: The state of the debt, the work of mourning, & the new international* (P. Kamuf, Trans.). New York, NY: Routledge.

Donald, D. (2012). Indigenous Métissage: A decolonizing research sensibility. *International Journal of Qualitative Studies in Education, 25*(5), 533–555.

Erickson, G. (2000). Research programmes and the student science learning literature. In R. Millar, J. Leach, & J. Osborne (Eds.), *Improving science education: The contribution of research* (pp. 271–292). Philadelphia, PA: Open University Press.

Foucault, M. (1977). *Discipline and punish: The birth of the prison.* London, UK: Allen Lane.

Foucault, M. (1979). Truth and power. In M. Morris & P. Patton (Eds.), *Power, truth, strategy* (pp. 29–48). Sydney, NSW: Feral Publications.

Harding, S. (2008). *Sciences from below: Feminisms, postcolonialities, and modernities.* Durham, NC: Duke University Press.

Higgins, M. (2014a). De/colonizing pedagogy and pedagogue: Science education through participatory and reflexive videography. *Canadian Journal of Science, Mathematics and Technology Education, 14*(2), 154–171.

Higgins, M. (2014b). Rebraiding photovoice: Putting to work Indigenous conceptions of praxis and standpoint theory. *Australian Journal of Indigenous Education, 43*(2), 208–217.

Higgins, M. (2016a). Decolonizing school science: Pedagogically enacting agential literacy and ecologies of relationships. In C. Taylor & C. Hughes (Eds.), *Posthuman Research Practices* (pp. 267–289). Basingstoke, UK: Palgrave Macmillan.

Higgins, M. (2016b). Placing photovoice under erasure: A critical and complicit engagement with what it theoretically is (not). *International Journal of Qualitative Studies in Education, 29*(5), 670–685.

Higgins, M., & Kim, E. J. A. (2019). De/colonizing methodologies in science education: rebraiding research theory–practice–ethics with Indigenous theories and theorists. *Cultural Studies of Science Education, 14*(1), 111–127.

Higgins, M., Madden, B., Berard, M. F., Lenz Kothe, E., & Nordstrom, S. (2017). De/signing research in education: Patchwork (ing) methodologies with theory. *Educational Studies, 43*(1), 16–39.

Higgins, M., Madden, B., & Korteweg, L. (2015). Witnessing (the lack of) deconstruction: White teachers' "perfect stranger" position in urban Indigenous education. *Race Ethnicity and Education, 18*(2), 251–276.

Higgins, M., Wallace, M. F., & Bazzul, J. (2018). Disrupting and displacing methodologies in STEM education: From engineering to tinkering with theory for eco-social justice. *Canadian Journal of Science, Mathematics and Technology Education, 18*(3), 187–192.

Holbrook, J., & Rannikmae, M. (2007). The nature of science education for enhancing scientific literacy. *International Journal of Science Education, 29*(11), 1347–1362.

Holbrook, J., & Rannikmae, M. (2009). The meaning of scientific literacy. *International Journal of Environmental and Science Education, 4*(3), 275–288.

Jackson, A. Y., & Mazzei, L. A. (2012). *Thinking with theory in qualitative research: Viewing data across multiple perspectives*. New York, NY: Routledge.

Kawagley, A. O. (2006). *A Yupiaq worldview: A pathway to ecology and spirit*. Long Grove, IL: Waveland Press.

Kirby, V. (2011). *Quantum anthropologies: Life at large*. Durham, NC: Duke University Press.

Kuokkanen, R. J. (2007). *Reshaping the university: Responsibility, Indigenous epistemes, and the logic of the gift*. Vancouver, BC: UBC Press.

Kuokkanen, R. (2010). The responsibility of the academy: A call for doing homework. *Journal of Curriculum Theorizing, 26*(3), 61–74.

Latour, B. (1993). *We have never been modern*. Cambridge, MA: Harvard University Press.

Lenz Taguchi, H. (2010). *Going beyond the theory/practice divide in early childhood education: Introducing an intra-active pedagogy*. London, UK: Routledge.

Little Bear, L. (2000). *Jagged worldviews* colliding. In M. Battiste (Ed.), *Reclaiming Indigenous voice and vision* (pp. 77–85). Vancouver, BC: University of British Columbia.

Little Bear, L. (2016, June). *Big Thinking and rethinking: Blackfoot metaphysics 'waiting in the wings'*. Keynote address at 2016 Congress of the Humanities and Social Sciences, Calgary, Alberta.

Marker, M. (2015). Borders and the borderless Coast Salish: Decolonising historiographies of Indigenous schooling. *History of Education, 44*(4), 480–502.

Mazzei, L. (2007). *Inhabited silence in qualitative research: Putting poststructural theory to work*. New York, NY: Peter Lang.

McComas, W. F. (1998). The principal elements of the nature of science: Dispelling the myths. In W. F. McComas (Ed.), *The nature of science in science education* (pp. 53–70). Dordrecht, NL: Springer, Netherlands.

McGloin, C. (2009). Considering the work of Martin Nakata's "cultural interface": A reflection on theory and practice by a non-Indigenous academic. *The Australian Journal of Indigenous Education, 38*(1), 36–41.

McKinley, E. (2001). Cultural diversity: Masking power with innocence. *Science Education, 85*(1), 74–76.

McKinley, E. (2007). Postcolonialism, Indigenous students, and science education. In S. K. Abell & N. G. Lederman (Eds.), *Handbook of research on science education* (pp. 199–226). Mahwah, NJ: Lawrence Erlbaum.

McKinley, E., & Stewart, G. (2012). Out of place: Indigenous knowledge in the science curriculum. In B. Fraser, K. Tobin, & C. J. McRobbie (Eds.), *Second international handbook of science education* (pp. 541–554). Dordrecht, NL: Springer, Netherlands.

Nakata, M. (2006). Australian Indigenous studies: A question of discipline. *The Australian Journal of Anthropology, 17*(3), 265–275.

Nakata, M. (2007a). *Disciplining the savages: Savaging the disciplines. Exploring inscriptions of Islanders in Western systems of thought*. Canberra, Australia: Aboriginal Studies Press.

Nakata, M. (2007b). The cultural interface. *The Australian Journal of Indigenous Education, 36*(S1), 7–14.

Patel, L. (2016). *Decolonizing educational research: From ownership to answerability*. New York, NY: Routledge.

Peat, D. (2002). *Blackfoot physics: A new journey into the Native American universe*. Newbury Port, MA: Weiser Books.

Roth, W. M. (2003). Scientific literacy as an emergent feature of collective human praxis. *Journal of Curriculum Studies, 35*(1), 9–23.

Sammel, A. (2009). Turning the focus from 'other' to science education: Exploring the invisibility of whiteness. *Cultural Studies of Science Education, 4*, 649–656.

Snively, G., & Corsiglia, J. (2001). Discovering Indigenous science: Implications for science education. *Science Education, 85*, 6–34.

Spivak, G. C. (1976). Translator's preface. In J. Derrida, *Of grammatology* (G. C. Spivak, Trans.) (pp. ix-lxxxvii). Baltimore, MD: Johns Hopkins University Press.

Spivak, G. C. (1988). Can the subaltern speak? In C. Nelson & L. Grossberg (Eds.), *Marxism and the interpretation of culture* (pp. 271–313). Urbana, IL: University of Illinois Press.

Spivak, G. C. (1993/2009). *Outside in the teaching machine*. New York, NY: Routledge.

Spivak, G. C. (1994). Responsibility. *Boundary 2, 21*(3), 19–64.

Spivak, G. C. (1999). *A critique of postcolonial reason*. Cambridge, MA: Harvard University Press.

St. Pierre, E. A. (2011a). Post qualitative research: The critique and the coming after. In N. K. Denzin & Y. S. Lincoln (Eds.), *The SAGE Handbook of Qualitative Research (4th Ed.)* (pp. 611–626). Thousand Oaks, CA: SAGE Publications.

Stengers, I. (2018). *Another science is possible: A manifesto for slow science*. Cambridge, UK: Polity.

van Eijck, M., & Roth, W. M. (2007). Keeping the local local: Recalibrating the status of science and traditional ecological knowledge (TEK) in education. *Science Education, 91*(6), 926–947.

van Eijck, M., & Roth, W. M. (2009). Authentic science experiences as a vehicle to change students' orientations toward science and scientific career choices: Learning from the path followed by Brad. *Cultural Studies of Science Education, 4*(3), 611–638.

CHAPTER 8

Towards Being Wounded by Thought: Indigenous Metaphysics Is (Still) Waiting in the Wings of Science Education

Any act of reading is besieged and delivered by the precariousness of intertextuality. And translation is, after all, one version of intertextuality.... heavy-handedness cannot punctuate an entire text where "penser" (to think) carries within itself and points at "panser" (to dress a wound); for does not thinking seek forever to clamp a dressing over the gaping and violent wound of the impossibility of thought? (Spivak, 1976, p. lxxxvi)

While the purpose of this chapter is to conclude this book, how does one go about "closing" a book whose primary task is to unsettle the meta-physics of *clôture* (i.e., as double(d) *closure* [verb] and *enclosure* [noun])? When "each act of reading the 'text' is a preface to the next" (Spivak, 1976, p. xii), *a* conclusion (as with *an* introduction; see Chapter 1) must always be open at both ends, provisional, and contingent as "any act of reading is besieged and delivered by the precariousness of inter-textuality" (Spivak, 1976, p. lxxxvi). *Translating* text into the *context* of (a) conclusion does not necessitate its (whole) surrender to the context: a conclusion remains open to (its own) intertextuality, never (fully) achieving closure.[1]

As Spivak (1976) implicitly asks, how could a conclusion ever achieve closure? The heavy-handedness of the metaphysics of *clôture* can never fully foreclose the possibility of thought (*penser*) that confronts from outside the concepts, categories, and constructs that we hold and that

© The Author(s) 2021
M. Higgins, *Unsettling Responsibility in Science Education*,
Palgrave Studies in Educational Futures,
https://doi.org/10.1007/978-3-030-61299-3_8

hold us. This confrontation *always* retains the possibility of rupturing the clôture of metaphysics, leaving us *wounded by thought* (see also Britzman, 2003; Lather, 2007) and (re)opened by that which is *to-come*. However, this potential is never fully achieved, "for does not thinking [(i.e., *penser*)] seek forever to clamp a dressing over the gaping and violent wound [(i.e., *panser*)] of the impossibility of thought" (Spivak, 1976, p. lxxxvi)? In other words, not all thought (re)opens: thought is, at once, both the possibility of thinking *anew* and thinking *again*. As *poison* and *panacea*, it is nonetheless a necessary and unavoidable *pharmakon*: "thinking about science is part of doing science" (Barad, 2000, p. 245). Thinking, both *penser* and *panser*, are never disentangled from relationships (e.g., Indigenous science) to-come that vacillate between *becoming* and *unbecoming*, *being* and *non-being* beyond the *clôture* of metaphysics (see Barad, 2012a).

Again, even if the metaphysics of *clôture* is totalizing, it is never totalized: the possibility of being wounded by thought, such as Indigenous science to-come, is *always already* present. Also, such can also be said for the entangled (neo-)colonial logics by which it is entangled and (co-)constituted. Importantly, such a radical openness is at the core of both Indigenous science and quantum mechanics. It is how relationships are (re)generated through the world's ongoing becoming (see also Apffel-Marglin, 2011; Cajete, 1994, 2000; Little Bear, 2016). Further, as Barad (2012a) states, "*ontological indeterminacy*, a radical openness, an infinity of possibilities, is at the core of *mattering*" (p. 16, emphasis mine), in both senses of the word (i.e., materializing and coming-to-significance). Accordingly, Kuokkanen (2007) reminds that disrupting and displacing the violence of metaphysical *clôture* as both process and product, as well as the ontological indeterminacy (rather than epistemological uncertainty; see Barad, 2011) generated by this task are the necessary conditions for the spirit of the gift to circulate, for *new* relations to be formed and for relations to be formed *anew* (see also Apffel-Marglin, 2011). For Kuokkanen (2007), such rupturing is in itself a gift, albeit a difficult one. As Indigenous science continues to be, in many ways, considered *to-come*,[2] this book centres the invitation to (re)open the spaces of science education (through a double(d) inversion of the gaze back onto itself; see Lather, 2007) so that science education might be wounded by thought and open(ed) to new and renewed relationships to Indigenous science (to-come).

Thus, the task within this conclusion is not to strive for the "rage for unity" (Spivak, 1976); that is, the tidying up of untimely and disorderly loose ends to achieve the closure typically presumed of concluding. Rather than tuck, trim, and tidy away the partially extended relationalities(-to-come) with which we have journeyed, this chapter is presented as an open-ended conclusion whose possible possibilities also reside in (and beyond) what may have been and what may yet be actualized. Towards this end, this open-ended conclusion explores two parts and processes.

First, I offer a summary and synthesis of the book to highlight the ways in which Indigenous science is *to-come*. However, this potentiality is not transcendental. Rather, it occurs within relationship, where you— the reader—are entangled (recognizing that the "you" addressed here is subjectively (un)done in material-discursive relations as the language-practices you possess also possess you). You are encouraged to pay attention to the ways in which thought manifested throughout: *(How) Did it wound, allowing for the possibility of Indigenous science to-come (i.e., penser)? (How) Did thought work to cover the wound, leaving Indigenous science as yet-to-come (i.e., panser)?*

Second, the notion of being wounded by thought (i.e., *penser* and *panser*) is quickly revisited once more to conclude by turning the message of Indigenous metaphysics still waiting in the wings of science education (see Little Bear, 2016) on its head to (re)open this closing with a message of affirmation.

Summary of the Argument so Far (for Readers in a Hurry...)[3]: Mapping Pathways Travelled upon and Those (yet-)to-Come

Because we must begin *some-where* and *some-time*, let's begin by recapping the central premise of this book[4]: working towards the radical potentiality of thinking (*penser*) Indigenous science within the context of science education, rather than upholding thinking (*panser*) that, through foreclosure, excludes, differs, and defers Indigenous science. This matters as, currently and often within science education, Indigenous science is either excluded or included in ways that differ from or defer its intended meanings, as well as its pedagogical potentiality for *all* students. As the goal of science-education-as-usual is to bring students to know *what* scientists know (i.e., scientific facts) or enact *how* scientists know (i.e., scientific methods), science education often continues to naturalize and normalize the subject position of "scientist" which is emblematic of the masculine, Eurocentric, and anthropocentric subject of Western modernity. When presented as *the* (only or most valid) way-of-knowing-Nature, students' learnings are perpetually inflected and inflicted by the ways in which (neo-)colonial logics and Western modern metaphysics (fore)closes the possibility of being and becoming otherwise. The most significant consequence of this is that "school science" potentially produces experiences of cultural *assimilation* and *acculturation* rather than *enculturation* for the vast majority of students (i.e., ~90% of students, see Aikenhead & Elliot, 2010). However, for students whose daily lived experiences continue to be negatively impacted by Eurocentrism (re)produced with/in (and beyond) science education, learning with/in the cultural practice of "school science" largely continues to be a form of epistemic violence (Bang & Marin, 2015; McKinley, 2001, 2007).

As such, the questions engaged herein are not those of *empowering* those excluded from science to participate, particularly if the culture of science and school science remain unperturbed by the effort. As Kuokkanen (2007) reminds, such efforts can be "more patronizing than helpful and... do not lead to any real transformation" (p. 152). Rather, after Kayumova, McGuire, and Cardello (2019), this book embraces a recent move in science education "from empowerment to response-ability" (e.g., de Freitas, Lupinacci, & Pais, 2017; Higgins & Tolbert, 2018; Wallace, Higgins, & Bazzul, 2018). This is particularly significant given the stakes: science education makes palatable and possible the ongoing dispossession and devastation of Indigenous Land, as well as erasure of Indigenous peoples (Bang & Marin, 2015). One more time: science education has a responsibility, but is it able to respond?

Towards this end, the central question guiding this entire inquiry has been *how is Indigenous science to-come with/in the context of science education?* Recall that this central question is understood and has been explored in through three guiding inflections that are inseparably entangled. First, *to-come* continues to signal that Indigenous science, in the context of science education, has not yet (wholly) arrived. Secondly, *to-come* signals ethical indebtedness: *How might the structure, culture, and discipline of science education be (re)opened and re(con)figured to receive Indigenous science to-come, on its own terms, and in ethical relation?* Thirdly, *to-come* entails a responsibility (and response-ability) for and towards that which is *to-come*: *What types of practices might allow for and nurture the possibility of Indigenous science to-come?* While Indigenous science is still *to-come*—in all three inflections— upon conclusion of this book,[5] I continue to be driven by the goal of exploring what possibilities are made possible through the labour of attempting to be response-able and accountable (see Barad, 2010) to Indigenous metaphysics waiting in the wings. As a means of responding to this guiding question and goal, I *translate* the pathways[6] of science education journeyed upon in each chapter into the format of a rough sketch.[7] These sketches offer a synthesis of Indigenous science to-come in one or more of its inflections, as well as an overview of the chapter contributions to the field of science education. Further, the *deconstructive methodologies* employed throughout the book will be revisited for their significance with respect to the larger project of unsettling science education.[8] As these sketches are enactments, they are *mappings* that are meant to invite relation as living, breathing concepts rather than (re)*tracings* who are epistemologically and ontologically foreclosed through representationalism.

Chapter 1: Unsettling Metaphysics in Science Education

The trailhead of this journey introduces and frames the simultaneously co-constitutive and othering relation between Western modern science and Indigenous ways-of-knowing-in-being as it manifests within spaces of science education. This chapter asks: *What does metaphysics (i.e., the co-constitutive space of epistemology, ontology, ethics, among others) have to do with science education and the ways in which Indigenous science remains (yet-)to-come?* And provides multiple tools to equip oneself for the remainder of the journey. Most notably, unsettling science education is presented as a double(d) approach to address the ways in which settler-colonial logics linger and lurk within and as sedimented and stratified knowledge-practices (Bang & Marin, 2015; Bang, Warren, Rosebery, & Medin, 2012). As a more nascent approach to the question of Indigenous science within science education, this work is supplemented by two larger orientations that address its double(d) meaning. The first orientation addresses *settler colonialism* by providing an overview of similar yet different pathways explored with/in science education: decolonizing (e.g., Aikenhead & Elliot, 2010; Chinn, 2007) and post-colonial (e.g., Carter, 2004; McKinley, 2007) approaches to science education in response to the metaphysics of modernity. In particular, this overview maps the ways in which Eurocentrism operates (e.g., mis- and under-representation, deficit thinking, universalism) and how decolonizing and post-colonial science respond through curricular deconstruction and reconstruction. Accordingly, the second orientation addresses *settled meanings*. It unpacks deconstruction (e.g., Spivak, 1976) in relationship to decolonizing methodologies, as well as decolonizing science education, as a (meta-)methodological approach to (re)open the metaphysics of modernity. Of particular importance to this book is the displacement and disruption of the self/other, nature/culture, and ethical possibility/impossibility binaries. Importantly, this sets the stage for bearing witness to the ways in which settler coloniality often manifests as absent presence and to (re)open the space of response *within* science education towards Indigenous ways-of-knowing-in-being.

While this first chapter's primary purpose is to provide a general orientation for the work that is to come within the remainder of the book, a significant contribution that emerges is (re)positioning decolonizing science education at the ontological turn (e.g., Kayumova et al., 2019; Higgins & Tolbert, 2018). The critical call to decolonize has been primarily taken up by extending the openings produced through treating both science, and science as problematic cultural spaces, to be examined through *socio-cultural* approaches (e.g., sociology of science). As cultural and socio-cultural approaches offer both methodological possibility and problematic when they interface with the metaphysics of modernity (see Carter, 2010; Latour, 1993), moving towards ontological plurality[9] and deconstructing ontology-as-usual (i.e., Cartesianism) (re)opens the conversation beyond the question of which epistemology best fits *the* ontology of Nature.

While there is no specific deconstructive practice conceptually sketched out within this chapter, to introduce *deconstruction* into a field where it is under-employed and -explored bears significance. In turn, this contributes specifically to a small yet growing body of research that engages in (mis)reading science education texts for their settled (neo-)colonial referents (e.g., Carter, 2004, 2005; McKinley & Aikenhead, 2005). Further, these insights are employed to frame de/colonizing in a fulsome way within the field of science education[10] to make explicit the ways that ways in which decolonizing and colonizing discourses cannot be wholly framed in opposition as they are always already co-constitutive (e.g., (neo-)colonial logics even seep into decolonizing efforts).

Chapter 2: The Homework of Response-Ability in Science Education

Continuing to journey with the insight from the previous chapter that science education has a responsibility towards Indigenous science (to-come) as the result of its exclusionary practices rooted in Eurocentrism and (neo-)colonialism, this chapter engages with the differential question of the degree to which science education has the ability to respond towards Indigenous ways-of-living-with-nature (IWLN) and traditional ecological knowledge (TEK). Through revisiting a significant personal pedagogical encounter (see Higgins, 2014), I make the case that having a responsibility and being able to respond are not one and the same, particularly when the very systems that exclude Indigenous science (e.g., Eurocentrism) (fore)close the ability to respond as well as how we come to understand responsibility.

Thinking with the work of Sami scholar Rauna Kuokkanen (2007, 2008, 2010), attention is brought to unsettle responsibility: disrupting and displacing the ways in which responsibility is often and inadvertently overcoded by the (neo-)colonial logics that it sets out to refuse and resist as the result of epistemic ignorance.[11] Significantly, epistemic ignorance is not strictly a lack of knowledge—it can also be the result of knowledge that acts as resistance, which (fore)closes the possibility of knowing otherwise (Kuokkanen, 2008). For there to be responsibility, there must be an ability to respond: there is a "homework of response-ability" required to (re)open the norms of responsiveness towards the possibility of heeding the call of Indigenous science from within the structure of science education (see Kuokkanen, 2007, 2010). The "home" part of homework can take many meanings: home as cultural, disciplinary, geographical, historical, epistemological, and ontological, among others. The "work" part involves coming to understand how these multiple "homes" foreclose the ability to respond to Indigenous science to-come).

The most significant contribution to science education from this chapter is the translating of Kuokkanen's (2007) *homework of response-ability* for the context of science education. While it is not new to study the "home" of this field (be it historical, philosophical, cultural, or other), Bazzul (2017) suggests that this *homework* is rarely employed to put science education back into its contemporary context: "science education seems to ignore the social, historical, and political realities that give shape to its practices, knowledges, and literatures" (p. 67). Significantly, the homework of response-ability is not only a call for engaging with the ways in which science education is always already a situated knowledge (see van Eijck & Roth, 2007), but that its multiple and co-constitutive homes are rich sites for critical labour, such as (re)opening science education to Indigenous science to-come. In turn, as many of science education's multiple homes are absent presences which are often sedimented and stratified, as well as locations in which (neo-)colonial referents linger and lurk, the homework of response-ability is also the primary deconstructive tool put to work and offered by this chapter to unsettle science education.

Chapter 3: Serious Play: Inflecting the Multicultural Science Education Debate Through and for (Socratic) Dialogue

Wandering through the pathway(s) presented by the multicultural science education debate, the most glaring feature of the scholarly literature is strongly conflicting epistemic demands on "what counts" as science (and in turn school science curricula) and respectful education within multi-cultural science education classrooms (see Lewis & Aikenhead, 2001; McKinley & Stewart, 2012; van Eijck & Roth, 2007). At stake is the inclusion of TEK and IWLN alongside WMS. This highly contested commitment and practice nearly (wholly) polarizes the field, resulting in two seemingly diametrically opposed positions. *Cross-culturalists* advocate for the inclusion of TEK and IWLN on the basis that they are equally (but not similarly) valid; *universalists* do not consider these placed-based ways-of-knowing- nature as *equally valid* to the "universal" standard of WMS. These positions directly impact conceptions and enactments of respecting and including diverse knowledge traditions that students bring and, in turn, students themselves.

Mapping the contours of the path, the culture of the debate can be said to deploy both dialectic negation (see Bohm, 1996), as well as adversariality (as method; see Moulton, 1983). This regularly brings the potentiality and possibility of creative movement through the field of science education to a viscous stasis, although never achieving a grinding halt. On this note, Southerland (2000) commented early on that the multicultural science education debate was, "generating a lot of heat, but very little light" (p. 289). While the critical task of addressing conflicting values "head on" is a one of necessity and importance, the literature seems to indicate that changes through current modes of engagement have been, and continue to be, slow and partial. This invites an important question that is revisited in many ways throughout the book: *How might coming at these issues of metaphysical closure and seemingly sedimented knowledge-practices sideways and obliquely provide new (re)openings?*

A central contribution to science education that is produced through this chapter is (re)considering the multicultural science education debate obliquely through its antagonism. While I am not the first and likely will not be the last to consider the antagonistic nature of the multi-cultural science education debate (e.g., McKinley & Stewart, 2012; van Eijck & Roth, 2007), putting theories, strategies, and practices of conflict resolution to work are notable (e.g., dialogue; see Bohm, 1996). A multiplicity of tools, practices, and considerations are provided for attending to, and reconfiguring, the (re)production of adversariality. These include consideration of meaning-making as relational and embodied (e.g., paying attention to emotions and physical responses as indicators of the movement of thought), as well as seeking out shared meanings as productive pathways and paths of lesser resistance. Examples of shared meanings sought in Chapter 3 include knowledge as knowledge-practice (van Eick & Roth, 2007); and knowledge-practices as epistemologically situated (van Eick & Roth, 2007), ontologically situated (Cobern & Loving, 2008), and culturally hybrid (van Eijck & Roth, 2009).

The deconstructive tool put to work within Chapter 3 was the play of (re)signification. As the first of many deconstructive strategies employed and leveraged throughout this book, the play of (re)signification invites the reader to consider the ways in which textual realism (i.e., representational fidelity; see Carter, 2004) is a cultural myth that diverts and at times disallows the wandering of the pathways of science education otherwise.

Chapter 4: Mirrors, Prisms, and Diffraction Gratings: Placing the Optics of the Critical Gaze in Science Education Under Erasure (After the Critique of Critique)

Following the orientation provided by the trailhead of Chapters 3 and 4 asks if the metaphors through which the field of science education critically engages are (re)shaping the possible pathways for, and ways-of-navigating, inclusion of Indigenous science. (Re)considering critique within this chapter began with the notion that critique—the conventional mode through which the culture of dominance (e.g., Eurocentrism) within science education is challenged—has *run out of steam* (Latour, 2004a). As Latour (2004a) states, the very tools employed by critics have been appropriated by those who have been and continue to be its intended targets. As such, critique is easily reversed and brought to a standstill (as evidenced within Chapter 3). Thus, critique becomes protective (of positions, concepts, and categories) rather than productive (in seeking shared meanings) (see Spivak, 1976), unfortunately contributing to the (re)production of Indigenous science as yet-to-come within science education (see McKinley & Aikenhead, 2005; see also Kuokkanen, 2007; Spivak, 1994).

Critique (in the conventional sense) is meant to (re)open possibilities for knowing and being otherwise. However, it presents itself as already prescriptive as a result of proposing particular ways of navigating its pathways. Such operationalization of critique enacts and upholds the very metaphysical structures worked against within this book. Critique-as-usual (re)naturalizes many of the tools and strategies of (neo-)colonization and *clôture* (e.g., distance, separation, hierarchy, sameness/difference). Nonetheless, critique should not, and cannot, be dismissed. Its role continues to be central and critical in terms of (re)opening possibilities for Indigenous science to-come.

Considering critical engagement as epistemologically situated and metaphorically mediated allowed for the possibility of it to *become otherwise*. Approaching conventional critique as mirrored correspondence (see Barad, 2012b; Latour, 2004a) opened up space to differently journey the path of critique. This entailed contemplating optical metaphors whose differential metaphysics might allow: (a) Indigenous science (to-come) to be more intelligible from within spaces of science education; (b) for the potential to reveal, as well as displace, (neo-)colonial structures and strategies that (re) centre WMS (both implicitly and explicitly); and, (c) for the production of differential modes of critique that might be more productive in being response-able and accountable towards Indigenous science to-come.

Through deconstructive (mis)reading, what science education *is* (e.g., critique as mirrored correspondence) was substituted with what it *is (not)* (e.g., Foucaultian prismaticity, Baradian diffraction). Such deconstruction works to displace and disrupt, all the while upholding the structure one is working within. The re(con)figured practices of critique (i.e., prism and diffraction grating) differentially reveal otherwise taken-for-granted concepts, categories, knowings, and beings operating with/in multicultural science education.[12] These optical metaphors present unique contributions to the field. As Bazzul and Carter (2018) state in their literature review of Foucaultian application in science education, there are but a few examples of Foucaultian critique within science education in general and, I would suggest, much less with respect to the particular context of Indigenous science. Inclusion of diffraction grating as critical metaphor is all-the-rarer as Barad's (2012b) articulation of diffraction as critique is but recent, and the ontological turn only beginning to bear upon science education (see Milne & Scantlebury, 2019).

Chapter 5: Tinkering with/in the Multicultural Science Education Debate: Towards Positing an(Other) Ontology

The continued excursion through science education presented with/in Chapter 4 entails wandering the pathways of multicultural science askew and obliquely by considering uncommonly considered, but ever-present, ontology. Journeying with insights garnered from the previous chapters reveals that ontology as singular, naturalized, and taken-for-granted comes to bear on the ways in which Indigenous science is *to-come*. In short, ontology at once acts as the handmaiden of Eurocentric and the (cultural) meter stick through which epistemological realism (i.e., the epistemology of WMS) is upheld and (re) centred. Here, Cobern and Loving's (2008) call for ontological situatedness as a means of resolving the multicultural science education debate simply displaces the terms (e.g., presenting WMS as universal) rather than disrupting them: alignment assumes and presupposes separation and separability, as well as requires mirrored correspondence (extending Chapter 3's metaphor of the critical mirror).

By extending critical alternatives from the previous chapter, science education is (re)opened to a differential consideration of ontology. They include: (a) ontology as plural rather than singular; (b) ontology as a metaphysical choice rather than pre-supposed; and (c) the situation where the aforementioned ontological conditions are cuts one colludes in making, and such metaphysical choosing (re)opens ontological agency to one's (distributed) ethical responsibility (see Barad, 2007). This invites the positing of *an* ontology and may begin by naming ontology, and produce movement towards accounting for and being accountable to what is produced and producible within the (never-fully-stable) ontology of teaching, learning, and meaning-making with/in science education.

The deconstructive practice of tinkering (i.e., using tools intended for other tasks) utilized in Chapter 5 invites the reader to pay attention to the ways in which the ends and the means never fully coincide. One such consideration that is noteworthy is how tinkering asks for attention to practice, particularly those that get labelled as non-practices (see Spivak, 1976, 1988, 1993/2009). A highly significant example here is the enactment of ontology as *practice* (see Barad, 2007, 2010; Cajete, 1994, 2000), rather than pre-supposed.

Chapter 6: Positing Cartesianism as an Ontology Within Science Education: Towards a More Response-Able Inheritance with Dr. Frédérique Apffel-Marglin

Chapter 6 extends the wanderings presented in Chapter 5 through continuing the work of tinkering within what it means to posit *an* ontology. Particularly, it explores what it might mean to position Cartesianism—the ontology of WMS—as *an* ontology in order to account for and be accountable to what it produces and makes producible. A series of expert interviews with Dr. Frédérique Apffel-Marglin reveals that Cartesianism came to be operationalized in sixteenth and seventeenth century modern Europe as the result of a confluence of inseparable forces: economy, geography, military, politics, religion, as well as science. These inseparable forces come to constitute *a* metaphysics of clôture which, *there-then*, created multiple forms of separation and separability. Notably, for the status of Indigenous science to-come, is the separation (and distance) produced between people who worked the land as knowledge-practice, from the land (i.e., the commons) and the knowledge-practices that were enacted through the spatial-economic practice of the enclosure (see Apffel-Marglin, 2011). Furthermore, through the birth of the laboratory and its associated technologies (e.g., modest witness in concert with other forces), science comes to be constructed as *a-cultural* and *a-political*, etc; it appears to transcend bias. Considering diffraction invites attention to the ways in which these *there-thens* transposed into *here-nows*, as well as the (neo-)colonial knowledge-practices through which this multiplicity of entangled knowledge-practices have been diffused and distributed (see Battiste, 2005; Blaut, 1993; Spivak, 1999).

While the contributions of this chapter are multiple given Dr. Apffel-Marglin's expertise on the "birth of modernity", as well as why the metaphysics of modernity is a significant location to labour in accounting for and being accountable to Indigenous science to-come, I wish to high-light two *orientations* to decolonizing and post-colonial science education that diffractively emerge from this series of expert interviews. The first is the importance of privileging transdisciplinarity (see also Battiste, 2013; Smith, Maxwell, Puke, & Temara, 2016). Science, science education, and Indigenous science (to-come) are irreducible, always failing to fit within *a* single discipline. While this is a well-recognized quality of Indigenous ways-of-knowing-in-being (e.g., Battiste, 2013; Cajete, 2000; Smith, 1999/2012; Smith et al., 2016), science and science education are more likely to be framed with/in disciplinary logics (as a form *of* disci-pline, through the logics of self-sameness and not-otherness). Remaining open to knowledge outside of one's disciplinary tradition is not only an ethical move, but it is also one that is productive as knowledge-practice (see Bazzul & Carter, 2018). To follow the lived life of concepts and categories into other spaces allows one to consider their differential being and becoming (e.g., Battiste, Bell, Findlay, Findlay, & Henderson, 2005). Tinkering, as leveraging the un/common, is a remarkable practice of pursuing this productive supplement.

The second orientation, related to the productive supplementarity of transdisciplinarity, is considering the multiplicity of ways in which the metaphysics of individualism work to separate and make separable ecolo-gies of relationships, as well as the ways in which they (inseparably) complement one another. One of the ways that this can be put into prac-tice is, as signalled on the first page of the introduction, by considering entanglements and ecologies of relationships as extending beyond a *here-now* to a multiplicity of *there-thens*. In turn, the move towards Indigenous science to-come cannot and should not be considered a response to curricular *place-less-ness*, but rather *place-full-ness*. While place may not considered through an Indigenous sense of place, or even actively consid-ered at all, place (i.e., proximal Nature-Culture, space-time-mattering) is never absent: place has its own agency that always comes to bear (see Barad, 2010; Cajete, 2000). The question of place becomes *which* place comes to bear (as well as when, where, and how)? As demonstrated within Chapter 6 (and introduced in Chapter 3), dominant conceptions of knowing Nature cannot be so easily disentangled from a laboratory *there-then,* which always already comes to bear on a *here-now*. To (too easily) suggest a conception of place without attending to the ways in which Nature-Culture already comes to manifest runs the risk of masking the workings of power (see McKinley, 2001).

While there are a few deconstructive tools put to work within this chapter above and beyond tinkering, *diffraction* as methodology is worth highlighting. As means of "reading insights through one another" (Barad, 2007, p. 30) to account for and be accountable to patterns of difference that are produced, diffraction was an invaluable tool to reveal the ways in which seemingly disparate historicities are always already entangled within contemporary science education: it is to do the (home)work of inheritance (see Barad, 2010; Derrida, 1994/2006). Again, this bears significance as concepts and practices from the ontological turn, particularly those that are methodological rather than substantive, are only beginning to gain traction within science education (see Milne & Scantlebury, 2019).

Chapter 7: Response-Ability Revisited: Towards Re(con)figuring Scientific Literacy

As the logics through which the metaphysics of modernity are (re)produced are circular, walking the pathways of science education may often feel as though it is a form of déjà-vu—although what seems as sameness is always already differential. In turn, breaking from modernist orbits requires sustained and focused efforts to work against slippage into processes through which (neo-)coloniality is maintained (e.g., the dialectic negation of othernesss). Nakata's (2007a, 2007b) concept of the cultural interface provides a rich lived-conceptual location from which to consider the multiplicity of conflicting, contradictory, and co-constitutive interactions that occur between Indigenous metaphysics and the metaphysics of modernity. While such circularity is a boon, it is also a gift: resistance to dominance can happen in (almost) any location through which power circulates. This is, of course, with the cautionary note that not all locations are equally productive, and that some locations are a privilege to inhabit even if inhabitation is critical (see Spivak, 1988). Furthermore, not all locations are equally pliable and pry-able.

To be responsible for the complex nature of the culture interface, one needs to be able to respond, to be response-able (see Kuokkanen, 2007; Spivak, 1994). However, to be able to respond to that which is *to-come* requires that one be able to respond to that which one does not already know (Barad, 2010). Stated otherwise, one must be wounded by (the impossible) thought (i.e., *penser*) and forego the urge to clamp a dressing over the wound (i.e., *panser*). To work the possibility of getting lost beyond the circularity of the pathways of science and to be wounded by thought, I targeted a location that is central to science education; one which displayed indeterminacy, thus making them pliable and pry-able and open to response-ability as deconstruction: scientific literacy. Significantly, scientific literacy is a sedimented location which (re)produces settler colonialism (see Sammel, 2009): unsettling is indispensable. (Re)opening scientific literacy to other possibilities of "literacy" that work towards plurality and dialogue (i.e., not needing to negate other forms, but rather to situate them): agential literacy and ecologies of relationships (i.e., an Indigenous "sense of place"). Both of these productive (mis)readings allow for the possibility of responding to forms of Indigenous science that are often *to-come:* community as more than a collection of individuals (which extends beyond humans), other-than-human agency, and non-linear temporalities (see Cajete, 1994, 2000).

The primary deconstructive tool employed within this chapter is *response-able reconstruction,* which takes seriously the notion that deconstruction and reconstruction are neither separate nor separable (see Jackson & Mazzei, 2012). Rather, drawing from Barad (2010), response-able reconstruction is a recursive, iterative, and co-constitutive process which stems from the possibilities that arise from the relationality between the two. Notably, it is a practice of re(con)figuring which can productively leverage the openings generated by homework of response-ability towards enacting science education otherwise.

An Open-Ended Conclusion: Indigenous Metaphysics Is (Still) Waiting in the Wings of Science Education

This book is now reaching its close. Indigenous metaphysics is *still* waiting in the wings of science education, and is neither wholly nor hospitably received on the main stage of science education. Nonetheless, I conclude with a message of affirmation. What appears as a partial absence is not lack, a mere criticism, nor a critical negation. Rather, it is a potential relationality, a co-constitutive moment, and movement *to-come.*

As Barad (2012a) suggests us in *What is the measure of nothingness?*, even nothingness is itself is a doing that is rife with possibility and potentiality. This is even the case for what is classically understood as "pure" nothingness: the vacuum of space. As she states,

> From the point of view of classical physics, the vacuum has no matter and no energy. But the quantum principle of ontological indeterminacy calls the existence of such a zero-energy, zero-matter state into question, or rather, makes it into a question with no decidable answer. Not a settled matter, or rather, no matter. And if the energy of the vacuum is not determinably zero, it isn't determinably empty. (Barad, 2012a, pp. 8–9)

Rather than this determinability being a question of epistemological uncertainty, it is one of ontological indeterminacy, "the indeterminacy of *being/non-being, a ghostly non/existence*" (Barad, 2012a, p. 12, emphasis in original). Even the smallest of particles vacillates between being a something and a nothing. It is rife with potentiality as it is never "just itself"; it is co-constituted and co-constituting a plurality of other particles and particles-to-come. As she states, "even the smallest bits of matter are an enormous multitude. Each 'individual' is more up of all possible histories of virtual intra-actions with all Others. Indeterminacy is an un/doing of identity that unsettles the very foundation of non/being" (p. 15). This has consequences for ethics (as co-constituted by epistemology and ontology, of course): "individuals are infinitely indebted to all Others, where indebtedness is not about a debt that follows or results from a trans/action, but rather, a debt that is the condition of possibility of giving/receiving" (pp. 15–16).

As Indigenous science already displays and has always deployed "ingenuity, creativity, resourcefulness, and ability of people to learn and to teach a harmonious way of existence with Nature" (Cajete, 2000, p. 78), the possibility of hospitably *receiving* Indigenous science is not only an ethical call; science education can learn much from Indigenous ways-of-knowing-in-being and its practices of relational balance, (re)generation, and renewal. Thus, wandering the pathways of science education to heed the call of Indigenous science to-come matters even if it is but a possible possibility: "matter is never a *settled* matter. It is always already radically open" (Barad, 2012a, p. 16, emphasis mine). Nature (i.e., space, time, matter) is and was never fully totalized within (neo-)coloniality. It is with hope and affirmation that I end: *Indigenous science is still yet-to-come.*[13]

NOTES

1. In other words, the task of a conclusion is "daring to repeat the book and reconstitute it in another register" (Spivak, 1976, p. xii); it is an audacious yet impossible act of translation. As Derrida (1976) reminds more generally, translation is never fully achieved or achievable: meaning is always differed and deferred. Just as there is no *originary* text there can also be no *terminal* form of textuality; "there is no exception to this rule" (Spivak, 1976, p. xii), even when presented otherwise. Importantly, this perpetual (re)opening relates explicitly to questions of science education's ability to receive Indigenous science to-come on its own terms: "the question of hospitality begins with the question of translation" (Kuokkanen, 2007, p. 137). Importantly, translation is both possibility and problematic: it is at once the grounds for dialoguing across difference and, yet, "hospitality is not possible when the guest is required to speak the language of the host" (Kuokkanen, 2007, p. 76). This becomes all-the-more significant when the (neo-)colonial demands of translation make it such that the effort of putting knowledge-practices into relation (re)produces the erasure of Indigenous peoples, places, and practices. Accordingly, there remains the need to be hyper-vigilant in the work of leveraging the productive im/possibilities of translation in creating spaces of dialogue between Indigenous science to-come and WMS within science education (see also Ahenakew, 2016; Higgins & Kim, 2019).

2. It bears repeating: the notion of *to-come* says more about the inhospitability of science education than the ways in which Indigenous peoples have been practicing ways-of-living-with-Nature since time immemorial. This is to say that Indigenous science remains "waiting in the wings" of science education.

3. A conclusion is conventionally (and often conveniently) the end or finish of a given task, event, or process: it often synthesizes what has already come. While conclusions hold the potentiality for a differential engagement with that which has preceded it, they can also work to (fore)close the possible possibilities of meaning-making. In turn, I borrow Latour's (2004b) cheeky formulation of a conclusion as a "summary of the argument (for readers in a hurry...)" (p. 231). While conclusions hold the potentiality for a differential engagement with that which has preceded it, they can also work to (fore)close the possible possibilities of meaning-making.

   However, given the contemporary pressures of educational institutions which are often *busy by design* (and all-the-more-so for those who individually do the work of being the bridge over the gaping disjuncture between commitments to diversity and how inclusion is enacted), the above is not an individual(istic) admonition. Rather it is an invitation: as the temporality of reading and writing is forever askew (see Spivak, 1976), it is an invitation to work backwards and outwards to the texts most relevant and pressing.

4. One final time (for those reading from front-to-back), the mantra of beginning some-where and some-time is not only a persistent reminder that we are *always already* within the question of Indigeneity within science education, but also an invitation to address it as such. To "begin" as if nothing has been done is a form of know-nothing-ism which serves to mask power with innocence, and obscures: (a) the multiple and ongoing responses articulated and practiced within science education for the last few decades; (b) that Indigenous peoples are the most researched peoples in the world; and, (c) that science's Othering of (and, in turn, co-constitution by) Indigeneity irreducibly places the two in relation of obligation. This is to state that science education has a responsibility: one that is *to* the Other rather than *for* the Other. The distinction here is not semantic: a responsibility *to* is to be accountable and answerable to the ways in which Othering has occurred, rather than enacting a form of responsibility-as-usual which leaves (neo-)colonial forms of power stratified and sedimented (see McKinley, 2001). Hence the question: science education has a response-ability, but is it able to respond?

For readers beginning at the end, Kuokkanen (2007) suggests that there is *home*work to be done:

> The academic responsibility for doing homework on [I]ndigenous epistemes must begin at an even more elemental level than examining one's beliefs, biases, and assumptions. It must start with the acknowledgement that the "[I]ndigenous" exists, be it in terms of peoples, their epistemes, or how they have been configured in the geopolitical past and present. (p. 115)

(Re)opening responsiveness requires doing a creative excavation of the multiple ways in which the *home* of science education (e.g., cultural, disciplinary, geographic, historical) differs and defers the possibility of hospitably receiving Indigenous science. However, as Kuokkanen (2007) emphasizes, this critical and creative work can only begin once we have accepted the existence of Indigenous science or ways-of-living-with Nature.

5. It can be, and has been stated that the possibility for Indigenous science to (wholly) arrive on its own terms and in ethical relation is an impossible possibility in this contemporary moment (see Carter, 2004; McKinley & Aikenhead, 2005).

6. After Cajete (1994), we are encouraged to (re)consider science education as a journey*ing* of pathways. This is subtle deconstructive prompt that is significant in three major ways. First, it addresses the (neo-)colonial desire for representation: what might appear as a sedimented and stratified *path* is inseparable from its enactment, its journeying, its *way*. Secondly, journeying *the* path denaturalizes it as it is and can never be the same times: a pathway of science education comes to reveal itself as but *a* path, (re)opening up a multiplicity of ethical possibilities. Lastly, such tactical insights provide direction for strategically straying off the beaten path or taking the pathway in unintended ways to lose sight of the prescriptive and often problematic ways in which the path is regularly travelled.

7. This partial account is not only a textual strategy to present future orientations and possibilities offered by the work within this book, but also whose incompleteness is meant as gift: presenting *an* account rather than *the* account invites orientations to continue wandering the pathways of science education rather than destinations which one accepts or rejects (i.e., science *is* or *is not* this).

8. The significance of deconstruction in science education cannot be understated. It is a form of methodological engagement that allows for the possibility of (re)opening science education towards Indigenous science to-come by identifying and tinkering with/in the structure of education, between what it *is*, *is not*, and *could be(come)*; as well as (mis)reading science education for its subtle and lingering (neo-)colonial referents and enactments. Despite its significance, deconstruction is rarely seen at this intersection, despite calls for its use in science education (e.g., Carter, 2004; McKinley & Aikenhead, 2005), science studies (e.g., Barad, 2010; Latour, 1993), as well as in spaces of Indigenous education that overlap with decolonizing and post-colonial science education (e.g., Battiste, 2013; Cajete, 1994; Donald, 2012).

9. Significantly, as there has been much learned from approaches that diligently study and critically interrogate the culture of science and school science, moving towards ontological plurality does not require the doing away of culture, nor its politics.

10. Notably, this framing significantly extends de/colonizing's introduction elsewhere for science education (i.e., Higgins, 2014), and situates it within a larger conversation which was already ongoing (e.g., Rhee & Subreenduth, 2006) and provides a richer theoretical context between the decolonizing and post-colonial theories that inform it.

11. For example, the taken-for-granted Western notion of responsibility doubly individualizes responsibility by making responsibility an individual affair and rendering the individual the adjudicator of whether responsible

action has occurred or not. Such becomes problematic when *epistemic ignorance* renders the possibility of accounting for and being accountable to one's own actions murky and diffuse (i.e., individuals who perceive their actions as not being oppressive are often also the final arbiter of such within conventional notions of responsibility; see Kuokkanen, 2007). In turn, unsettling response-ability entails decentering the individualistic subject of responsibility to engage with the larger questions of how this "I" is shaped (i.e., the homework of response-ability). Moving beyond the "I" as the ethical subject means considering the ability to respond as being more than an individual affair (without excusing the "I" from responsibility): responsibility must be more than a self-accounting for individually doing or having done the right thing.

12. Furthermore, such metaphors invite a critical suspension (see Bohm, 1996; Foucault, 1997). As some of the issues facing the place of Indigenous science within school-based science education has been differentially (re)produced for decades now (McKinley & Stewart, 2012), it may be worth considering momentary suspension towards displacing (rather than attempting the impossible and never achieved destruction) of dominant logics (despite seeming counter-intuitive to pause amidst this critical contemporary moment in which Indigenous science is yet-to-come) (see also Patel, 2016; Stengers, 2018).

13. As this conclusion, like all conclusions, is always already an introduction, a supplement that escapes the structure which may be unintelligible as such and yet-to-come, I subtly offer an invitation here, in the margins. As Derrida (1976) suggests, "*thought* is ... the blank part of the text" (Derrida, 1976, p. 93, emphasis in original); this "closing" is also an opening. You are invited to critically create conditions for students and for yourself to pedagogically be wounded by thought (i.e., *penser*): to think Nature beyond the metaphysics of modernity through which the wound of knowing nature is bandaged closed (i.e., *panser*). However, it bears repeating one last time that one cannot be wounded by thought (*penser*) if the wound is already (ad)dressed (*panser*): responding to world beyond oneself requires that the world not be (fore)closed prior its encounter (see also Spivak, 1999). To engage in the homework of response-ability is to engage in the necessary (re)opening the possibility of being wounded by thought by addressing the ways in which the multiplicity of "home" in homework is (fore)closed to its (co-)constitutive otherness. All of this, to engage with and tell stories that Nature (i.e., space, time, matter) might tell (with) us as a community of learners, were we able to listen anew through different attunement (see Barad, 2007; Cajete, 2015).

REFERENCES

Ahenakew, C. R. (2016). Grafting Indigenous ways of knowing onto non-Indigenous ways of being. *International Review of Qualitative Research, 9*(3), 323–340.

Aikenhead, G. S., & Elliot, D. (2010). An emerging decolonizing science education in Canada. *Canadian Journal of Science, Mathematics and Technology Education, 10*(4), 321–338.

Apffel-Marglin, F. (2011). *Subversive spiritualities: How rituals enact the world.* New York, NY: Oxford University Press.

Bang, M., & Marin, A. (2015). Nature–culture constructs in science learning: Human/non-human agency and intentionality. *Journal of Research in Science Teaching, 52*(4), 530–544.

Bang, M., Warren, B., Rosebery, A. S., & Medin, D. (2012). Desettling expectations in science education. *Human Development, 55*(5–6), 302–318.

Barad, K. (2000). Reconceiving scientific literacy as agential literacy. In R. Reed & S. Traweek (Eds.), *Doing Science + Culture* (pp. 221–258). New York, NY: Routledge.

Barad, K. (2007). *Meeting the universe halfway: Quantum physics and the entanglement of matter and meaning.* Durham, NC: Duke University Press.

Barad, K. (2010). Quantum entanglements and hauntological relations of inheritance: Dis/continuities, spacetime enfoldings, and justice-to-come. *Derrida Today, 3*(2), 240–268.

Barad, K. (2011). Erasers and erasures: Pinch's unfortunate 'uncertainty principle'. *Social Studies of Science.* https://doi.org/10.1177/030631271140 6317.

Barad, K. (2012a). What is the measure of nothingness? Infinity, virtuality, justice. *Documenta, 13*, 1–17.

Barad, K. (2012b). Interview with Karen Barad. In R. Dolphijn & I. van der Tuin (Eds.), *New materialism: Interviews & cartographies* (pp. 48–70). Ann Arbor, MI: Open Humanities Press.

Battiste, M. (2005). You can't be the global doctor if you're the colonial disease. In P. Tripp & L. J. Muzzin (Eds.), *Teaching as activism* (pp. 121–133). Montreal, QC: Queen's University Press.

Battiste, M. (2013). *Decolonizing education: Nourishing the learning spirit.* Saskatoon, SK: Purich Publishing.

Battiste, M., Bell, L., Findlay, I., Findlay, L., & Henderson, J. (2005). Thinking place: Animating the Indigenous humanities in education. *The Australian Journal of Indigenous Education, 34*, 7–18.

Bazzul, J. (2017). From orthodoxy to plurality in the nature of science (NOS) and science education: A metacommentary. *Canadian Journal of Science, Mathematics and Technology Education, 17*(1), 66–71.

Bazzul, J., & Carter, L. (2018). (Re)considering Foucault for science education research: Considerations of truth, power and governance. *Cultural Studies in Science Education, 12,* 435–452.

Blaut, J. (1993). *The colonizer's model of the world: Geographical diffusionism and Eurocentric history.* New York, NY: Guilford Press.

Bohm, D. (1996). *On dialogue.* New York, NY: Routledge.

Britzman, D. P. (2003). *Practice makes practice: A critical study of learning to teach.* Albany, NY: SUNY Press.

Cajete, G. (1994). *Look to the mountain: An ecology of indigenous education.* Durango, CO: Kivaki Press.

Cajete, G. (2000). *Native science: Natural laws of interdependence.* Santa Fe, NM: Clear Light Books.

Cajete, G. (2015). *Indigenous community: Rekindling the teachings of the seventh fire.* St Paul, MN: Living Justice Press.

Carter, L. (2004). Thinking differently about cultural diversity: Using postcolonial theory to (re)read science education. *Science Education, 88*(6), 819–836.

Carter, L. (2005). A place for alternative readings: Can they be of use? Responding to comments on "Thinking differently about cultural diversity: Using postcolonial theory to (re)read science education.". *Science Education, 89*(6), 913–919.

Carter, L. (2010). The armchair at the borders: The 'messy' ideas of borders, border zones and epistemological diversity in multicultural science education. *Science Education, 94,* 1–20.

Chinn, P. (2007). Decolonizing methodologies and Indigenous knowledge: The role of culture, place and personal experience in professional development. *Journal of Research in Science Teaching, 44*(9), 1247–1268.

Cobern, W. W., & Loving, C. C. (2008). An essay for educators: Epistemological realism really is common sense. *Science & Education, 17,* 425–447.

de Freitas, E., Lupinacci, J., & Pais, A. (2017). Science and technology studies × educational studies: Critical and creative perspectives on the future of STEM education. *Educational Studies, 56*(6), 551–559.

Derrida, J. (1976). *Of grammatology* (G. C. Spivak, Trans.). Baltimore, MD: John Hopkins University Press.

Derrida, J. (1994/2006). *Specters of Marx: The state of the debt, the work of mourning, & the new international* (P. Kamuf, Trans.). New York, NY: Routledge.

Donald, D. (2012). Indigenous Métissage: A decolonizing research sensibility. *International Journal of Qualitative Studies in Education, 25*(5), 533–555.

Foucault, M. (1997). *The politics of truth.* New York, NY: Semiotext(e).

Higgins, M. (2014). De/colonizing pedagogy and pedagogue: Science education through participatory and reflexive videography. *Canadian Journal of Science, Mathematics and Technology Education, 14*(2), 154–171.

Higgins, M., & Kim, E. J. (2019). De/colonizing methodologies in science education: Rebraiding research theory-practice-ethics with Indigenous theories and theorists. *Cultural Studies of Science Education, 14*(1), 111–127.

Higgins, M., & Tolbert, S. (2018). A syllabus for response-able inheritance in science education. *Parallax, 24*(3), 273–294.

Jackson, A. Y., & Mazzei, L. A. (2012). *Thinking with theory in qualitative research: Viewing data across multiple perspectives.* New York, NY: Routledge.

Kayumova, S., McGuire, C. J., & Cardello, S. (2019). From empowerment to response-ability: Rethinking socio-spatial, environmental justice, and nature-culture binaries in the context of STEM education. *Cultural Studies of Science Education, 14*(1), 205–229.

Kuokkanen, R. J. (2007). *Reshaping the university: Responsibility, Indigenous epistemes, and the logic of the gift.* Vancouver, BC: UBC Press.

Kuokkanen, R. J. (2008). What is hospitality in the academy? Epistemic ignorance and the (im)possible gift. *Review of Education, Pedagogy, and Cultural Studies, 30*(1), 60–82.

Kuokkanen, R. (2010). The responsibility of the academy: A call for doing homework. *Journal of Curriculum Theorizing, 26*(3), 61–74.

Lather, P. (2007). *Getting lost: Feminist efforts toward a double(d) science.* New York, NY: State University of New York.

Latour, B. (1993). *We have never been modern.* Cambridge, MA: Harvard University Press.

Latour, B. (2004a). Why has critique run out of steam? From matters of fact to matters of concern. *Critical Inquiry, 30*(2), 225–248.

Latour, B. (2004b). *Politics of nature: How to bring the sciences into democracy.* Cambridge, MA: Harvard University Press.

Lewis, B., & Aikenhead, G. (2001). Introduction: Shifting perspectives from universalism to cross-culturalism. *Science Education, 85*, 3–5.

Little Bear, L. (2016, June). *Big Thinking and rethinking: Blackfoot metaphysics 'waiting in the wings'.* Keynote address at 2016 Congress of the Humanities and Social Sciences, Calgary, Alberta.

McKinley, E. (2001). Cultural diversity: Masking power with innocence. *Science Education, 85*(1), 74–76.

McKinley, E. (2007). Postcolonialism, Indigenous students, and science education. In S. K. Abell & N. G. Lederman (Eds.), *Handbook of research on science education* (pp. 199–226). Mahwah, NJ: Lawrence Erlbaum.

McKinley, E., & Aikenhead, G. (2005). Comments on "Thinking differently about cultural diversity: Using postcolonial theory to (re)read science education". *Science Education, 89*(6), 901–906.

McKinley, E., & Stewart, G. (2012). Out of place: Indigenous knowledge in the science curriculum. In B. Fraser, K. Tobin, & C. J. McRobbie (Eds.), *Second international handbook of science education* (pp. 541–554). Dordrecht, NL: Springer, Netherlands.

Milne, C., & Scantlebury, K. (Eds.). (2019). *Material practice and materiality: Too long ignored in science education* (Vol. 18). Switzerland: Springer.

Moulton, J. (1983). A paradigm of philosophy: The adversary method. In S. Harding & M. B. Hintikka (Eds.), *Discovering reality: Feminist perspectives on epistemology, metaphysics, methodology, and philosophy of science* (pp. 149–164). Dordrecht, NL: Reidel.

Nakata, M. (2007a). *Disciplining the savages: Savaging the disciplines. Exploring inscriptions of Islanders in Western systems of thought.* Canberra, Australia: Aboriginal Studies Press.

Nakata, M. (2007b). The cultural interface. *The Australian Journal of Indigenous Education, 36*(S1), 7–14.

Patel, L. (2016). *Decolonizing educational research: From ownership to answerability.* New York, NY: Routledge.

Rhee, J. E., & Subreenduth, S. (2006). De/colonizing education: Examining transnational localities. *International Journal of Qualitative Studies in Education, 19*(5), 545–548.

Sammel, A. (2009). Turning the focus from 'other' to science education: Exploring the invisibility of whiteness. *Cultural Studies of Science Education, 4,* 649–656.

Smith, L. T. (1999/2012). *Decolonizing methodologies: Research and Indigenous People* (2nd ed.). London, UK: Zed Books.

Smith, L. T., Maxwell, T. K., Puke, H., & Temara, P. (2016). Indigenous knowledge, methodology and mayhem: What is the role of methodology in producing Indigenous insights? A discussion from mātauranga Māori. *Knowledge Cultures, 4*(3), 131–156.

Southerland, S. A. (2000). Epistemic universalism and the shortcomings of curricular multicultural science education. *Science & Education, 9,* 289–307.

Spivak, G. C. (1976). Translator's preface. In J. Derrida (Ed.), *Of grammatology* (G. C. Spivak, Trans.) (pp. ix–lxxxvii). Baltimore, MD: Johns Hopkins University Press.

Spivak, G. C. (1988). Can the subaltern speak? In C. Nelson & L. Grossberg (Eds.), *Marxism and the interpretation of culture* (pp. 271–313). Urbana, IL: University of Illinois Press.

Spivak, G. C. (1993/2009). *Outside in the teaching machine.* New York, NY: Routledge.

Spivak, G. C. (1994). Responsibility. *boundary 2, 21*(3), 19–64.

Spivak, G. C. (1999). *A critique of postcolonial reason.* Cambridge, MA: Harvard University Press.

Stengers, I. (2018). *Another science is possible: A manifesto for slow science.* Cambridge, MA: Polity.

van Eijck, M., & Roth, W. M. (2007). Keeping the local local: Recalibrating the status of science and traditional ecological knowledge (TEK) in education". *Science Education, 91*(6), 926–947.

van Eijck, M., & Roth, W. M. (2009). Authentic science experiences as a vehicle to change students' orientations toward science and scientific career choices: Learning from the path followed by Brad. *Cultural Studies of Science Education, 4*(3), 611–638.

Wallace, M., Higgins, M., & Bazul, J. (2018). Thinking with Nature: Following the contours of minor concepts for ethico-political response-ability in science education. *Canadian Journal of Science, Mathematics and Technology Education, 18*(3), 199–209.

INDEX

© The Editor(s) (if applicable) and The Author(s) 2021   343
M. Higgins, *Unsettling Responsibility in Science Education*,
Palgrave Studies in Educational Futures,
https://doi.org/10.1007/978-3-030-61299-3

Printed in the United States
by Baker & Taylor Publisher Services